Chromatography Theory

Chromatography Theory

Jack Cazes
Florida Atlantic University
Boca Raton, Florida

Raymond P. W. Scott
Georgetown University
Washington, D.C., and
Birkbeck College, University of London
London, England

MARCEL DEKKER, INC.　　　NEW YORK · BASEL

ISBN: 0-8247-0778-8

This book is printed on acid-free paper.

Headquarters
Marcel Dekker, Inc.
270 Madison Avenue, New York, NY 10016
tel: 212-696-9000; fax: 212-685-4540

Eastern Hemisphere Distribution
Marcel Dekker AG
Hutgasse 4, Postfach 812, CH-4001 Basel, Switzerland
tel: 41-61-261-8482; fax: 41-61-261-8896

World Wide Web
http://www.dekker.com

The publisher offers discounts on this book when ordered in bulk quantities. For more information, write to Special Sales/Professional Marketing at the headquarters address above.

Copyright © 2002 by Marcel Dekker, Inc. All Rights Reserved.

Neither this book nor any part may be reproduced or transmitted in any form or by any means, electronic or mechanical, including photocopying, microfilming, and recording, or by any information storage and retrieval system, without permission in writing from the publisher.

Current printing (last digit):
10 9 8 7 6 5 4 3 2 1

PRINTED IN THE UNITED STATES OF AMERICA

Preface

Chromatographic separations involve a large number of interacting variables that must be optimized in order to achieve the maximum resolution and the minimum analysis time for any given separation. To select the best conditions it is ncessary to understand chromatography theory. A rational theory that quantitatively described a chromatographic separation was first put forward in the mid-1940s. It was further developed during the 1950s and, except for some minor additions, was complete by 1960. Since that time, much work has been reported supporting the fundamental relationships that were established. This book provides the basic theory of gas and liquid chromatography together with the foundations of thin layer chromatography. The thermodynamic and dynamic principles of chromatographic retention are considered in detail and the kinetic processes that lead to band dispersion are quantitatively developed for all three major chromatographic techniques. The factors effecting the efficacy of preparative chromatography are reviewed and a discourse on moving bed and simulated moving bed chromatography is included. In addition, the column design and the design of chromatographic equipment are also discussed.

The book has been written in such a way that the mathematical skills required to understand the arguments are reduced to a minimum. Although, the algebra is occasionally lengthy, the contents of this book will be easily understood by undergraduate scientists who have taken basic courses in algebra and calculus.

This book is directed to analysts who utilize chromatographic techniques on a routine basis, scientists interested in designing chromatographic equipment, graduate students and postgraduate research fellows, and all who wish to have a fundamental understanding of the processes involved in chromatographic separation.

Jack Cazes
Raymond P. W. Scott

Contents

Preface iii

Part 1

The Mechanism of Retention 1

Molecular Interactions, the Thermodynamics of Distribution,
the Plate Theory, and Extensions of the Plate Theory

Chapter 1   Introduction to Chromatography Theory 3

Brief History of Chromatography
The Development Process 7
  Displacement Development 7
  Frontal Analysis 8
  Elution Development 9
  Elution Development in Thin Layer Chromatography 12
Chromatographic Terminology 14
Synopsis 16
References 17

Chapter 2   The Control of Chromatographic Retention and Selectivity (The Plate Theory) 19

The Plate Theory 20
The Retention Volume of a Solute 24
The Conditions for Chromatographic Separation 25
  The Capacity Ratio of a Solute 26
  The Separation Ratio (Selectivity) of a Solute 27
The Compressibility of the Mobile Phase 28
The Column Dead Volume 34
  Chromatographic Dead Volumes 38
  Experimental Examination of Dead Volume Measurement 39
Synopsis 45
References 45

Chapter 3   The Distribution Coefficient and Its Control of Retention Solute 47

The Thermodynamic Treatment of Solute Distribution in Chromatography 47
The Distribution of Standard Free Energy 54
  Thermodynamic Analysis of Some Homologous Series 54
Molecular Interactions and their Influence on the Magnitude of the
Distribution Coefficient (K) 62
Molecular Interactions 63
  Dispersion Forces 63
  Polar Forces 65
  Dipole-Dipole Interactions 66

Dipole-Induced-Dipole Interactions 67
Ionic Forces 69
Hydrophobic and Hydrophilic Interactions 71
The Distribution of Standard Free Energy Between Different Types of
Molecular Interactions 75
   The Concept of Complex Formation 77
Applied Thermodynamics 80
Synopsis 83
References 85

Chapter 4 The Theory of Mixed Phases in Chromatography 87

Introduction 87
Effect of Low Concentrations of Moderator 88
   Langmuir's Adsorption Isotherm for a Single Layer 88
   Effect of Bonding Characteristics 92
   Langmuir's Adsorption Isotherm for a Double Layer 93
   Mono-layer Adsorption 94
   Bi-layer Adsorption 95
Interactive Mechanisms with a Stationary Phase Surface in LC 98
   Interactions of a Solute with a Stationary Phase Surface 99
   Experimental Support for the Sorption and Displacement Processes 102
High Concentrations of Moderator 106
Mixed Mobile Phases in LC mixed solvents 109
   Ternary Mixtures 115
The Combined Effect of Temperature and Solvent Composition on Solute
Retention 118
Validation of Temperature/Solvent-Composition Independence 123
Aqueous Solvent Mixtures 124
The Thermodynamic Properties of Water/Methanol Association 133
Solute Interactions with Associated Solvents 135
   General Comments 139
Synopsis 140
References 142

Chapter 5 Programming Techniques 143

Flow Programming 144
   Flow Programming with a Compressible Mobile Phase 146
Temperature Programming 149
Gradient Elution or Solvent Programming 157
Synopsis 163
References 164

Chapter 6 Extensions of the Plate Theory 165

The Gaussian Form of the Elution Equation 165
Retention Measurements on Closely Eluting Peaks 167
Quantitative Analysis from Retention Measurements 171
Peak Asymmetry 175
Column Efficiency 179
The Position of the Points of Inflection 182
Resolving Power of a Column 183
Effective Plate Number 187
Elution Curve of a Finite Charge 190
Summation of Variances 193

Contents

The Maximum Sample Volume that Can Be Placed on a Chromatographic
Column 194
Vacancy Chromatography 195
The Peak Capacity of a Chromatographic Column 202
Temperature Changes During the Passage of a Solute through a Theoretical
Plate in Gas Chromatography...209
A Theoretical Treatment of the Heat of Absorption Detector 218
Synopsis...231
References...233

Part 2

The Mechanism of Dispersion

Dispersion in Columns and Mobile Phase Conduits, the Dynamics of Chromatography, the Rate Theory and Experimental Support of the Rate Theory........................235

Chapter 7   The Dynamics of Peak Dispersion 237

The Random Walk Model 240
The Diffusion Process...243
Sources of Dispersion in a Packed Column 245
  Dispersion from the Multi-Path Effect 245
  Dispersion from Longitudinal Diffusion..247
  Longitudinal Diffusion in the Stationary Phase 248
Dispersion Due to Resistance to Mass Transfer...250
  Resistance to Mass Transfer in the Mobile Phase...250
  Resistance to Mass Transfer in the Stationary Phase..251
Quantitative Treatment of Resistance to Mass Transfer Dispersion....................252
  Diffusion Controlled Dispersion in the Stationary Phase 254
  Diffusion Controlled Dispersion in the Mobile Phase......................................255
Fudge Factors..257
Synopsis...258
References...259

Chapter 8   The Rate Theory Equations..261

The Giddings Equation 261
The Huber Equation..262
The Knox Equation...264
The Horvath and Lin Equation...265
The Golay Equation..266
Effect of Mobile Phase Compressibility on the HETP Equation for a Packed
GC Column 267
Mobile Phase Compressibility: Its Effect on the Interpretation of Chromatographic Data
in LC...273
Effect of Mobile Phase Compressibility on the HETP Equation for a Packed LC Column...275
Extensions of the HETP Equation..276

Packed GC Columns..278
Packed LC Columns 279
Open Tubular GC Columns...281
Synopsis...283
References...284

Chapter 9 Extra-column Dispersion...287

Dispersion Generated by Different Parts of the Chromatographic System 290
Dispersion Due to Sample Volume 290
Dispersion in Sample Valves..293
Dispersion in Unions and Stainless Steel Frits...294
Dispersion in Open Tubular Conduits..295
Low Dispersion Connecting Tubes..300
  Serpentine Tubes 302
Dispersion in the Detector Sensor Volume 305
  Dispersion in Detector Sensors Resulting from Newtonian Flow...................305
  Apparent Dispersion from Detector Sensor Volume.....................................306
Dispersion Resulting from the Overall Detector Time Constant.......................310
Synopsis...311
References...312

Chapter 10 Experimental Validation of the Van Deemter Equation............315

The Accurate Measurement of Column Dispersion..316
The Multi-path, or (A), Term 321
The Longitudinal Diffusion, or (B), Term..324
The Optimum Velocity and the Minimum Plate Height 325
Dispersion Due to Resistance to the Mass Transfer of the Solute between the
Two Phases 328
The Effect of Particle Size on the Magnitude of the Van Deemter (C) Term......329
The Effect of the Function of (k') on Peak Dispersion....................................330
Synopsis...333
References...333

Chapter 11 The Measurement of Solute Diffusivity and Molecular Weight..335

The Relationship between Dispersion in a Packed Column to Solute Molecular Weight 343
Synopsis...356
References...358

Chapter 12 Chromatography Column Design...................................359

The Design of Packed Columns for GC and LC...359
Performance Criteria 361
  The Reduced Chromatogram..361
Instrument Constraints..363
Elective Variables..364
Column Specifications and Operating Conditions 365
Analytical Specifications..366
The Column Design Process for Packed Columns..367
The Optimum Particle Diameter..370
  Packed GC Columns..373
  Packed LC Columns 376
The Optimum Column Radius 379
The Optimum Flow Rate..381

Contents ix

The Minimum Solvent Consumption..382
Maximum Sample Volume 383
Synopsis..383
Reference 394

Chapter 13   Chromatography Column Design: The Design of Open Tubular Columns for GC..385

The Optimum Column Radius 388
The Minimum Length of an Open Tubular Column 390
Minimum Analysis Time..391
The Optimum Flow-Rate 391
Maximum Sample Volume and Maximum Extra Column Dispersion................392
Synopsis..393

Chapter 14   Chromatography Column Design: Application of the Design Equations to Packed Liquid Chromatography Columns and Open Tubular Gas Chromatography Columns 395

Optimized Packed Columns for LC 395
The Optimum Particle Diameter...396
The Optimum Velocity 398
The Minimum Plate Height...400
The Minimum Column Length..401
The Minimum Analysis Time 402
The Optimum Column Radius 403
The Optimum Flow Rate...404
Solvent Consumption 405
The Maximum Sample Volume 406
Gradient Elution 407
Optimized Open Tubular Columns for GC 409
The Optimum Column Radius 409
The Optimum Mobile Phase Velocity 411
The Calculation of (H_{min})..412
The Optimum Column Length 413
The Analysis Time..414
The Optimum Flow Rate...415
The Maximum Sample Volume 415
Synopsis..417
Reference 418

Chapter 15   Preparative Chromatography..419

Column Overload..419
Column Overload Due to Excess Sample Volume..420
Column Overload Due to Excess Sample Mass..427
The Control of Sample Size for Normal Preparative Column Operation............431
The Moving Bed Continuous Chromatography System 433
Synopsis..439
References..441

Chapter 16   Thin Layer Chromatography Theory 443

TLC Measurements...447
Resolution 449
Measurement of TLC Efficiency...451

Dispersion on a Thin Layer Plate 452
Synopsis..453
References..454

Appendix 1 The integration of the differential equation that describes the rate of change of solute concentration within a plate to the volume flow of mobile phase through it..455

Appendix 2 Accurate and Precise Dispersion Data for LC Columns............................457

Appendix 3   List of Symbols..463

Author Index   467

Subject Index   469

Part 1

The Mechanism of Retention

Molecular Interactions, the Thermodynamics of Distribution, the Plate Theory and Extensions of the Plate Theory

Chapter 1

Introduction to Chromatography Theory

The reader new to chromatography theory should first understand that the dynamic and thermodynamic effects that result in a chromatographic separation are logical and easy to understand. It is wise to always bear in mind the comment made by Einstein: *"first order effects are simple"*. In other words, in any physical chemical process, the phenomenon that accounts for the major effect will be elementary in nature and easy to understand. The overall simplicity of the chromatographic process and the factors that control it will be continually stressed throughout this book. Only when second order effects are considered, and dealt with quantitatively, does the theory and accompanying mathematics become more complex. Thus, the reader is challenged. As a balance, an excerpt from a lecture given in the early 1940s at the University of London by a physicist, G. W. Poole, *viz, "those who really know speak in words that everyone can understand"* offers an equivalent challenge to the authors. To fully understand the chromatography theory presented in this book, effort will be required by both the authors and the reader.

There are two essential theories needed to explain the processes involved in a chromatographic separation; perhaps it will be a surprise to the reader to discover that both were postulated and experimentally confirmed before 1960. There have been a number of contributions to chromatography theory over the intervening 40 years, but most of these have been largely involved in the extension and confirmation of existing concepts and a more detailed examination of second order effects. The more recent and original contributions to chromatography theory have come largely from developments in the field of preparative chromatography and electrochromatography. Chromatography theory is not merely an abstract study of the separation process; rather it is directly related to the practice of the technique. In fact, the remarkable advances and improvements that have taken place in chromatographic performance over the years past have been directly predicted and pre-empted from theoretical studies. In addition, without fully understanding the physical processes involved in a chromatographic separation, the optimum use of a chromatograph would be virtually impossible or, at the very least, depend heavily on serendipity.

Chromatography has been defined in the classical manner as,

"A separation process that is achieved by the distribution of the substances to be separated between two phases, a stationary phase and a mobile phase. Those solutes, distributed preferentially in the mobile phase, will move more rapidly through the system than those distributed preferentially in the stationary phase. Thus, the solutes will elute in order of their increasing distribution coefficients with respect to the stationary phase."

It follows that during the development of a chromatographic separation, two processes will occur simultaneously and to large extent, independently. Firstly, the individual solutes in the sample are moved apart in the distribution system as a result of their different affinities for the stationary phase. Secondly, as the bands are moved apart, their tendency to spread or disperse is constrained to ensure that the separation that has been achieved is maintained. Thus, the phase system must be chosen to provide the necessary relative retention of the solutes, and the distribution system must be appropriately designed to minimize this dispersion and permit the components of the mixture to be eluted discretely. Chromatography theory provides a basis for these choices; it discloses the mechanisms that control retention, it explains the different processes that can cause band dispersion, and it shows how solute retention and band dispersion are controlled by the operating variables of the chromatographic system. In addition, chromatography theory reveals how the separation is affected by the properties of those parts of the chromatographic system that are not directly associated with solute distribution (*i.e.,* sampling devices, detector sensors, etc.).

Brief History of Chromatography Theory

There are two fundamental chromatography theories that deal with solute retention and solute dispersion and these are the *Plate Theory* and the *Rate Theory*, respectively. It is essential to be familiar with both these theories in order to understand the chromatographic process, the function of the column, and column design. The first effective theory to be developed was the plate theory, which revealed those factors that controlled *chromatographic retention* and allowed the

Introduction to Chromatography Theory

efficiency of the column, that is, its capacity for containing peak dispersion, to be measured.

The original plate theory was first applied to chromatography by Martin and Synge [1] in 1941, who borrowed the plate concept from distillation theory. The equation which they derived took the less useful exponential form, but the work established the *plate concept* as a valid approach for the mathematical examination of elution processes in chromatography. It will be seen throughout this book that the theory can also be used very effectively to investigate a wide range of other types of chromatographic phenomena. Largely as a result of ignorance, the plate theory has come under some unjust criticism. This is partly due to its age (in the modern idiom, *new* is good, *old* is bad) and partly because it is thought that the theory assumes that the solute is in continuous equilibrium with the two phases which, in fact, can never happen in a chromatographic system. However, it will be clear, when the plate theory is considered, that equilibrium is *not* assumed and, in fact, the *plate* concept itself is specifically introduced to contend with the *non*-equilibrium condition that exists between the two phases. The reader is well advised to study the plate theory carefully as it has a wide field of application. The theory is easy to understand and requires the minimum knowledge of algebra and calculus. It can be used to derive, in a very simple manner, the essential retention equations, the equation from which the efficiency of the system can be calculated, and the different equations used to describe resolution.

After its initial introduction, the plate theory was further elaborated by Mayer and Tomkins (1947) [2] and Glueckauf (1955) [3] but it was not until 1956 that Said [4] generated the more precise and useful form that is generally used today. Said's treatment furnished a simple differential equation that could be integrated to give a Poisson function which, under limiting conditions, simplified to a Gaussian or Error function. It will be seen that the Gaussian or Error function curve is the generally accepted characteristic shape of an *ideal* chromatography elution curve. At the time of publication, the work of Said went almost unnoticed by the chromatography field and this more useful form of the plate theory was not generally known or appreciated until reported and discussed by Keulemans [5] in 1959.

The original Rate Theory which describes dispersion in packed beds evolved over a number of years, probably starting with the work of Lapidus and Amundson [6] in 1952, extended by that of Glueckauf [7] and Tunitski [8] in 1954. The final form of the equation that described dispersion in packed beds as a function of the linear velocity of the fluid passing through it, and which was directly applicable to packed gas chromatography (GC) columns, was described by Van Deemter *et al.* [9] in 1956. Several years later, in the 1960s, during the renaissance of liquid chromatography (LC), the equation of Van Deemter *et al.* was also found to be directly applicable to LC columns. In 1958, the rate theory was applied to open tubes by Golay [10,11], who developed a specific dispersion equation for capillary columns which has been used virtually unchanged and modified, since that time. The Golay equation can also be used to examine dispersion in parts of the chromatograph other than the column, *e.g.*, sample valves, connecting tubes, and detector cells, etc.

Neither the Golay equation nor the equation of Van Deemter *et al.* took into account mobile phase compressibility which, although it does not strongly affect the magnitude of the dispersion (even in GC columns), it does change the method of measurement of both the linear velocity and the resistance to mass transfer terms in the dispersion equation. The effect of pressure change along the column was investigated by Giddings [12, 13] and a modified form of both the Van Deemter equation and the Golay equation, that would accommodate mobile phase compressibility, was developed by Ogan and Scott [14].

In the late 1960s and early 1970s, solute dispersion occurring in LC columns was critically examined and compared to that predicted by the Van Deemter equation. Poor agreement between theory and experiment became apparent and, as a consequence, alternative dispersion equations (largely based on the original Van Deemter equation) were proposed by Huber and Hulsman [15], Kennedy and Knox [16], and Horvath and Lin [17]. In retrospect, it would appear that the poor agreement between theory and experiment was at least partly due to extra-column dispersion, the significance of which was not fully appreciated at that time. Later experiments in which measurements were made with high precision and with apparatus where extra-column dispersion was reduced to a minimum [18] confirmed that the Van Deemter equation could accurately describe solute dispersion in LC columns.

It appears that the equation introduced by Van Deemter is still the simplest and the most reliable for use in general column design. Nevertheless, all the equations helped to further understand the processes that occur in the column. In particular, in addition to describing dispersion, the Kennedy and Knox equation can also be employed to assess the efficiency of the packing procedure used in the preparation of a chromatography column.

As soon as it was shown by the plate theory that retention data could be used to calculate the magnitude of the distribution coefficient between the two phases, then many equilibrium systems began to be examined chromatographically. The standard entropies and enthalpies of distribution were measured for a wide range of solutes and phase systems from retention data taken over a range of temperatures. Some of the early pioneers in this work were Herington [19] in 1956, Kwantes and Rijnders [20] in 1958 and Adlard *et al*. [21]. The thermodynamic explanation of retention has been used extensively over the intervening years in attempts to describe various different retention mechanisms, including entropically driven retention such as exclusion chromatography and the unique selectivity of chiral stationary phases. The work continues in many laboratories and some of the more recent examples of this type of work are those of Martire and his group [22-26].

Contemporary development of chromatography theory has tended to concentrate on dispersion in electro-chromatography and the treatment of column overload in preparative columns. Under overload conditions, the adsorption isotherm of the solute with respect to the stationary phase can be grossly nonlinear. One of the prime contributors in this research has been Guiochon and his co-workers, [27-30]. The form of the isotherm must be experimentally determined and, from the equilibrium data, and by the use of appropriate computer programs, it has been shown possible to calculate the theoretical profile of an overloaded peak.

The Development Process

A chromatographic separation can be developed in three ways, by *displacement development*, by *frontal analysis*, and by *elution development*, the last being almost universally used in all analytical chromatography. Nevertheless, for the sake of completeness, and because in preparative chromatography (under certain conditions of mass overload) displacement effects occur to varying extents, all three development processes will be described.

Displacement Development

Displacement development is only really effective if the stationary phase is a solid and the solutes are adsorbed on its surface. The sample mixture is placed on the front of the distribution system, and the individual solutes compete for the immediately available adsorption sites. Initially, all the nearby adsorbent sites will be saturated with the most strongly held component. As the sample band moves through the system the next available adsorption sites will become saturated with the next most strongly adsorbed component. In this manner, the components will array themselves along the distribution system in order of their decreasing adsorption strength. Now, in displacement development, the sample components are held on the stationary phase so strongly that they cannot be eluted by the mobile phase or, at least, only very slowly. They can, however, be displaced by a substance more strongly held than any of the solutes (called the displacer). The displacer is passed through the system and first displaces the most strongly held component. In turn this component will displace the next and so on. Thus, the displacer forces the adsorbed components progressively through the distribution system, each component displacing the one in front until they are all displaced sequentially from the system.

The solutes will be characterized by the order in which they elute and the amount of each solute present will be proportional to the length of each band, not the height. The great disadvantage of this type of development is that the solutes are never actually *separated* from one another. The solutes leave the system sequentially and in contact with each other and, in practice, each somewhat mixed with its neighbor. As already stated, this type of development is almost unknown in analytical chromatography and only very rarely used in preparative LC. However, displacement effects can occur in overloaded distribution systems (which, in fact, can sometimes aid in large-scale separations) and in the development of thin layer plates with multicomponent solvents. The development of thin layer plates, using a multicomponent solvent mixture, is a somewhat complex elution process that involves the displacement separation of the mobile phase components and will be discussed in due course.

Frontal Analysis

This type of chromatographic development will only be briefly described as it is rarely used and probably is of academic interest only. This method of development can only be effectively employed in a column distribution system. The sample is fed continuously onto the column, usually as a dilute solution in the mobile phase. This is in contrast to displacement development and elution development, where discrete samples are placed on the system and the separation is subsequently processed. Frontal analysis only separates part of the first compound in a relatively pure state, each subsequent component being mixed with those previously eluted. Consider a three component mixture, containing solutes (A), (B) and (C) as a dilute solution in the mobile phase that is fed continuously onto a column. The first component to elute, (A), will be that solute held *least* strongly in the stationary phase. Then the second solute, (B), will elute but it will be mixed with the first solute. Finally, the third solute (C), will elute in conjunction with (A) and (B). It is clear that only solute (A) is eluted in a pure form and, thus, frontal analysis would be quite inappropriate for most practical applications of chromatography. This development technique has been completely superseded by elution development.

Elution Development

Elution development is by far the most common method of processing a chromatographic separation and is used in all types of chromatography. Elution development is best described as a series of absorption-extraction processes which are continuous from the time the sample is injected into the distribution system until the time the solutes exit from it. The elution process is depicted in Figure 1.

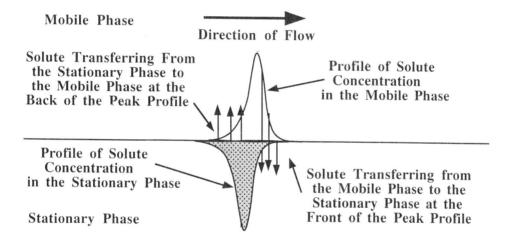

Figure 1. The Elution of a Solute Through a Chromatographic System

The concentration profiles of the solute in both the mobile and stationary phases are depicted as Gaussian in form. In due course, this assumption will be shown to be the ideal elution curve as predicted by the Plate Theory. Equilibrium occurs between the mobile phase and the stationary phase, when the probability of a solute molecule striking the boundary and entering the stationary phase is the same as the probability of a solute molecule randomly acquiring sufficient kinetic energy to leave the stationary phase and enter the mobile phase. The distribution system is continuously thermodynamically driven toward equilibrium. However, the *moving* phase will continuously displace the concentration profile of the solute in the mobile phase forward, relative to that in the stationary phase. This displacement, in a grossly exaggerated form, is depicted in Figure 1. As a result of this displacement, the concentration of solute in the mobile phase at the front of the peak *exceeds* the equilibrium concentration with respect to that in the stationary phase. Accordingly, a *net* quantity of solute in the front part of the peak is continually *entering* the stationary phase from the mobile phase in an attempt to re-establish equilibrium as the peak progresses along the column. At the rear of the peak, the reverse occurs. As the concentration profile moves forward, the concentration of solute in the stationary phase at the *rear* of the peak is now in *excess* of the equilibrium concentration. A net amount of solute must now *leave* the stationary phase and enters the mobile phase in an endeavor to re-establish equilibrium. Thus, the solute moves through the chromatographic system as a result of solute entering the mobile phase at the rear of the peak and returning to the stationary phase at the front of the peak. It should be emphasized, however, that solute is always transferring between the two phases over the whole of the peak in an attempt to attain or maintain thermodynamic equilibrium. However, the solute band progresses through the system as a result of a *net* transfer of solute from the mobile phase to the stationary phase in the *front half* of the peak, which is compensated by a *net* transfer of solute from the stationary phase to the mobile phase at the *rear half* of the peak.

The processes involved in establishing solute equilibrium between two phases is quite complicated, but a simplified explanation will be given here. The distribution of kinetic energy of the solute molecules contained in the stationary phase and mobile phase is depicted in Figure 2A and 2B. Solute molecules can only leave the stationary phase when their kinetic energy is equal to or greater than the potential energy of their association with the molecules of stationary phase. The distribution of kinetic energy between the molecules dissolved in the stationary phase at any specific temperature T, is considered to take the form of a Gaussian curve as shown in Figure 2A. (In fact, other distribution functions might be more appropriate to describe the energy distribution, but the specific function used will not affect the following explanation and so, for simplicity, the Gaussian function will be assumed.) The number of molecules at the boundary surface (N_1) that have a kinetic energy in excess of the potential energy associated with their molecular interactions with the stationary phase (E_A), (*i.e.,* those molecules represented by the shaded area of the distribution curve) will leave the stationary phase and enter the mobile phase. Those with an energy less than (E_A) will remain in the stationary phase. The distribution of energy of the solute molecules in the mobile phase is depicted in Figure 2B. The distribution is again taken as Gaussian in form and it is seen that the number of molecules (N_2) striking the surface that have an energy less than (E_A) (*i.e.*, the shaded area in figure 2B) will remain in the stationary phase after entering the liquid, whereas the others having energies above (E_A) will collide with the surface and 'bounce back'. 'Bounce back' is, perhaps, a somewhat picturesque term in this context. In fact, some may bounce back, others may communicate their excess energy to a another solute molecule which will give it sufficient energy to enter the other phase.

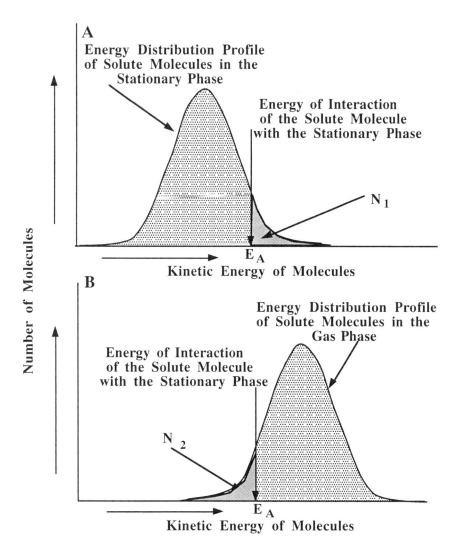

Figure 2. Energy Distribution of Solute Molecules in the Stationary and Mobile Phase

In either case, the net effect is the same; there will be no net molecule transfer if its energy is too great.

Under equilibrium conditions,

$$N_1 = N_2$$

This description of the dynamics of solute equilibrium is oversimplified, but is sufficiently accurate for the reader to understand the basic principles of solute distribution between two phases. For a more detailed explanation of dynamic equilibrium between immiscible phases the reader is referred to the kinetic theory of gases and liquids.

Consider a distribution system that consists of a gaseous mobile phase and a liquid stationary phase. As the temperature is raised the energy distribution curve in the gas moves to embrace a higher range of energies. Thus, if the column temperature is increased, an increasing number of the solute molecules in the stationary phase will randomly acquire sufficient energy (E_A) to leave the stationary phase and enter the gas phase. As a result, the distribution coefficient with respect to the stationary phase of all solutes will be reduced as the temperature rises and it will be seen in due course that this will cause the band velocity of all the solutes to be increased.

Elution Development in Thin Layer Chromatography

Before moving on to chromatographic terminology it is pertinent to discuss the development processes that take place on a thin layer plate as, although it is basically an elution technique, it is complicated by the frontal analysis of the mobile phase itself. The mobile phases used to elute the solutes in TLC are usually multi-component, containing at least three individual solvents, if not more. If the plate is not pre-conditioned with solvent, there is an elaborate modification of the plate surface which is depicted, for a ternary solvent mixture, in Figure 3. The edge of the plate is dipped into a tray of the solvent mixture which begins to migrate along the plate, driven by surface tension forces. The different solvents array themselves on the surface in the manner shown in Figure 3. The solvent that interacts most strongly with the stationary phase is extracted from the mixture and forms an adsorbed layer on the surface that corresponds to the area (X) in the diagram. The now binary mixture continues to migrate along the plate and the next solvent component that interacts most strongly with the stationary phase (solvent B) is adsorbed as a layer on the surface corresponding to the area (Y) in the diagram. Finally, the remaining solvent (C) with the weakest interactions with the stationary phase continues to migrate and cover the surface with a layer of solvent (C) in the area (Z). It is seen that the distribution system, which has resulted from the frontal analysis of the three components of the mobile phase, is now quite complex.

Introduction to Chromatography Theory

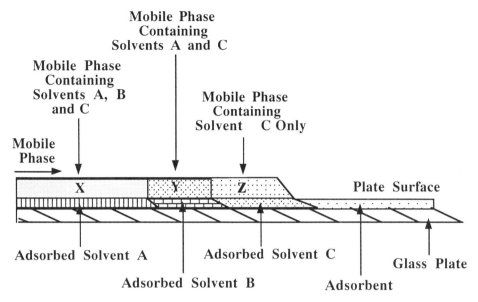

Figure 3. The Development of a Thin Layer Plate

Consider now how the *solutes* will interact during the separation process. In the first section (X) solutes will be distributed between the ternary solvent mixture (A), (B) and (C) and the surface covered with solvent (A). In the next section (Y) the solutes will be distributed between a binary solvent mixture of (B) and (C) and a surface covered with solvent (B). Finally, distribution will take place in section (Z) between pure solvent (C) and a surface covered with solvent (C). Even this is an over-simplification, as the composition of the mobile phase in each section will not be constant but will decrease along the plate. Furthermore, as the separation progresses, the lengths of sections (X), (Y) and (Z) will continually increase. Such a system is extremely difficult to treat theoretically in a precise manner, particularly as the boundaries are not as sharp as those drawn in Figure 3. In fact, the overall effect is as though the separation was carried out sequentially on three separate sections of a plate, each section having a different stationary phase and mobile phase. In each section, the separation will then be achieved by elution development, but the overall effect will be a form of gradient elution.

The complexity of the system increases with the number of solvents used and, of course, their relative concentrations. The process can be simplified considerably by pre-conditioning the plate with solvent vapor from the mobile phase before the separation is started. Unfortunately, this only partly reduces the adsorption effect, as the equilibrium between the *solvent vapor* and the adsorbent surface will not be the same as that between the *liquid solvent* and the surface. It is clear that by forming a gradient by the frontal analysis of the mobile phase and carefully choosing the solvent mixture, very delicate *pseudo*-gradients can be created, which, in no small measure, accounts for the great versatility, popularity, and success of TLC.

Chromatographic Terminology

Before starting to discuss chromatographic theory, it would be advantageous to define some of the terms that will be used in the subsequent treatment. Much of the terminology involved in chromatography theory will be defined as the pertinent arguments develop, but some basic terms are essential to initiate the discussion. Chromatography nomenclature has evolved over the years, but it was not until the late 1950s that an attempt was made to rationalize the various terms used to describe the characteristics of a chromatogram.

The British Chromatography Discussion Group initially nominated R. G. Primevesi, N. G. McTaggart, C. G. Scott, F. Snelson and M. M. Wirth to classify the various chromatography terms existing at that time into a formal nomenclature. Their first report was published in the *Journal of the Institute of Petroleum* in 1967 [31]. A summary of the nomenclature that was recommended by the group is shown diagramatically in Figure 4 which will apply to both GC and LC. Since that time a number of groups, including IUPAC, have been involved in classifying and defining chromatography terms and, for the most part, the terms recommended by these groups will be employed throughout this book. Some of the basic terms are defined as follows.

The *baseline* is any part of the chromatogram where only mobile phase is emerging from the column.

The *injection point* is that point when the sample is placed on the column. If the sample has a finite volume, then the injection point corresponds the start of the sampling process.

The *dead point* is the position of the peak maximum of an unretained solute. It is *not* the initial part of the dead volume peak as this represents a retarded portion of the peak that is caused by dispersion processes. The importance of employing the peak maximum for such measurements as dead volume and retention volume will be discussed in later chapters of the book that deal with peak dispersion.

The *peak maximum* is the highest point (the apex) of the peak.

Introduction to Chromatography Theory 15

The *dead time* (t_0) is the time elapsed between the *injection point* and the *dead point*.

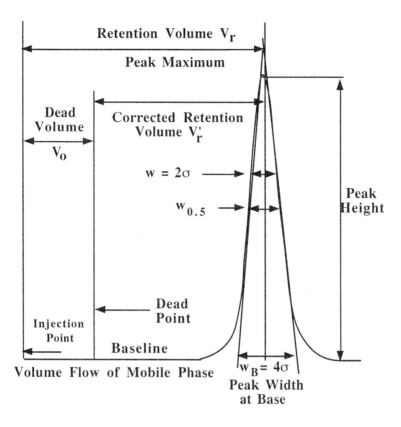

Figure 3. The Nomenclature of a Chromatogram.

The *dead volume* (V_0) is the volume of mobile phase passed through the column between the *injection point* and the *dead point*.

Thus, $$V_0 = Qt_0$$

where (Q) is the flow rate in ml/min. (If the mobile phase is compressible then (Q) must be the *mean* flow rate.)

The *retention time* (t_r) is the time elapsed between the *injection point* and the *peak maximum*. Each solute will have a characteristic retention time.

The *retention volume* (V_r) is the volume of mobile phase passed through the column between the *injection point* and the *peak maximum*.

Thus, $$V_r = Qt_r$$

Each solute will also have a characteristic retention volume.

The *corrected retention time* (t'_r) is the time elapsed between the *dead point* and the *peak maximum*.

The *corrected retention volume* (V'_r) is the volume of mobile phase passed through the column between the *dead point* and the *peak maximum*. It will also be the *retention volume* minus the *dead volume*.

Thus, $$V'_r = V_r - V_0 = Q(t_r - t_0)$$

The *peak height* (h) is the distance between the *peak maximum* and the *baseline* geometrically produced beneath the peak.

The *peak width* (w) is the distance between each side of a peak measured at 0.6065 of the peak height. The peak width measured at this height is equivalent to two standard deviations (2σ) of the Gaussian curve and, thus, has significance when dealing with chromatography theory.

The *peak width at half height* ($w_{0.5}$) is the distance between each side of a peak, measured at half the peak height. The peak width measured at half height has no significance with respect to chromatography theory.

The *peak width at the base* (w_B) is the distance between the intersections of the tangents drawn to the sides of the peak and the *peak base* geometrically produced. The peak width at the base is equivalent to four standard deviations (4σ) of the Gaussian curve and, thus, also has significance when dealing with chromatography theory.

Synopsis

In a chromatographic separation, the individual components of a mixture are moved apart in the column due to their different affinities for the stationary phase and, as their dispersion is contained by appropriate system design, the individual solutes can be eluted discretely and resolution is achieved. Chromatography theory has been developed over the last half century, but the two critical theories, the Plate Theory and the Rate Theory, were both well established by 1960. There have been many contributors to chromatography theory over the intervening years but, with the exception of the development of preparative column theory, some theory of retention mechanisms and electro-chromatography, the majority of the contributions have been largely confined to the extension, and confirmation, of existing concepts and a more detailed examination of second order effects. There are three basic methods of chromatographic development, displacement development, frontal analysis and elution development. The latter is almost the only development procedure used in contemporary analytical chromatography. Elution development is best described as a series of absorption-extraction processes between the two phases which are continuous from the time the sample is injected into the chromatographic system until the time the solutes exit from it. The solute band progresses through the system as a result of a *net* transfer of solute from the mobile phase to the stationary phase in the front half of the peak which is compensated by a *net* transfer of solute from the stationary phase to the mobile phase at the rear half of the peak. Chromatographic development in thin layer chromatography (TLC) can be more complex if a multicomponent solvent mixture is used as the mobile phase. This is due to the frontal analysis effect of the mobile phase on the plate producing different distribution systems along the plate surface. In practice, TLC development is actually a gradient elution process that gives the technique a high degree of versatility. A basic understanding of chromatography terminology is necessary to discuss theoretical aspects of the technique and, as new concepts are discussed, so will their terminology be explained.

References

1. A. J. P. Martin and R. L. M. Synge, *Biochem. J.*, **35**(1941)1358.
2. S. W. Mayer and E. R. Tompkins, *J. Am. Chem. Soc.*, **69**(1947)2866.
3. E. Glueckauf, *Trans. Faraday Soc.*, **51**(1955)34.
4. A. S. Said, *Am. Inst. Chem. Eng. J.*, **2**(1956)477.
5. A. I .M. Keulemans, "*Gas Chromatography*"(2nd Ed.), Reinhold Publishing Company, Amsterdam, (1959)108.
6. L. Lapidus and N. R. Amundson *J. Phys. Chem.*, **56**(1952)984.
7. E. Glueckauf, *Principles of Operation of the Ion Exchange Column*, Conference on Ion Exchange, London, (1954).
8. N. N. Tunitski, *N. N. Doklady Acad. Nauk,* (Comp. rend. acad. sci.U.R.S.S.) **9**(1954)577.
9. J. J. Van Deemter, F. J. Zuiderweg and A. Klinkenberg, *Chem. Eng. Sci.,* **5**(1956)24.

10. M. J. E. Golay, in *"Gas Chromatography"*, (Ed. V. J. Coates, H. J. Nobels and I. S.Fagerson), Academic Press, New York (1957).
11. M. J. E Golay, in *"Gas Chromatography 1958"* (Ed. D. H. Desty), Butterworths Scientific Publications, London (1958)36.
12. J. C. Giddings, *Anal. Chem.*, **36**(4)(1964)741.
13. J. C. Giddings, *Dynamics of Chromatography*, Marcel Dekker, New York, (1965)56.
14. K. Ogan and R. P. W. Scott, *J. High Res. Chromatogr.*, **7 July** (1984)382.
15. J. F. K. Huber and J. A. R. J. Hulsman, *Anal. Chim. Acta.*, **38**(1967)305.
16. G. J. Kennedy and J. H. Knox, *J. Chromatogr. Sci.*, **10**(1972)606.
17. Cs. Horvath and H. J. Lin, *J. Chromatogr.*, **149**(1976)401.
18. E. Katz, K. L. Ogan and R. P. W. Scott, *J. Chromatogr.*, **270**(1983)51.
19. E. F. G. Herington in *"Vapor Phase Chromatography"* (Ed. D. H. Desty), Butterworths Scientific Publications Ltd., London (1956)5.
20. A. Kwantes and G. W. A. Rijnders," *Gas Chromatography 1958"*, Butterworths Scientific Publications Ltd., London (1958)125.
21. E. R. Adlard, M. A. Kahn and B. T. Whitham in, *Gas Chromatography 1960"*, Butterworths Scientific Publications Ltd., London(1960)57.
22. D. E. Martire and P. Reidl, *J. Phys. Chem.*, **72**(1968)3478.
23. Y. B. Tewari, J. P. Sheriden and D. E. Martire, *J. Phys. Chem.*, **74**(1970)3263.
24. Y. B. Tewari, D. E. Martire and J. P. Sheriden, *J. Phys. Chem.*, **74**(1970)2345.
25. J. P. Sheriden, D. E. Martire and Y. B. Tewari, *J. Am. Chem. Soc.*, **94**(1972)3294.
26. H. L. Liao and D. E. Martire, *J. Am. Chem. Soc.*, **96**(1974)2058.
27. G. Guiochon and A. Katti, *Chromatographia*, **24**(1987)165.
28. S. Golshan-Shirazi, A. Jaulmes and G. Guiochon, *Anal. Chem.*, **60**(1988)1856.
29. S. Golshan-Shirazi, A. Jaulmes and G. Guiochon, *Anal. Chem.*, **61**(1989)1276.
30. S. Golshan-Shirazi, A. Jaulmes and G. Guiochon, *Anal. Chem.*, **61**(1988)1368.
31. R. G. Primevesi, N. G. McTaggart, C. G. Scott, F. Snelson and M. M. Wirth, *J. Inst. Petrol.*, **53**(1967)367.

Chapter 2

The Control of Chromatographic Retention and Selectivity (The Plate Theory)

It is stated in the definition of chromatography that *"the solutes will elute in order of their increasing distribution coefficients with respect to the stationary phase"*. It follows, that the relative retention of two substances in a chromatographic system (the time interval between their elution) will determine how well they are separated. The greater the retention difference between any pair of solutes, the better will be the resolution and the farther apart they will appear on the chromatogram. Consequently, an algebraic expression for the retention volume of a solute will display those factors that control retention, how the retention can be increased, and how the separation can be improved. The chromatogram that depicts the elution of a solute from a column is actually a graph relating the concentration of the solute in the mobile phase leaving the column to elapsed time. As the flow rate is constant, the chromatogram will also be a curve relating the concentration of solute in the exiting mobile phase to the volume of mobile phase passed through the column. Thus, an equation is required that will relate the concentration of the solute in the mobile phase leaving the column to the volume of mobile phase that has passed through it. In Figure 1, a simple chromatogram shows the elution of a single peak. The expression, f(v), is the elution curve equation and will be derived using the plate theory.

Once the elution-curve equation is derived, and the nature of f(v) identified, then by differentiating f(v) and equating to zero, the position of the peak maximum can be determined and an expression for the retention volume (V_r) obtained. The expression for (V_r) will disclose those factors that control solute retention.

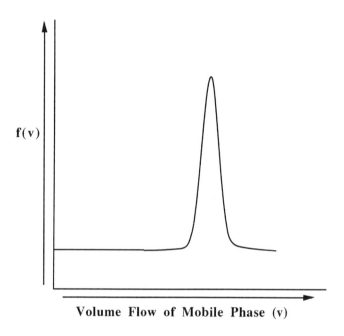

Figure 1. The Elution Curve of a Single Peak

The Plate Theory

The plate theory assumes that the solute, during its passage through the column, is always in equilibrium with the mobile and stationary phases. However, as the solute is continuously passing from one phase to the other, equilibrium between the phases *never* actually occurs. To accommodate this non-equilibrium condition, a technique originally introduced in distillation theory is adopted, where the column is considered to be divided into a number of cells or plates. Each cell is allowed a specific length and, as a consequence, the solute will spend a finite time in each cell. The cell is chosen to be of such size as to give the solute sufficient residence time to establish equilibrium with the two phases. Consequently, the smaller the plate is found to be, the faster will equilibrium be achieved and the more plates there will be in the column. It follows that the number of theoretical plates contained by a column will be directly related to equilibrium rate and, for this reason, has been termed the *column efficiency*. It will be shown, in due course, that the peak width (the dispersion or peak spreading) is inversely proportional to the square root of the efficiency and, thus, the higher the efficiency, the more narrow the peak.

Said [1] developed the Martin concept [2] to derive the elution curve equation in the following way.

Consider the equilibrium conditions that are assumed to exist in each plate, then

$$X_S = KX_m \qquad (1)$$

where (X_m) is the concentration of solute in the mobile phase,

(X_S) is the concentration of solute in the stationary phase,

and (K) is the distribution coefficient of the solute between the two phases.

Note that (K) is defined with *reference to the stationary phase*; i.e.,

$$K = X_S/X_m$$

Thus, the larger the value of (K), the more the solute will be distributed in the *stationary* phase. (K) is a dimensionless constant and, in gas/liquid and liquid/liquid systems, (X_S) and (X_m) can be measured as *mass of solute per unit volume of phase*. In gas/solid and liquid/solid systems, (X_S) and (X_m) can be measured as *mass of solute per unit mass of phase*.

Equation (1) reiterates the general distribution law and represents the adsorption isotherm as linear. In both gas/solid chromatography (GSC) and liquid chromatography (LC), virtually all the solutes exhibit Langmuir type isotherms between the two phases which, over a wide concentration range, will certainly *not* be linear. However, at the extremely low solute concentrations employed in most chromatographic separations, that portion of the isotherm that is pertinent, and over which the chromatographic process is operating, will indeed be very close to linear. It will be shown later that the operating portion of the adsorption isotherms must always be very close to linear if the system is to have practical use, since nonlinear isotherms produce asymmetrical peaks

Differentiating equation (1),

$$dX_S = KdX_m \qquad (2)$$

Consider three consecutive plates in a column, the (p-1), the (p) and the (p+1) plates and let there be a total of (n) plates in the column. The three plates are depicted in Figure 2. Let the volumes of mobile phase and stationary phase in each plate be (v_m) and (v_s) respectively, and the concentrations of solute in the mobile and stationary phase in each plate be $X_m(p-1)$, $X_s(p-1)$, $X_m(p)$, $X_s(p)$, $X_m(p+1)$, and $X_s(p+1)$, respectively. Let a volume of mobile phase, dV, pass from plate (p-1) into plate (p) at the same time displacing the same volume of mobile phase from plate (p) to plate (p+1). As a result, there will be a change of mass (dm) of solute in plate (p) which will equal the difference in the mass entering plate (p) from plate (p-1) and the mass of solute leaving plate (p) and entering plate (p+1). It is now possible to apply a simple mass balance procedure to the plate (p).

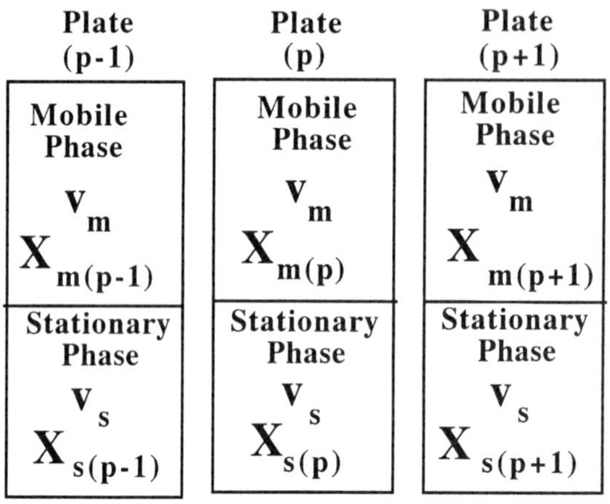

Plate (p-1)	Plate (p)	Plate (p+1)
Mobile Phase v_m $X_{m(p-1)}$	Mobile Phase v_m $X_{m(p)}$	Mobile Phase v_m $X_{m(p+1)}$
Stationary Phase v_s $X_{s(p-1)}$	Stationary Phase v_s $X_{s(p)}$	Stationary Phase v_s $X_{s(p+1)}$

Figure 2. Three Consecutive Theoretical Plates in a Column

Now, the change in mass will be the product of concentration and change in volume; thus, the change of mass of solute (dm) in plate (p) will be

$$dm = (X_{m(p-1)} - X_{m(p)})dV \qquad (3)$$

Now, as equilibrium is maintained in the plate (p) by definition, the mass (dm) will be distributed between the two phases, resulting in a solute concentration change of $dX_{m(p)}$ in the mobile phase and $dX_{s(p)}$ in the stationary phase. Then,

$$dm = v_s \, dX_{s(p)} + v_m \, dX_{m(p)} \qquad (4)$$

Substituting for $dX_{s(p)}$ from equation (2),

$$dm = (v_m + Kv_s) \, dX_{m(p)} \qquad (5)$$

Equating equations (3) and (5) and rearranging,

$$\frac{dX_{m(p)}}{dV} = \frac{X_{m(p-1)} - X_{(p)}}{v_m + Kv_s} \qquad (6)$$

It is mathematically convenient to change the variable, so instead of measuring the volume flow of mobile phase in ml, it will be measured in units of $(v_m + Kv_s)$.

Thus, the new variable (v) will be given by

$$v = \frac{V}{(v_m + Kv_s)} \tag{7}$$

The function $(v_m + Kv_s)$ is termed the 'plate volume' and so the flow through the column will be measured in 'plate volumes' instead of milliliters. The 'plate volume' is defined as that volume of mobile phase that can contain all the solute in the plate at the equilibrium concentration of the solute in the mobile phase. The meaning of 'plate volume' must be understood, as it is an important concept and is extensively used in different aspects of chromatography theory.

Differentiating equation (7),
$$dv = \frac{dV}{(v_m + Kv_s)} \tag{8}$$

Substituting for dV from (8) in (6)
$$\frac{dX_{m(p)}}{dV} = X_{m(p-1)} \; X_{(p)} \tag{9}$$

Equation (9) describes the rate of change of concentration of solute in the mobile phase in plate (p) with the volume flow of mobile phase through it. The integration of equation (9) will provide the elution curve equation for any solute eluted from any plate in the column. A simple method for the integration of equation (9) is given in Appendix 1, where the solution, the elution curve equation for plate (p), is shown to be

$$X_{m(p)} = \frac{X_o e^{-v} v^p}{p!}$$

where $(X_{m(p)})$ is the concentration of the solute in the mobile phase
      leaving the (p)th plate,
and (X_o) is the initial concentration of solute placed on the 1st
      plate of the column.

Thus, the elution curve equation for the last plate in the column, the (n) th plate *(that is, the equation relating the concentration of solute in the mobile phase entering the detector to volume of mobile phase passed through the column)* is given by

$$X_{m(n)} = \frac{X_o e^{-v} v^n}{n!} \tag{10}$$

Equation (10) is the basic elution curve equation from which much information concerning the chromatographic process can be educed. Equation (10) is a Poisson function but, in due course, it will be shown that if (n) is large, the function tends to the normal Error function or Gaussian function. In most chromatography systems, (n) >>100 and, thus, most chromatography peaks will be Gaussian or nearly Gaussian in shape. However, the shape of a peak can be distorted in a number of ways which will be discussed in another chapter.

The Retention Volume of a Solute

The retention volume of a solute is that volume of mobile phase that passes through the column between the injection point and the peak maximum. Consequently, by differentiating equation (10), equating to zero and solving for (v), an expression for the retention volume (V_r) can be obtained.

Restating equation (10), $$X_{m(n)} = X_o \frac{e^{-v} v^n}{n!}$$

$$\frac{dX_{m(n)}}{dv} = X_o \frac{-e^{-v} v^n + e^{-v} n v^{(n-1)}}{n!}$$

$$= X_o \frac{-e^{-v} v^{(n-1)}}{n!}(n - v)$$

Equating to zero and solving for (v), then at the peak maximum,

$$n - v = 0 \quad \text{or} \quad v = n$$

It is seen that the peak maximum is reached after (n) *plate volumes* of mobile phase have passed through the column. Thus, the retention volume in **ml of mobile phase** will be obtained by multiplying by the 'plate volume', ($v_m + Kv_s$).

Thus, $$V_r = n(v_m + Kv_s)$$

$$= nv_m + nKv_s$$

The volume of mobile phase per plate (v_m), multiplied by the number of plates (n), will give the total volume of mobile phase in the column (V_m). Similarly, the volume of stationary phase per plate (v_s), multiplied by the number of plates (n), will give the total volume of stationary phase in the column (V_s).

Thus, $$V_r = V_m + KV_s \qquad (11)$$

Control of Chromatographic Retention

It is seen that, as predicted, the function for the retention volume is indeed simple and depends solely on the distribution coefficient and the volumes of the two phases that are present in the column.

In practice, the retention volume of an unretained peak eluted at the dead volume (V_o), will be made up of the volume of mobile phase in the column (V_m) and extra-column volumes, from sample valves, connecting tubes, unions, etc. (V_E).

Thus,
$$V_o = V_{r(o)} + V_E$$
$$= V_m + V_E \qquad (12)$$

If $(V_E) \ll (V_m)$, (V_E) can be ignored, but for accurate measurements, (V_E) should be measured and taken into account.

Thus,
$$V_{r(measured)} = V_m + KV_S + V_E \qquad (13)$$

Again it is seen that only when *second order effects* need to be considered does the relationship become more complicated. The dead volume is made up of many components, and they need not be identified and understood, particularly if the thermodynamic properties of a distribution system are to be examined. As a consequence, the subject of the column dead volume and its measurement in chromatography systems will need to be extensively investigated. Initially, however, the retention volume equation will be examined in more detail.

Returning to equation (13), it is now possible to derive an equation for the adjusted retention volume, (V'_r),
$$V'_r = V_r - V_o$$

Thus, from equations (12) and (13),
$$V'_r = V_m + KV_S + V_E - (V_m + V_E)$$

and
$$V'_r = KV_S \qquad (14)$$

The Conditions for Chromatographic Separation

For two solutes (A) and (B),

$$V_{r(A)} = V_m + K_{(A)}V_S + V_E \qquad (15)$$

$$V_{r(B)} = V_m + K_{(B)}V_S + V_E \qquad (16)$$

The corrected retention volumes will be

$$V'_{r(A)} = K_{(A)}V_S \quad (17)$$

$$V'_{r(B)} = K_{(B)}V_S \quad (18)$$

It follows that for two solutes (A) and (B) to be separated if their retention volumes differ,

$$V_{r(A)} <> V_{r(B)} \quad \text{and} \quad V_{r(A)} \neq V_{r(B)}$$

or

$$V_m + K_{(A)}V_S <> V_m + K_{(B)}V_S \quad \text{and} \quad V_m + K_{(A)}V_S \neq V_m + K_{(B)}V_S$$

i.e., $\quad K_{(A)}V_S <> K_{(B)}V_S \quad \text{and} \quad K_{(A)}V_S \neq K_{(B)}V_S$

Consequently, either $\quad K_{(A)} <> K_{(B)} \quad$ or $\quad V_{S(A)} <> V_{S(B)}$

The extent to which two solutes are separated depends exclusively on the relative magnitudes of their individual distribution coefficients ($K_{(A)}$) and ($K_{(B)}$) and the amount of stationary phase with which they can interact, ($V_{(A)}$) and ($V_{(B)}$). Consequently, for them to be separated, either their distribution coefficients must differ (choose appropriate phase systems), the amount of stationary phase with which they interact must differ (choose a stationary phase with exclusion properties), or a subtle combination of both.

The Capacity Ratio of a Solute

An eluted solute was originally identified from its corrected retention volume which was calculated from its corrected retention time. It follows that the accuracy of the measurement depended on the measurement and constancy of the mobile phase flow rate. To eliminate the errors involved in flow rate measurement, particularly for mobile phases that were compressible, the *capacity ratio* of a solute (k') was introduced. The capacity ratio of a solute is defined as the ratio of its distribution coefficient to the phase ratio (a) of the column, where

$$a = \frac{V_m}{V_s} = \frac{v_m}{v_s}$$

Thus, $\quad k' = \dfrac{KV_s}{V_m} \quad$ and, as $\quad V'_r = KV_S$

Then $\quad k' = \dfrac{V'_r}{V_m} \quad$ In practice, $\quad k' = \dfrac{V'_r}{V_o - V_E} \quad$ (19)

The value of (k'), if not calculated by the data processing computer, is usually measured from the chart as,

$$k' = \frac{\text{Distance between the dead volume peak and the peak maxima}}{\text{Distance between the injection point and the dead volume peak}}$$

This calculation, however, assumes that the extra-column volume (V_E) is negligible. It will be seen later that there are two dead volumes that are effective in chromatography and, thus, there will also be *two* values for the *capacity ratio* of a solute. The two dead volumes, called the thermodynamic and the dynamic dead volumes are significantly disparate and, thus, the corresponding capacity ratios are also quite different. The expression given above gives the thermodynamic dead volume of the solute. The two dead volumes and capacity ratios are used under different circumstances in measuring chromatographic properties. The use of the thermodynamic dead volume and thermodynamic capacity ratio (k') will be discussed later in Part 1 of this book. The use of the dynamic dead volume and the dynamic capacity ratio (k") will be discussed in Part 2 of this book.

In qualitative work, it is clear that some caution must be shown in comparing (k') values for the same solute from different columns and for different solutes on the same column. Both (V_m) and (V_S) will vary between different columns and may, due to the exclusion properties of solid stationary phases and supports, even vary between different solutes. To obtain a parameter that would be largely independent of (V_m) and (V_S) for solute identification, the *separation ratio* was introduced.

The Separation Ratio (Selectivity) of a Solute

The separation ratio of two solutes (A) and (B), ($\alpha_{A/B}$), is taken as the ratio of their corrected retention volumes,*i.e.*,

$$\alpha_{A/B} = \frac{V'_{r(A)}}{V'_{r(B)}} = \frac{K_{(A)} V_s}{K_{(B)} V_s} = \frac{K_{(A)}}{K_{(B)}}$$

It is clear that the *separation ratio* is simply the ratio of the distribution coefficients of the two solutes, which only depend on the operating temperature and the nature of the two phases. More importantly, they are independent of the mobile phase flow rate and the phase ratio of the column. This means, for example, that the same separation ratios will be obtained for two solutes chromatographed on either a packed column or a capillary column, providing the temperature is the same and the same phase system is employed. This does, however, assume that there are no exclusion effects from the support or stationary phase. If the support or stationary phase is porous, as, for example, silica gel or silica gel based materials, and a pair of solutes differ in size, then the stationary phase available to one solute may not be available to the other. In which case, unless both stationary phases have exactly the same pore distribution, if separated on another column, the separation ratios may not be the same, even if the same phase system and temperature are employed. This will become more evident when the measurement of dead volume is discussed and the importance of pore distribution is considered.

To aid in solute identification, a standard substance is usually added to a mixture and the separation ratio of the solutes of interest to the standard is used for identification purposes. In practice, separation ratios are calculated as the ratio of the distances in centimeters between the dead point and the maximum of each peak. If the flow rate is sufficiently constant and data processing is employed, then the corresponding retention times can be used.

The Compressibility of the Mobile Phase

By measuring the retention volume of a solute, the distribution coefficient can be obtained. The distribution coefficient, determined over a range of temperatures, is often used to determine the thermodynamic properties of the system; this will be discussed later. From a chromatography point of view, thermodynamic studies are also employed as a diagnostic tool to examine the actual nature of the distribution. The use of thermodynamics for this purpose will be a subject of discussion in the next chapter. It follows that the accurate measurement of (V'_r) can be extremely important and that those factors that impinge on the accurate measurement of retention volume need to be considered. The retention volume of a solute (from equation (11)) is given by

$$V_r = V_m + KV_S \qquad \text{Thus,} \qquad K = \frac{V_r - V_m}{V_S}$$

Control of Chromatographic Retention

If the mobile phase is a liquid, and can be considered incompressible, then the volume of the mobile phase eluted from the column, between the injection and the peak maximum, can be easily obtained from the product of the flow rate and the retention time. For more precise measurements, the volume of eluent can be directly measured volumetrically by means of a burette or other suitable volume measuring vessel that is placed at the end of the column. If the mobile phase is compressible, however, the volume of mobile phase that passes through the column, measured at the exit, will no longer represent the true retention volume, as the volume flow will increase continuously along the column as the pressure falls. This problem was solved by James and Martin [3], who derived a correction factor that allowed the actual retention volume to be calculated from the retention volume measured at the column outlet at atmospheric pressure, and a function of the inlet/outlet pressure ratio. This correction factor can be derived as follows.

Consider the column depicted in Figure 3.

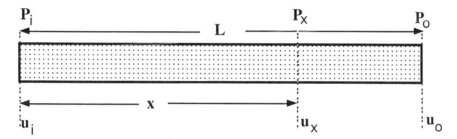

Figure 3. Pressure and Velocity Distribution Along a Packed Column

The column may be packed or it may be an open tube but in this example, a packed column will be specifically considered. The column is considered to have a length (L) and inlet and outlet pressures and inlet and outlet velocities of (P_i), (P_o), (u_i) and (u_o), respectively. The pressure and velocity at a distance (x) from the front of the column is (P_x) and (u_x), respectively. According to D'Arcy's equation for fluid flow through a packed bed, at any point in the column,

$$u_x = -\frac{K}{\eta}\frac{dp}{dx} \qquad (20)$$

where (η) is the viscosity of the gas,

and (K) is a constant.

(The D'Arcy equation is the accepted equation for the flow of fluid through a packed bed and its derivation will be found in most physics textbooks. If a capillary column is considered, then the Poiseuille's equation for fluid flow through an open tube would be used). Now, the mass of mobile phase passing any point in the column must be conserved (*i.e.,* the mass of fluid passing any point in the column per unit time is constant) and, thus, the product of pressure and flow rate will be constant at all points along the length of the column. Consequently, under isothermal conditions,

$$P_i Q_i = P_o Q_o$$

where (Q_i) is the volume flow of mobile phase into the column at (P_i),
and (Q_o) is the flow of mobile phase from the column at (P_o).

Now, at a position (x) along the column,

$$Q_x = a_f u_x$$

where (a_f) is the cross-sectional area of the column available for gas flow and is assumed constant for a well-packed column.

Thus, $\quad a_f P_i u_i = a_f P_x u_x = a_f P_o u_o \quad$ and $\quad P_o Q_o = a_f P_x u_x$

Substituting for (u_x) from equation (20),

$$P_o Q_o = -a_f P_x \frac{K}{\eta} \frac{dp}{dx}$$

or

$$P_o Q_o dx = -a_f P_x \frac{K}{\eta} dp$$

Integrating from x = 0 to x = x and (P_i) to (P_x), $\quad P_o Q_o \int_0^x dx = -a_f \frac{K}{\eta} \int_{P_i}^{P_x} P\, dp$

$$P_o Q_o x = a_f \frac{K}{\eta} (P_i^2 - P_x^2)$$

When x = L, then $P_x = P_o$. Thus,

$$P_o Q_o L = a_f \frac{K}{\eta} (P_i^2 - P_o^2)$$

Then by simple ratio, $\quad \dfrac{x}{L} = \dfrac{(P_i^2 - P_x^2)}{(P_i^2 - P_o^2)} \quad$ or $\quad \dfrac{x}{L} = \dfrac{(\gamma^2 - \dfrac{P_x^2}{P_o^2})}{(\gamma^2 - 1)}$

Rearranging, $\quad \dfrac{P_x}{P_o} = \left[\gamma^2 - (\gamma^2 - 1)\dfrac{x}{L} \right]^{0.5}$ (21)

where (γ) is $\left(\dfrac{P_i}{P_o}\right)$ the inlet/outlet pressure ratio of the column.

Equation (21) allows the value of the ratio $\left(\dfrac{P_x}{P_p}\right)$ to be calculated along the length of the column for different inlet/outlet pressure ratios. Using equation (21), the values of $\left(\dfrac{P_x}{P_p}\right)$ at different points along the column were calculated and the results shown as curves relating $\left(\dfrac{P_x}{P_p}\right)$ to fraction of the column length are shown in Figure 4.

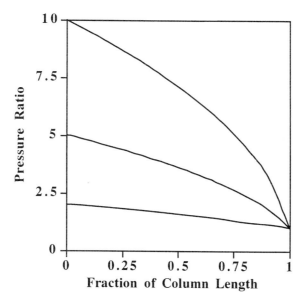

Figure 4. Graph Showing Change of Pressure Along a GC Column for Different Inlet/Outlet Pressure Ratios

It is seen that, at low inlet/outlet pressure ratios (*i.e.*, $\gamma < 2$), the pressure falls almost linearly from the inlet to the outlet of the column. At higher inlet/outlet pressure ratios, (*i.e.*, $\gamma > 2$), however, the relationship is no longer linear and the pressure falls very rapidly in the latter half of the column. It is clear that the average flow rate can not be used in measuring retention volumes, but a special correction factor must be employed. If the retention volume of a solute at the outlet pressure (P_o) is $(V_{r(o)})$ and the true retention volume is (V_r) measured at the *mean column pressure* (P_r), then

$$P_r V_r = P_o V_{r(o)}$$

Thus,
$$V_r = \frac{V_{r(o)} P_o}{P_r} \qquad (22)$$

It is now necessary to derive an expression for (P_r) and then the correction factor $\frac{P_o}{P_r}$ can be evaluated. The average pressure in the column (P_r) is defined as, $P_r = \frac{\int P_x dx}{\int dx}$.

Now it has been shown that
$$dx = -\frac{a_f P_x}{P_o Q_o} \frac{K}{\eta} dp$$

Thus,
$$P_r = \frac{\int -\frac{a_f P_x^2}{P_o Q_o} \frac{K}{\eta} dp}{\int -\frac{a_f P_x}{P_o Q_o} \frac{K}{\eta} dp} = \frac{\int P_x^2 dp}{\int P_x dp}$$

Integrating between (P_i) and (P_o)

$$P_r = \frac{2}{3} P_o \frac{\left(\frac{P_i}{P_o}\right)^3 - 1}{\left(\frac{P_i}{P_o}\right)^2 - 1} \qquad (23)$$

Substituting for (P_r) in equation (22) from equation (23),

$$V_r = \frac{3}{2} V_{r(o)} \frac{\left(\frac{P_i}{P_o}\right)^2 - 1}{\left(\frac{P_i}{P_o}\right)^3 - 1} = V_{r(o)} \frac{3}{2} \left(\frac{\gamma^2 - 1}{\gamma^3 - 1}\right)$$

Thus, the correction factor for retention data in GC is $\frac{3}{2}\left(\frac{\gamma^2 - 1}{\gamma^3 - 1}\right)$. This factor must be used in all accurate retention volume measurements in GC, particularly when using the chromatography data to evaluate the thermodynamic properties of a distribution

Control of Chromatographic Retention

system. It should be noted that the factor depends on the packing being *homogeneous* and the *flow impedance constant* along the column.

Noting the inverse relationship $\frac{u_o}{u} = \frac{P}{P_o}$, equation (21) can also be used to demonstrate the change in mobile phase velocity along a column for different inlet/outlet pressure ratios. A set of such curves calculated using equation (21) is shown in Figure 5. It is seen that the mobile phase velocity increases very rapidly towards the end of the column, even at an inlet/outlet pressure ratio of only two. At inlet/outlet pressures of 10 and 15, the acceleration of the mobile phase in the last quarter of the column length is very great indeed.

In the case of the column with an inlet/outlet pressure ratio of 15, the mobile phase velocity increases by over an order of magnitude in the last 15% of the column length. At first sight, this would appear to indicate that there would be a high degree of dispersion (peak spreading) in the final stages of elution. It will be seen in Part 2 of this book, however, that as a result of a compensating change in *solute diffusivity* at the higher pressures, the increase in dispersion is not nearly as great as the curves in Figure 5 might suggest.

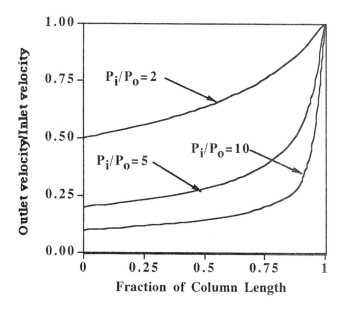

Figure 5. Graph Showing the Change in Velocity Along a Column for Different Inlet/Outlet Pressure Ratios

The Column Dead Volume

The accurate determination of the column dead volume is extremely important when measuring retention data, particularly if the actual corrected retention volume is relatively small. Under these circumstances, the magnitude of the retention volume will be of the same order of magnitude as the dead volume, or considerably less, and so any errors involved in dead volume measurement will cause serious errors in the values taken for the corrected retention volume. The dead volume is made up of a number of different parts of the actual column volume, and the distribution of the total column volume into those parts of chromatographic significance can be very complex. The composition and measurement of the dead volume have been the subject of a number of investigations [4-10]. In the early years of chromatography, the column volumes that were considered important were the interstitial volume (the volume between the particles occupied by the mobile phase), the column pore volume (the volume within the particle occupied by the mobile phase) and the volume of stationary phase in the column. However, in using these basic chromatographic column volumes, certain assumptions were made that were not valid. It was assumed that all the mobile phase in the interstitial volume of the column was moving and none was static, which, although possibly true in GC (particularly if a capillary column is used), is certainly not true in LC when a packed column is employed. In addition, the contents of the pores were assumed to have the same composition as the mobile phase which, when solvent mixtures are employed as the mobile phase in LC, has also been proved not to be so [10]. Finally, it was assumed that all the stationary phase present in the column was available to the solute and, thus, exclusion effects were ignored. The distribution of the total column volume between the different chromatographically critical volumes can be understood by a simple logical breakdown of the column into the individual volumes involved. The column contains three materials, the mobile phase, the stationary phase and the support. However, conventionally it has been assumed that all the mobile phase is mobile, which is obviously not true for that proportion of the mobile phase that is contained in the pores of the support (if there is one) which will be static. However the term *mobile phase* is so well established that it will be used to designate all the phase that is not part of the stationary phase or support and will include the *moving phase* which will be considered as that part of the mobile phase that actually moves.

If the column volume is (V_c) and contains a volume of mobile phase (V_M), a volume of stationary phase (V_S) and a volume of support (*e.g.*,silica) (V_{Si}), then

Control of Chromatographic Retention

$$V_c = V_M + V_S + V_{Si} \tag{25}$$

Recalling the plate theory, it must be emphasized that (V_M) is not the same as (V_m). (V_m) is the moving phase and a significant amount of (V_M) will be static (*e.g.*, that contained in the pores). It should also be pointed out that the same applies to the volume of stationary phase, (V_S), which is not the same as (V_s), which may include material that is unavailable to the solute due to exclusion.

Now, the column volume, assuming it has a circular cross-section, will be given by

$$V_c = \pi r^2 l \tag{26}$$

where (r) is the column radius,
and (l) is the column length.

Hence,
$$V_M + V_S + V_{Si} = \pi r^2 l \tag{27}$$

The individual volumes, (V_M), (V_S) and (V_{Si}), can now be apportioned to the different chromatographic activities.

(V_M) can be initially divided into two parts, that contained in the pores (V_p) and that contained in the interstitial volume between the particles (V_I), thus,

$$V_M = V_I + V_p \tag{28}$$

In addition, the interstitial volume can also be divided into two parts, the interstitial volume that is actually moving $(V_{I(m)})$ and that part of the interstitial volume around the points of contact of the particles that is static $(V_{I(s)})$.

Hence,
$$V_I = V_{I(m)} + V_{I(s)} \tag{29}$$

If, in LC, the mobile phase is a mixture of solvents, the pore contents will not be homogeneous. One solvent component, the one with stronger interactions with the stationary phase, will be preferentially adsorbed on the surface [10] relative to the other. Consequently, although the bulk of the contents the pores, $(V_{p(1)})$, will have the same composition as that of the mobile phase, the pore contents close to the surface, $(V_{p(2)})$, where adsorption is taking place, will have a composition that differs from that of the bulk mobile phase.

Hence,
$$V_p = V_{p(1)} + V_{p(2)} \tag{30}$$

Now substituting for (V_i) and (V_p) from (29) and (30) in (28), a more informative distribution of the total volume of mobile phase becomes apparent,

$$V_M = V_{I(m)} + V_{I(s)} + V_{p(1)} + V_{p(2)} \qquad (31)$$

The stationary phase can be apportioned in a similar manner. For example, with a bonded phase, due to the porous nature of the support, some of the pores will become blocked with stationary phase and so the total amount of stationary phase can be divided into that which is chromatographically *available* ($V_{S(A)}$) and that which is chromatographically *unavailable* ($V_{S(U)}$).

It follows that,
$$V_S = V_{S(A)} + V_{S(U)} \qquad (32)$$

Substituting for (V_M) and (V_S) from equations (31) and (32) in equation (25), an expression for the total column volume can be obtained,

$$V_c = V_{I(m)} + V_{I(s)} + V_{p(1)} + V_{p(2)} + V_{S(A)} + V_{S(U)} + V_{Si} \qquad (33)$$

Now the retention volume ($V_{r(A)}$) of a solute (A) is

$$V_{r(A)} = V_0 + K_A V_S \qquad (34)$$

Equation (34) is generally quite correct and useful. However, if highly accurate retention measurements are important, then second order effects must be taken into account and equation (33) indicates that, for accurate data, equation (34) is grossly over simplified. From equation (33), a more accurate expression for solute retention would be

$$V_r = V_{I(m)} + K V_{I(s)} + K_1 V_{p(1)} + K_2 V_{p(2)} + K_3 V_{S(A)} \qquad (35)$$

where, (K) is the distribution coefficient of the solute between the moving phase and the static portion of the interstitial volume,

(K_1) is the distribution coefficient of the solute between the moving phase and the static pore contents, ($V_{p(1)}$),

(K_2) is the distribution coefficient of the solute between the moving phase and the static pore contents, ($V_{p(2)}$),

and (K_3) is the distribution coefficient of the solute between the moving phase and the available stationary phase ($V_{S(A)}$).

Control of Chromatographic Retention

It is important to realize that all static phases will contribute to retention and, as a result, a number of different distribution coefficients will control the retention of the solute. Nevertheless, the situation can be simplified to some extent. The static interstitial volume ($V_{I(s)}$) and the pore volume fraction ($V_{p(1)}$) will contain mobile phase that has the same composition as the moving phase and, consequently,

$$K = K_1 = 1$$

Thus, equation (35) will be reduced to

$$V_r = V_{I(m)} + V_{I(s)} + V_{p(1)} + K_2 V_{p(2)} + K_3 V_{S(A)} \qquad (36)$$

Equation (36) is certainly more accurate than equation (34) but still does not take into account any exclusion properties that the support may have. In addition, the particles are close-packed and touching so, in the interstitial volume, around the points of contact, some additional solute exclusion will almost certainly take place. The pore size of most silicas can range from 1-3 Å to several thousand angstrom and so can easily partially exclude some of the components of a mixture that cover a wide range of molecular size. It follows that equation (36) must be further modified,

$$V_r = V_{I(m)} + \Psi V_{I(s)} + \Omega V_{p(1)} + \Omega K_2 V_{p(2)} + \xi K_3 V_{S(A)} \qquad (37)$$

> where (Ψ) is that fraction of the static interstitial volume accessible to the solute,
> (Ω) is that fraction of the pore volume accessible to the solute,
> and (ξ) is the fraction of the stationary phase accessible to the solute.

In general, (Ω) and (ξ) will be equal, but the general case is assumed, where they are not. Equation (37) gives an explicit and accurate expression for the retention volume of a solute. The importance of each function in the expression will depend on the physical properties of the chromatographic system. At one extreme, using an open tubular column in GC, then

$$V_{I(s)} = V_p = 0$$

Thus, in this case the simple form of equation (34) is quite adequate. Alternatively, employing a wide pore silica base in LC for separating small molecular weight materials using a single solvent mobile phase, the contents of the pores will be homogeneous so $K_2 = 1$, there will be little or no exclusion from the pores,

Thus, $$\Omega = \xi = 1$$

and $$V_r = V_{I(m)} + \Psi V_{I(s)} + V_p + K_3 V_{S(A)}$$

It is seen that equation (37) must be modified on the basis of the experimental conditions chosen for measurement.

Chromatographic Dead Volumes

As already mentioned, there are two so called "dead volumes" that are important in both theoretical studies and practical chromatographic measurements, namely, the *kinetic dead volume* and the *thermodynamic dead volume*. The kinetic dead volume is used to calculate linear mobile phase velocities and capacity ratios in studies of peak variance. The thermodynamic dead volume is relevant in the collection of retention data and, in particular, data for constructing vant Hoff curves.

In equation (37), for an incompressible mobile phase, the kinetic dead volume is ($V_{i(m)}$) which is the volume of moving phase only. Consequently, at a flow rate of (Q) ml/s, the dead time (to) would be given by,

$$t_o = \frac{V_{i(m)}}{Q}$$

and, thus, the linear velocity (u) of the mobile phase is,

$$u = \frac{1}{t_o} = \frac{lQ}{V_{i(m)}}$$

where (l) is the column length

In equation (37), the thermodynamic dead volume is given by

$$V_{I(m)} + \Psi V_{I(s)} + \Omega V_p(l)$$

It is seen that the expression for the thermodynamic dead volume is more complex than the kinetic dead volume and depends, to a significant extent, on the size of the solute molecule. In common with the kinetic dead volume, it includes the volume of moving phase $V_{I(m)}$, but it also includes that volume of the interstitial volume that is size dependent (Ψ), as well as the volume of pores available to the solute which is also size dependent (Ω). Examination of equation (37) also shows that the major retention factor, ($\xi K_3 V_{S(A)}$), is likewise strongly dependent on the molecular size of the solute, ((ξ) is not unity), and, thus, unless the actual values of (Ψ), (Ω) and (ξ) are known or can be determined, it is not possible to accurately determine the difference between the retention volume of one solute and another. This is particularly true in LC, when using porous stationary phases supported on silica, if the molecular size of the two solutes differ significantly. As an added difficulty, the experimental determination of (Ψ), (Ω) and (ξ), although theoretically possible, can be extremely lengthy and tedious. Equation (37) has important implications for the measurement of the capacity factor (k') of a solute. Using equation (37) the equation for (k') can be seen to be

$$k' = \frac{\left(\begin{array}{c} V_{I(m)} + \psi V_{I(S)} + \Omega V_{p(1)} + \Omega K_2 V_{p(2)} + \\ \xi K_3 V_{S(A)} - V_{I(m)} + \psi V_{I(S)} + \Omega V_{p(1)} \end{array} \right)}{\left(V_{I(m)} + \psi V_{I(S)} + \Omega V_{p(1)} \right)}$$

which simplifies to

$$k' = \frac{\Omega K_2 V_{p(2)} + \xi K_3 V_{S(A)}}{V_{I(m)} + \psi V_{I(S)} + \Omega V_{p(1)}} \quad (38)$$

Equation (38) shows that the measurement of (k') incorporates the same errors as those met in trying to measure the thermodynamic dead volume. However, providing the solute is well retained, *i.e.*,

$$\Omega K_2 V_{p(2)} + \xi K_3 V_{S(A)} \gg V_{I(m)} + \Psi V_{I(s)} + \Omega V_{p(1)}$$

then the corrected retention volume can be used for thermodynamic calculations with greater confidence.

Experimental Examination of Dead Volume Measurement

Alhedai *et al.* [11] carried out a critical examination of a commercially available reverse phase column packing, Zorbax C_8, and measured some of the individual volumes included in equation (37). The exclusion characteristics of the interstitial cavities in the column were determined by measuring the retention volumes of salts having different molar volumes. The salts were ionically excluded from the pores of the packing and, so, only penetrated the interstitial cavities as they passed through the column. The results obtained are shown as a curve relating retention against ion volume in Figure 6.

40   Chromatography Theory

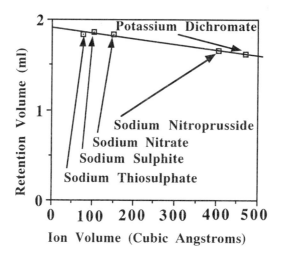

Courtesy of the Analyst (ref.11)

Figure 6. Graph of Retention Volume of a Series of Ions against Their Ionic Volume

It is seen that retention decreases linearly as the volume of the ion increases. It is also interesting to note that the retention, although a function of ion volume, is in no way related to the charge on the ion. The intercept of the curve on the retention volume axis gives a value for the total interstitial volume of the column, which differs only slightly from the retention volume of sodium nitrate. Consequently, the retention volume of sodium nitrate would give a close approximation to the interstitial volume of the column. The slope of the curve shown in Figure 1 clearly indicates the exclusion properties of the interstitial volume are significant.

The authors also investigated the effect of solvent composition on the retention of a series of solutes including a dispersion of silica smoke (mean particle diameter 0.002 μm). The silica smoke was used to simulate a solute of very large molecular size which would be completely excluded in the interstitial volume. The mobile phases used were a series of methanol/water mixtures The results they obtained are shown in Figure 7. The curves show that the retention volumes of the solutes are almost unaffected by the composition of the mobile phase. However, it must be emphasized that as the methanol composition was not examined below 10%v/v, the effect of the adsorption of methanol on the surface of the reverse phase was not evident. Below

Control of Chromatographic Retention

10%v/v the retention volume will be inversely proportional to the methanol concentration in accordance with the Langmuir adsorption isotherm.

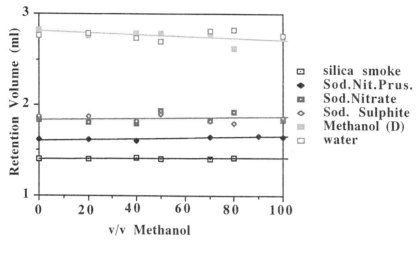

Courtesy of the Analyst (ref.11)

Figure 7. Graph of Retention Volume of a Number of Different Solutes against Composition of the Mobile Phase

The silica 'dispersion' showed the smallest retention volume. It should be noted, however, that the authors reported that the silica dispersion required sonicating for 5 hours before the silica was sufficiently dispersed to be used as "pseudo-solute". The retention volume of the silica dispersion gave the value of the kinetic dead volume, *i.e.*, the volume of the moving portion of the mobile phase. It is clear that the difference between the retention volume of sodium nitroprusside and that of the silica dispersion is very small, and so the sodium nitroprusside can be used to measure the kinetic dead volume of a packed column. From such data, the mean kinetic linear velocity and the kinetic capacity ratio can be calculated for use with the Van Deemter equation [12] or the Golay equation [13].

The thermodynamic dead volume would be that of a small molecule that could enter the pores but not be retained by differential interactive forces. The maximum retention volume was recorded for methanol and water which, for concentrations of methanol above 10%v/v, would be equivalent to the thermodynamic dead volume for small molecules (*viz.* about 2.8 ml). It is interesting to note that there is no significant difference between the retention volume of water and that of methanol over the complete range of solvent compositions examined, which confirms the validity of this method for measuring the thermodynamic dead volume. Bear in mind, however, that the lower concentrations of methanol were *not* examined where the surface area of the stationary phase was not completely covered with methanol and the Langmuir adsorption isotherm would apply; so, the method would not be appropriate for low concentrations of organic solvent. In addition, this method of measuring thermodynamic dead volume will only be valid for small molecules. Larger molecules will be partially excluded and, thus, their dead volumes will be commensurably smaller.

Alhedai *et al.* also examined the exclusion properties of a reversed phase material The stationary phase chosen was a C_8 hydrocarbon bonded to the silica, and the mobile phase chosen was *n*-octane. As the solutes, solvent and stationary phase were all dispersive (hydrophobic in character) and both the stationary phase and the mobile phase contained C_8 interacting moieties, the solute would experience the same interactions in both phases. Thus, any differential retention would be solely due to exclusion and not due to molecular interactions. This could be confirmed by carrying out the experiments at two different temperatures. If any interactive mechanism was present that caused retention, then different retention volumes would be obtained for the same solute at different temperatures. Solutes ranging from *n*-hexane to *n*-hexatriacontane were chromatographed at 30°C and 50°C respectively. The results obtained are shown in Figure 8.

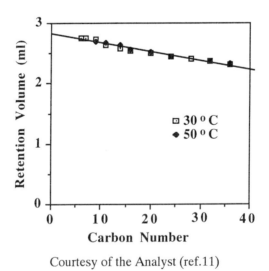

Courtesy of the Analyst (ref.11)

Figure 8. Graph of Retention Volume of n-Alkanes against Carbon Number

Table 2 Summary of the Physical and Chemical Properties of the Zorbax Column

Column Length	25 cm
Column Radius	2.3 mm
Column Packing	Zorbax C_8 Reverse Phase
Carbon Content of Reverse Phase	10.0%w/w
Equivalent Decane Content (Dimethyl Octane)	11.8%w/w
Total Column Volume	4.15 ml
Total Volume of Silica in the Column	0.96 ml
Total Volume of Stationary Phase in the Column	0.40 ml
Volume of Chromatographically Available Stationary Phase	0.49 ml

Volume of Chromatographically Unavailable Stationary Phase (by diff.) – total mobile phase in the column,

1. By weighing	2.79 ml
2. From the retention of the alkanes in n-octane extrapolated to an alkane carbon number of 3	2.78 ml
3. From the volume fraction average of the isotopes retention volume	2.77 ml

Total Interstitial Volume, value extrapolated from the retention volumes of ions of different size	1.91 ml
Interstitial Moving Phase Volume from the retention of 'silica smoke'	1.41 ml
Interstitial Static Phase volume, by difference	0.50 ml
Total Pore Volume. By Difference	0.87 ml
Pore Volume Containing Components of the Mobile Phase Having Composition differing from that of the Moving Phase, by difference	0.18 ml
Pore Volume Containing Mobile Phase of the Same Composition as the Moving Phase, by difference	0.70 ml

It is seen that all the points lie on the same straight line, irrespective of the operating temperature and, thus, the enthalpy term is close to zero and the solutes are not retained by differential molecular forces. Thus, the curve shows the effect of exclusion on the retention of the hydrocarbon solutes. The exclusion effect is significant and, by extrapolating to a carbon number of zero, the total volume of mobile phase in the column is obtained. In this case, the total interstitial volume plus the total pore volume is about 2.9 ml, which agrees well with the data from the retention of water and methanol. The problems associated with exclusion and dead volume measurement cannot be avoided but they can be minimized.

Probably the best compromise for silica based stationary phases is to use corrected retention volume data for solutes eluted at a (k') of greater than 5 and only compare chromatographic data for solutes of approximately the same molecular size.

A summary of the data for the Zorbax column obtained by Alhedai *et al.* [11] is shown in Table 2. It is seen that the distribution of the various chromatographically important volumes within a column is neither simple nor obvious. It would seem that about 70% of the column volume is occupied by mobile phase but only about 50% of that mobile phase is actually moving.

About 18% of the mobile phase is interstitial, but static, and about 31% of the mobile phase is contained within the pores and is also static. Just over 6% of the mobile phase in the pores has a different composition to that of the mobile phase proper and thus constitutes a second stationary phase.

There is good agreement between the different methods of measuring the thermodynamic dead volume. Weighing the column filled with mobile phase and then weighing dry, measuring the retention volume of one pure component of the mobile phase and measuring the retention volume of a completely permeating solute by extrapolation, both give very closely similar values.

The stationary phase constitutes about 12% of the column volume, which is equivalent to only about 17% of the mobile phase content of the column. The values given in Table 2 are probably representative of most reverse phase columns but will differ significantly with extremes of pore size and pore volume.

The primary factors that govern retention are the distribution coefficient (K) and the volume of stationary phase (V_S)). It is now necessary to identify those parameters that control the magnitude of the distribution coefficient itself and the volume of available stationary phase in a column. A study of these factors will be the subject of the next chapter.

Synopsis

In order to obtain an expression for the retention volume of a solute, the equation for the elution curve of the substance must be derived. The elution curve equation is obtained from the plate theory, which assumes the column consists of a number of theoretical plates which are of such a size that equilibrium can be assumed to occur between the solute and the two phases in each plate. A mass balance is applied to a plate and, from this, the differential equation for the change of concentration of the solute in the plate, with the flow of mobile phase through it, is obtained. The integration of this differential equation provides the elution curve equation. Differentiating the elution equation and equating to zero discloses the expression for the retention volume of a solute, which is shown to depend on the distribution coefficient of the solute between the two phases and the volumes of stationary and mobile phase in the column. Having obtained the retention volume of a pair of solutes, their separation will depend on the relative magnitudes of their distribution coefficients with respect to the stationary phase and the relative amount of stationary phase available to the two solutes. If the mobile phase is compressible, then a factor must be applied to the retention volume as measured at the column outlet and this correction factor involves the inlet/outlet pressure ratio. The column dead volume is complex, involving a number of chromatographically pertinent volumes; those of the mobile, stationary, moving and static phases, the interstitial and pore volumes and volumes of available and unavailable stationary phase. It follows that for accurate determination of retention volume, these different volumes must be measured and taken into account.

References

1. A. S. Said, *Am. Inst. Chem. Eng. J.*, **2**(1956)477.
2. A. J. P. Martin and R. L. M. Synge, *Biochem. J.*, **35**(1941)1358.
3. A. T. James and A. J. P. Martin, *Biochem. J.*, **50**(1952)579.
4. H. Engelhardt, H. Muller and B. Dreyer, *Chromatographia*, **19**(1984)240.
5. P. L. Zhu, *Chromatographia*, **20**(1985)425.
6. J. H. Knox and R. J. Kaliszan, *J. Chromatogr.*, **349**(1985)211.
7. R. J. Smith and C. S. Nieass, *J. Liq. Chromatogr.*, **9**(1986)1387.
8. R. P. W. Scott and P. Kucera, *J. Chromatogr.*, **149**(1978)93.
9. R. M. McCormick and B. L. Karger, *Anal. Chem.*, **52**(1980)2249.
10. R. P. W. Scott and C. F. Simpson, *Faraday Symp. Chem. Soc.*, **15**(1980)69.
11. A. Alhedai, D. E. Martire and R. P. W. Scott, *Analyst*, **114**(1989)869.

12. J. J. Van Deemter, F. J. Zuiderweg and A. Klinkenberg, *Chem. Eng. Sci.*, **5**(1956)271.
13. M. J. E. Golay, in *"Gas Chromatography 1958,"* (Ed. D. H. Desty), Butterworths, London (1958)36.

Chapter 3

The Distribution Coefficient and Its Control of Solute Retention

The distribution coefficient is an equilibrium constant and, therefore, is subject to the usual thermodynamic treatment of equilibrium systems. By expressing the distribution coefficient in terms of the standard free energy of solute exchange between the phases, the nature of the distribution can be understood and the influence of temperature on the coefficient revealed. However, the distribution of a solute between two phases can also be considered at the molecular level. It is clear that if a solute is distributed more extensively in one phase than the other, then the interactive forces that occur between the solute molecules and the molecules of that phase will be greater than the complementary forces between the solute molecules and those of the other phase. Thus, distribution can be considered to be as a result of differential molecular forces and the magnitude and nature of those intermolecular forces will determine the magnitude of the respective distribution coefficients. Both these explanations of solute distribution will be considered in this chapter, but the classical thermodynamic explanation of distribution will be treated first.

The Thermodynamic Treatment of Solute Distribution in Chromatography

Classical thermodynamics gives an expression that relates the equilibrium constant (the distribution coefficient (K)) to the change in *free energy* of a solute when transferring from one phase to the other. The derivation of this relationship is fairly straightforward, but will not be given here, as it is well explained in virtually all books on classical physical chemistry [1,2].

The expression is as follows,

$$RT \ln(K) = -\Delta G^o$$

where (R) is the gas constant,
(T) is the absolute temperature,
and (ΔG^o) is the standard free energy change.

Now, classical thermodynamics gives another expression for the standard free energy which separates it into two parts, the standard free enthalpy and the standard free entropy.
Thus,

$$\Delta G^o = \Delta H^o - T\Delta S^o$$

where (ΔH^o) is the standard enthalpy change,
and (ΔS^o) is the standard entropy change.

The physical significance of the enthalpy and entropy terms in chromatography needs to be explained. The enthalpy term represents the energy involved when the solute molecule interacts, by electrical forces, with the stationary phase (the nature of molecular interactions will be discussed later). However, when the solute interacts with the stationary phase it also suffers a change in freedom of movement and, thus, can no longer move in the same random manner. This change in entropy is more apparent on a molecular scale when the distribution of a solute between a gas and a liquid is considered, as in GC. In the gas, the solute molecules have high velocities and can travel in any direction. However, when in the liquid phase, they are held by interacting molecular forces to the molecules of stationary phase and can no longer travel through the phase at high velocities or with the same directional freedom of movement. This new *motion restriction* is measured as an *entropy change*. Thus, the free energy change is made up of an actual energy change resulting from the intermolecular forces between solute and stationary phase and an entropy change that reflects the resulting restricted movement, or loss in randomness, of the solute while preferentially interacting with the stationary phase.

Continuing,
$$\ln K = -\left(\frac{\Delta H^o}{RT} - \frac{\Delta S^o}{R}\right) \qquad (1)$$

or
$$K = e^{-\left[\frac{\Delta H^o}{RT} - \frac{\Delta S^o}{R}\right]}$$

The Distribution Coefficient

Equation (1) can be viewed in an over-simplistic manner and it might be assumed that it would be relatively easy to calculate the retention volume of a solute from the distribution coefficient, which, in turn, could be calculated from a knowledge of the standard enthalpy and standard entropy of distribution. Unfortunately, these properties of a distribution system are *bulk* properties. They represent, in a single measurement, the net effect of a large number of different types of molecular interactions which, individually, are almost impossible to separately identify and assess quantitatively.

It follows that although the thermodynamic functions can be measured for a given distribution system, they can not be predicted before the fact. Nevertheless, the thermodynamic properties of the distribution system can help explain the characteristics of the distribution and to predict, quite accurately, the effect of temperature on the separation.

Rearranging equation (1),
$$\ln K = -\frac{\Delta H^o}{RT} + \frac{\Delta S^o}{R}$$

Bearing in mind,
$$V'_r = KV_s \quad \text{and} \quad k = \frac{KV_s}{V_m} = \frac{K}{a}$$

$$\ln(V'_r) = -\frac{\Delta H^o}{RT} + \frac{\Delta S^o}{R} + \ln(V_s)$$

$$\ln(k') = -\frac{\Delta H^o}{RT} + \frac{\Delta S^o}{R} - \ln(a)$$

It is clear that a graph of $\ln(V'_r)$ or $\ln(k')$ against $1/T$ will give straight line. This line will provide actual values for the standard enthalpy (ΔH^o), which can be calculated from the slope of the graph and the standard entropy (ΔS^o), which can be calculated from the intercept of the graph. These types of curves are called van't Hoff curves and their important characteristic is that they will always give a linear relationship between $\ln(V'_r)$ and $(1/T)$. However, it is crucial to understand that the distribution system must remain *unchanged* throughout the temperature range. If the distribution system changes, then a graph relating $\ln(V'_r)$ against $(1/T)$ will *not be linear* and the curves *will not be* van't Hoff curves. This situation will be discussed in more detail later. It is also seen that there can be basically two types of graph. The first type, grossly exaggerated, is depicted in Figure 1.

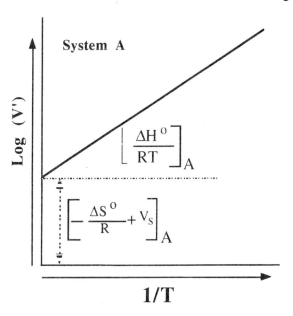

Figure 1 An Energy Driven Distribution

It is seen that distribution system (A) has a very large enthalpy value $\left[\dfrac{\Delta H^\circ}{RT}\right]_A$ (the slope of the curve is steep) but conversely a very low entropy contribution $\left[-\dfrac{\Delta S^\circ}{R}+V_s\right]_A$ (the intercept is relatively small). The large value of $\left[\dfrac{\Delta H^\circ}{RT}\right]_A$ means that the distribution in favor of the stationary phase is dominated by *molecular forces*. The solute is preferentially distributed in the stationary phase because the interactions of the solute molecules with those of the stationary phase are much greater than the interactive forces between the solute molecules and those of the mobile phase. It follows that the change in standard enthalpy is the major contribution to the change in standard free energy and, thus,

The distribution, in thermodynamic terms, is said to be "energy driven."

The second type of van't Hoff curve, also grossly exaggerated, is depicted in **Figure 2**.

It is seen that the distribution system (B), shown in Figure 2, is a completely different type. In this distribution system, there is only a very small enthalpy change $\left[\dfrac{\Delta H^\circ}{RT}\right]_B$ but, in this case, a very high entropy contribution $\left[-\dfrac{\Delta S^\circ}{R}+V_s\right]_B$.

The Distribution Coefficient

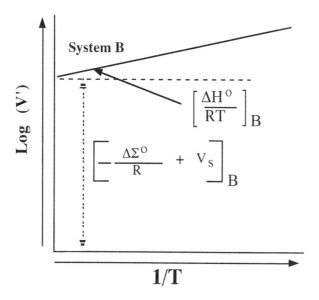

Figure 2. An Entropically Driven Distribution

This means that the distribution is *no longer* dominated by molecular forces. The entropy change is a measure of the loss of randomness or freedom that a solute molecule possesses when transferring from one phase to the other. The more random and 'more free' the solute molecule is to move in a particular environment, the greater its entropy in that environment. A large entropy change, similar to that shown in system (B), indicates that the solute molecules are more restrained with less random movement in the stationary phase than they were in the mobile phase. This loss of freedom is responsible for the greater distribution of the solute in the stationary phase and, consequently, greater solute retention. Because the change in entropy in system (B) is the major contribution to the change in free energy,

The distribution, in thermodynamic terms, is said to be "entropically driven".

Examples of entropically driven separations are chiral separations and separations that are dominated by size exclusion. However, it must be emphasized that chromatographic separations can not be exclusively "energetically driven" or "entropically driven"; but will always contain both components. It is by the careful adjustment of both "energetic" and "entropic" components of the distribution that very difficult and subtle separations can be accomplished.

Returning to the nature of van't Hoff curves, there are some cases in the literature that purport to show nonlinear van't Hoff curves. This, in fact, is a contradiction in terms. For example a graph relating log(V'$_r$) against 1/T for solutes eluted by a mixed solvent from a reverse bonded phase will sometimes be nonlinear. However, graphs relating log(V'$_r$) against 1/T can only be termed van't Hoff curves if they apply to an *established and constant distribution system,* where the nature of both phases does not change with temperature. If a mixed solvent system is employed as the mobile phase in liquid chromatography, the *solvents themselves* are distributed between the two phases *as well as* the solute. Depending on the actual concentration of solvent, as the temperature changes, so may the relative amount of solvent adsorbed on the stationary phase surface, and so the *nature of the distribution system* will also *change.* Consequently, the curves relating log(V'$_r$) against 1/T may not be linear and, as the distribution system is varying, the curves will *not constitute* van't Hoff curves. This effect has been known for many years and an early example is afforded in some work carried out by Scott and Lawrence [3] in 1969. Scott and Lawrence investigated the effect of water vapor as a moderator on the surface of alumina in the gas/solid separations of some *n*-alkanes. Examples of the results obtained by those authors are shown in Figure 3.

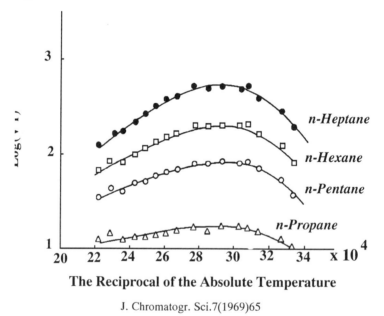

J. Chromatogr. Sci.7(1969)65

Figure 3. Graphs of Log(V'r) against 1/T for Some *n*-Alkanes Separated on Water-Vapor-Moderated Alumina

The alumina column was moderated by a constant concentration of water vapor in the carrier gas. As the temperature of the distribution system was increased, less of the water moderator was adsorbed on the surface. As a consequence, the alumina became

The Distribution Coefficient

more active. Thus, initially as the temperature was raised, the retention volume of each solute increased. When all the water was desorbed and the alumina surface assumed a constant interactive character, the retention volume began to fall again in the expected manner. This effect resulted in the curves shown in Figure 3. The same situation can occur in LC but the explanation is somewhat more complicated. Consider an ethanol/water mixture as the mobile phase in equilibrium with a hydrocarbon reversed phase. Ethanol is adsorbed onto the surface of a reverse phase according to the Langmuir isotherm and the surface is saturated at an ethanol concentration of about 25%v/v. Thus, for solvent mixtures in excess of 25%v/v of ethanol, the surface is covered by the solvent and constitutes a stable uniform stationary phase with respect to temperature. It follows that a plot of log(V'r) against 1/T for any solute would probably be linear if the solvent concentration is in excess 25%v/v. However, if the ethanol concentration is less than 25%v/v, then the surface will be only partly covered with solvent. Consequently, the surface coverage will change with temperature as the distribution coefficient of the *solvent* between the mobile phase and the reverse phase surface changes with temperature. Thus, as the properties of the surface will be temperature dependent, the distribution system changes with temperature and a graph relating log(V'r) with 1/T for any given solute will *not* be linear. It follows that thermodynamic measurements made with solvent mixtures in LC must be undertaken with caution and only made under conditions where the *physical nature* of the distribution system is proved to be independent of temperature.

In the majority of distribution systems met in gas chromatography, the slopes of the van't Hoff curves are positive and the intercept negative. The negative value of the intercept indicates that the standard entropy change of the solute results from the production of a less random and more orderly system during the process of distribution. However, from a chromatography point of view, this means that the entropy change *reduces* the magnitude of the distribution coefficient and thus also reduces the retention.

Summarizing, the greater the forces between the molecules, the greater the energy (enthalpy) contribution, the larger the distribution coefficient, and the **greater** the retention. Conversely, any reduction in the random nature of the molecules or any increase in the amount of order in the system reduces the distribution coefficient and *attenuates* the retention. In chromatography, the standard enthalpy and standard entropy oppose one another in their effects on solute retention. Experimentally it has been shown that there is considerable parallelism between the magnitude of the standard entropy and standard enthalpy for a series of solutes chromatographed on a given distribution system. This relationship is to be expected, but will be discussed in more detail later in this chapter.

The Distribution of Standard Free Energy

The standard free energy can be divided up in two ways to explain the mechanism of retention. First, the portions of free energy can be allotted to specific types of molecular interaction that can occur between the solute molecules and the two phases. This approach will be considered later after the subject of molecular interactions has been discussed. The second requires that the molecule is divided into different parts and each part allotted a portion of the standard free energy. With this approach, the contributions made by different parts of the solvent molecule to retention can often be explained. This concept was suggested by Martin [4] many years ago, and can be used to relate molecular structure to solute retention. Initially, it is necessary to choose a molecular group that would be fairly ubiquitous and that could be used as the first building block to develop the correlation. The methylene group (CH_2) is the first and obvious choice for simple aliphatic substances, as it is a common group and (as will be seen later) can only interact dispersively with any stationary phase. This choice is substantiated by the relationship between the $\ln(V'r)$ and number of methylene groups in the molecule for a range of quite different homologous series as shown below.

Thermodynamic Analysis of Some Homologous Series

Extending the concept that the standard free energy (ΔG^o) can be divided into different portions that represent the standard free energy of different parts of a molecule, consider the distribution of the standard free energy throughout an *n*-alkane molecule, allotting a portion to each methylene group and to the two methyl groups. Then, algebraically, this concept can be put in the following form.

$$RT\ln(V'_r(T)) = -n\Delta G^o(\text{Methylene Group}) - m\Delta G^o(\text{Methyl Group})$$

ΔG^o(Methylene Group) is standard free energy of the methylene group,
ΔG^o(Methyl group) is standard free energy of the methyl group,
(n) is the number of methylene groups,
and, (m) is the number of methyl groups (m=2 for an *n*-alkane).

The Distribution Coefficient

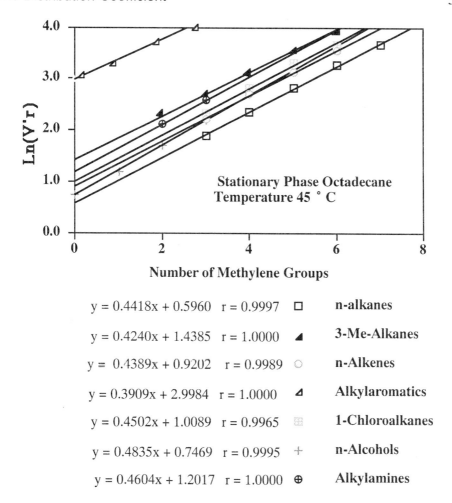

y = 0.4418x + 0.5960	r = 0.9997	□ n-alkanes
y = 0.4240x + 1.4385	r = 1.0000	▲ 3-Me-Alkanes
y = 0.4389x + 0.9202	r = 0.9989	○ n-Alkenes
y = 0.3909x + 2.9984	r = 1.0000	△ Alkylaromatics
y = 0.4502x + 1.0089	r = 0.9965	▨ 1-Chloroalkanes
y = 0.4835x + 0.7469	r = 0.9995	+ n-Alcohols
y = 0.4604x + 1.2017	r = 1.0000	⊕ Alkylamines

Figure 4. Graph of Log(V'r) against Number of Methylene Groups for Different Aliphatic Series

In fact, this procedure can be used for any aliphatic series such as alcohols, amines, etc. Consequently, before dealing with a specific homologous series, the validity of using the methylene group as the reference group needs to be established. The source of retention data that will be used to demonstrate this procedure is that published by Martire and his group [5-10] at Georgetown University and are included in the thesis of many of his students. The stationary phases used were all *n*-alkanes and there was extensive data available from the stationary phase *n*-octadecane. The specific data included the specific retention volumes of the different solutes at 0°C ($V'_{r(T_0)}$) and thus, ($V'_{r(T)}$) was calculated for any temperature (T_1) as follows,

$$V'_{r(T_1)} = V'_{r(T_0)} \rho_{T_1} \frac{T_1}{T_0}$$ where (ρ_{T_1}) is the density of the stationary phase at (T_1)

The relationship between $\log(V'_r(T))$ and the number of methylene groups in a molecule for a range of different solute types is shown in Figure 4. It is seen that the slopes (which are proportional to the contribution of each methylene group to the total standard free energy) are very similar for all the series, whereas the intercepts (standard free energy contributions from other groups and atoms) differ considerably. If the *slopes* are averaged and the *average* values taken, together with the appropriate *number of methylene groups* and these values are used in conjunction with the *actual* values for the *intercepts*, theoretical values for $\log(V'r(T))$ can be calculated.

These calculated values for $\log(V'r(T))$ are shown plotted against the experimentally measured values in Figure 5.

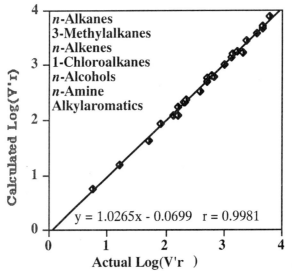

A constant average value is taken for the contribution of each methylene group

Figure 5. Graph of Log(V'r) against Number of Methylene Groups for Different Aliphatic Series

It is seen that by taking a mean value for the slope, there is very little divergence between the calculated and experimental values. Consequently, the methylene groups can, indeed, be taken as a reference group for assessing the effect of molecular structure on solute retention. The concept will now be applied to a simple *n*-alkanes series as discussed above, the data for which was obtained on the stationary phase *n*-heptadecane.

The Distribution Coefficient

This procedure will show how it is possible to identify the difference between the contribution of the methylene group and methyl group to solute retention and to show how any differences that occur might be explained. The curves for log(V'r(T)) against the number of methylene groups in each of the three *n*-alkanes for seven different temperatures are shown in Figure 6.

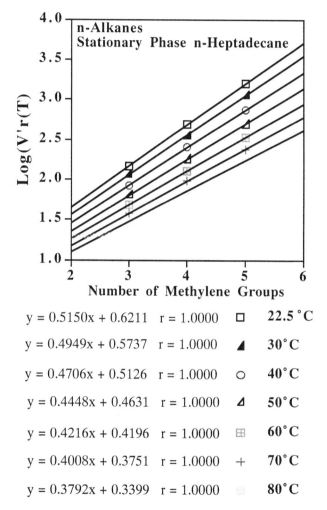

y = 0.5150x + 0.6211	r = 1.0000	□	22.5 °C
y = 0.4949x + 0.5737	r = 1.0000	▲	30 °C
y = 0.4706x + 0.5126	r = 1.0000	○	40 °C
y = 0.4448x + 0.4631	r = 1.0000	△	50 °C
y = 0.4216x + 0.4196	r = 1.0000	▨	60 °C
y = 0.4008x + 0.3751	r = 1.0000	+	70 °C
y = 0.3792x + 0.3399	r = 1.0000		80 °C

Figure 6. Graph of Log(V'r(T)) against Number of Methylene Groups for a Series of *n*-Alkanes

The expected straight lines are produced at each temperature with the index of determination very close to unity. Bearing in mind that the *slope* represents the portion of the standard free energy from each methylene group and each *intercept* represents the portion from *two* methyl groups, the contribution of a single methyl group is seen to be significantly less than that from one methylene group. The slopes (the free energy contribution of one methylene group, $\dfrac{\Delta G°_{\text{(Methylene Group)}}}{RT}$ and half the intercept (the contribution from one terminal methyl groups, $\dfrac{\Delta G°_{\text{(Methyl Group)}}}{RT}$) can now be plotted against the reciprocal of the absolute temperature to obtain the standard enthalpy and standard entropy of the interaction of the methyl and methylene groups with the stationary phase. The curves obtained are shown in Figure 7.

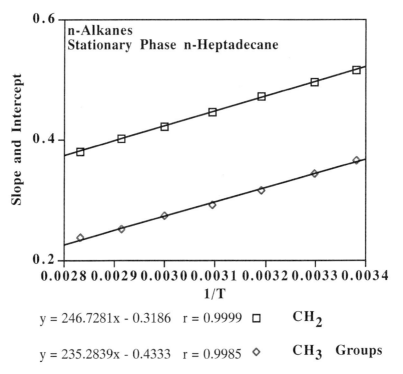

Figure 7 Graph of Intercept and Slope from [Log(V'$_r$(T))/Number of CH$_2$ Groups Curves] for a Series of *n*-Alkanes/ the Reciprocal of the Absolute Temperature

The curves provide relative values for the standard enthalpy ($\dfrac{\Delta H°_{\text{(Methylene Group)}}}{R}$ and $\dfrac{\Delta H°_{\text{(Methyl Group)}}}{R}$) and standard entropy ($\dfrac{\Delta S°_{\text{(Methylene Group)}}}{R}$ and $\dfrac{\Delta S°_{\text{(Methyl Group)}}}{R}$) of distribution for each group and the magnitude of which indicates the manner in which they interact with the stationary phase.

It is seen that despite the contribution of a methyl group to the free energy being much less than that of the methylene group, the energies of interaction of the two

The Distribution Coefficient

groups are very similar. However, the entropy term for the methyl group is nearly 50% greater than that of the methylene group and, as this acts in opposition to standard enthalpy contribution, it reduces the free energy associated with the methyl group by about 30% relative to that of the methylene group. This entropy difference between the two groups is due to the methylene group being *situated in a chain* (more rigidly held) and has, initially, a much lower entropy before solution in the stationary phase. In contrast, the methyl group, *situated at the end of the chain*, is much less restricted and thus, on interaction with the stationary phase molecules, the entropy change is much greater. It follows that the introduction of a methylene group into a solute molecule will increase its retention more than the introduction of a methyl group due to the greater change in entropy associated with the methyl group.

The alkanes are usually ideal in their general physical chemical behavior and it is important to examine other homologous series to ensure the validity of the argument. Data from the same source is available for a homologous series of aliphatic amines, alcohols, halogenated hydrocarbons etc. that can all be treated in the same way. Too many examples given here would be tedious, but perhaps one further example of the chlorinated hydrocarbons might be both interesting and convincing.

Retention data for the *n*-chloroalkanes was available over a limited temperature range of 76°C to 88°C on a *n*-C30 alkane stationary phase. The same procedure is used values for $Log(V'_r(T))$ for the solutes *n*-chlorobutane, *n*-chloropentane, *n*-chlorohexane and *n*-chloroheptane are plotted against the number of methylene groups at each temperature. The slope (representing the methylene contribution $(\frac{\Delta G°_{(CH_2)}}{RT})$, the intercept for the methyl group $(\frac{\Delta G°_{(CH_3)}}{RT})$ (taken from data for the *n*-alkane series which was also available on the same stationary phase and temperature range), and the intercept due to the interaction with the chlorine atom, $\left(\frac{\Delta G°_{(Cl)}}{RT}\right)$ (taken as the difference between the actual intercept and that of the methyl group) were each plotted against the reciprocal of the absolute temperature. The results are shown in Figure 8 and the difference between the methyl group, th methylene group, and the chlorine atom is clearly demonstrated. The enthalpy and entropy values for the methylene group are again very close to those obtained from the *n*-alkane series. As would be expected, the chlorine atom has both a higher enthalpy term and a higher entropy term than the methylene group.

60 Chromatography Theory

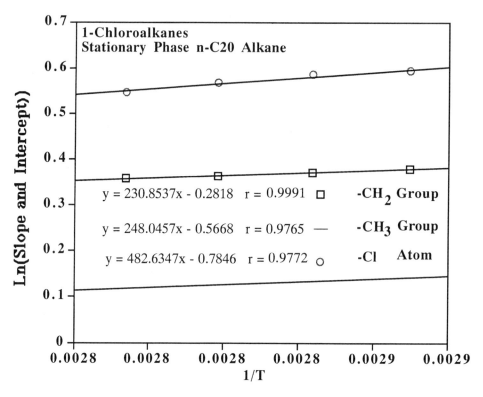

Figure 8. Graph of Intercept and Slope from [Log(V'$_r$(T))/Number of CH$_2$ Groups Curves] for a Series of 1-Chlorohydrocarbons against the Reciprocal of the Absolute Temperature

The high enthalpy contribution results from its larger mass and size providing stronger interactions with the stationary phase molecules, and its increased entropy contribution arises from it being a terminal *atom*, thus prior to interaction with the stationary phase, it has greater freedom.

The contribution of the methylene group and the chlorine atom can be calculated from the enthalpy and entropy values given in Figure 8 (cf. $\frac{\Delta G^\circ_{(Cl)}}{RT} = 0.6084$, c.f. $\frac{\Delta G^\circ_{(CH_2)}}{RT} = 0.3789$, calculated at 76°C.) The free energy contribution of one chlorine atom being nearly equivalent to 2 methylene groups.

It would follow that halogenated hydrocarbons would provide much stronger dispersive interactions than a methylene group, a fact that has been well established in both GC and LC. In conclusion it can be said,

The Distribution Coefficient

Different portions of the standard free energy of distribution can be allotted to different parts of a molecule and, thus, their contribution to solute retention can be disclosed. In addition, from the relative values of the standard enthalpy and standard entropy of each portion or group, the manner in which the different groups interact with the stationary phase may also be revealed.

Finally, another important and interesting fact is established from the data treated in this manner. All the examples given confirm that the standard entropy term tends to increase with the standard enthalpy term. Consequently, the increase in retention is not as great as that which would be expected from the increase in standard enthalpy alone.

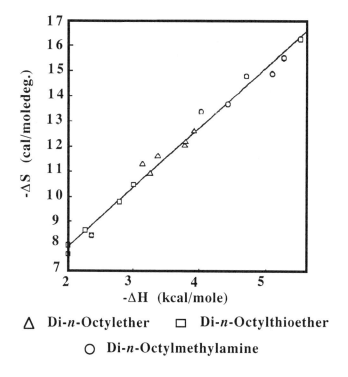

△ Di-*n*-Octylether   □ Di-*n*-Octylthioether

○ Di-*n*-Octylmethylamine

Figure 9. Graph of Standard Free Entropy against Standard Free Enthalpy for an Ether, Thioether and Amine

This relationship between entropy and enthalpy has been reported many times in the literature. An example of a graph relating (ΔH^o) to (ΔS^o), produced by Martire and his group [8], is shown in Figure 9. From a theoretical point of view, this relationship between standard enthalpy and standard entropy is to be expected. An increase in enthalpy indicates that more energy is used up in the association of the solute molecule with the molecules of the stationary phase. This means that the intermolecular forces are *stronger* and, thus, the stationary phase molecules hold the solute molecules *more tightly*. In turn, this indicates that the *freedom of movement* and the random nature of the solute molecule are also *more restricted*, which results in a larger change in standard entropy. It follows that, unless other significant retentive factors are present, any increase in standard enthalpy is usually accompanied by a corresponding increase in standard entropy.

Molecular Interactions and Their Influence on the Magnitude of the Distribution Coefficient (K)

The relative affinity of a solute for the two phases in a chromatographic system determines the magnitude of the distribution coefficient. As a result, for a solute to be retained, the *stationary phase* must be chosen to have *strong* interactions with the solute and, conversely, the *mobile phase* must be chosen to have relatively *weak* interactions with the solute (*i.e.*, the distribution coefficient must be large). In GC, the probability of collision between a solute molecule and a gas molecule is much less than that in a liquid and, in addition, when collisions do occur, any interactions that take place are relatively weak. Thus, the choice of carrier gas has minimal effect on solute retention or selectivity, but can be very important with respect to solute dispersion or the effective operation of the detector. In contrast, interactions between the solute molecules and those of the stationary phase will be much more frequent (a higher probability of collision due to the greater density, *i.e.*, a larger number of molecules per unit volume). In addition, the interactions with the stationary phase will be stronger because the molecules are much larger and can contain a variety of interacting groups. It is clear that, in GC, retention and selectivity are only determined by the choice of the *stationary phase*.

In LC, however, the mobile phase is a liquid, and can also provide strong interactions with the solute, which will reduce the magnitude of the distribution coefficient and, consequently, reduce retention. Thus, the choice of the phase system in LC can be more complex. The stronger interactions must still dominate in the stationary phase if the solute is to be retained and, so, the phases must be chosen on the basis of the type of interactions they can offer relative to the type of interactions that can occur with the solute. For example, the stationary phase would be chosen to have chemical groups that interact strongly with those of the solute, whereas the mobile phase would be chosen to contain groups that interact only weakly with the solute groups. This would allow the stronger interactions to reside in the stationary phase and the solute to be retained. The choice of phases on the basis of the types of

The Distribution Coefficient

interaction that can take place will become more apparent after the different types of interaction have been discussed.

However, at this time it is important to understand that there are three properties involved in a molecular interaction and they are the *type* of interaction, the *strength* of the interaction and the *probability* of it occurring. The first will now be discussed and the remainder will be considered in the next chapter.

Molecular Interactions

Molecular interactions are the result of *intermolecular forces* which are all electrical in nature. It is possible that other forces may be present, such as gravitational and magnetic forces, but these are many orders of magnitude weaker than the electrical forces and play little or no part in solute retention. It must be emphasized that there are three, and *only* three, different *basic* types of intermolecular forces, *dispersion forces*, *polar forces* and *ionic forces*. All molecular interactions must be composites of these three basic molecular forces although, individually, they can vary widely in strength. In some instances, different terms have been introduced to describe one particular force which is based not on the type of force but on the strength of the force. Fundamentally, however, there are only three basic types of molecular force.

Dispersion Forces

London [11] was the first to describe dispersion forces, which were originally termed 'London's dispersion forces'. Subsequently, London's name has been eschewed and replaced by the simpler term 'dispersion' forces. Dispersion forces ensue from charge fluctuations that occur throughout a molecule that arise from electron/nuclei vibrations. They are random in nature and are basically a statistical effect and, because of this, a little difficult to understand. Some years ago Glasstone [12] proffered a simple description of dispersion forces that is as informative now as it was then. He proposed that,

"although the physical significance probably cannot be clearly defined, it may be imagined that an instantaneous picture of a molecule would show various arrangements of nuclei and electrons having dipole moments. These rapidly varying dipoles, when averaged over a large number of configurations, would give a resultant of zero. However, at any instant, they would offer electrical interactions with another molecule, resulting in interactive forces".

Dispersion forces are ubiquitous and are present in all molecular interactions. They can occur in isolation, but are always present even when other types of interaction dominate. Typically, the interactions between hydrocarbons are exclusively dispersive and, because of them, hexane, at S.T.P., is a liquid boiling at 68.7°C and is not a gas. Dispersive interactions are sometimes referred to as 'hydrophobic' or 'lyophobic' particularly in the fields of biotechnology and biochemistry. These terms appear to have arisen because dispersive substances, *e.g.*, the aliphatic hydrocarbons, do not dissolve readily in water. Biochemical terms for molecular interactions in relation to the physical chemical terms will be discussed later.

The theory of molecular interactions can become extremely involved and the mathematical manipulations very unwieldy. To facilitate the discussion, certain simplifying assumptions will be made. These assumptions will be inexact and the expressions given for both dispersive and polar forces will not be precise. However, they will be reasonably accurate and sufficiently so, to reveal those variables that control the different types of interaction. At a first approximation, the interaction energy, (U_D), involved with dispersive forces has been calculated to be

$$U_D = \frac{3 h_p v_o \alpha_p^2}{4 l^6}$$

where (α_p) is the polarizability of the molecule,
(v_o) is the characteristic frequency of the molecule,
(h_p) is Planck's constant,
and (l) is the distance between the molecules.

The polarizability (α) is the crucial factor that controls the dispersive force acting on the molecule, which, for substances that have no dipoles, is given by

$$\frac{D-1}{D+2} = \frac{4}{3} \pi n_v \alpha_p$$

where (D) is the dielectric constant of the material
and (n_v) is the number of molecules per unit volume.

and
$$\frac{4}{3} \pi N_A \alpha_p = \frac{(D-1)}{(D+2)} \frac{M}{\rho} = P_M$$

where (ρ) is the density of the medium,
(M) is the molecular weight,
(N_A) is Avogadro's Number
and (P_M) is the molar polarizability, noting that the number of molecules per unit volume is Nρ/M.

The Distribution Coefficient

.It is clear from the equation that the molar polarizability is proportional to $\frac{M}{\rho}$, the molar volume; thus dispersive forces (and "hydrophobic" or "lyophobic forces") will be a function of the 'molar volume' of the interacting substances. The schematic representation of dispersive interactions is given in Figure 10. Dispersive interactions do not involve localized charges on any part of the molecule; they merely arise from a host of fluctuating, closely associated charges that, at any instant, can interact with instantaneous charges of an opposite kind situated on a neighboring molecule.

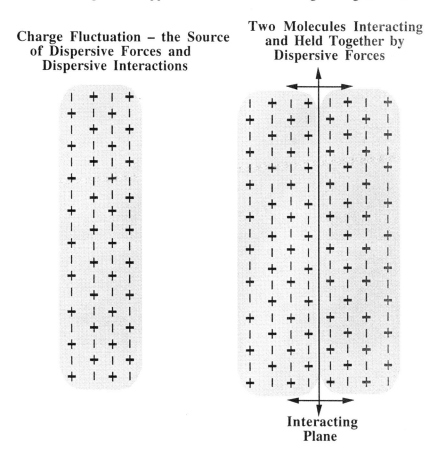

Figure 10. Dispersive Interactions

Polar Forces

Polar interactions can occur when a molecule contains a dipole or a number of dipoles which take the form of localized charges situated on different parts of the molecule. Each charge has an equal and opposite charge situated elsewhere on the molecule and, thus, the molecule has *no net charge* associated with it. Interactions occur between the charges on different molecules but are always accompanied by dispersive interactions as already discussed. The strength of the polar forces involved in polar interactions depend on the strength of the dipoles and can vary widely. Strong dipole-dipole interactions can simulate the energy of a chemical bond such as in hydrogen bonding. At the other extreme, interaction between dipoles and induced dipoles can be correspondingly weak as in (π)–(π) interactions.

Dipole-Dipole Interactions

To a first approximation, the energy (U_p) that arises during the interaction between two dipolar molecules is given by,

$$U_p = \frac{2\alpha_p \mu_p^2}{l^6}$$

where (α_p) is the polarizability of the molecule,
(μ_p) is the dipole moment of the molecule,
and (l) is the distance between the molecules.

The energy involved varies inversely with the sixth power of the distance between the molecules and as the square of the dipole moment, of which the latter can vary widely in strength. Dipole moments are often determined from bulk measurements of dielectric constant taken over a range of temperatures, but unfortunately the values so obtained do not always relate to the strength of any polar interactions that might occur with other molecules. For example, an extremely polar solvent, water, has a dipole moment of only 1.76 Debyes. Similarly, the dipole moment of another extremely polar substance, methanol, is only 2.9 Debyes. Strongly polar substances that appear to have unusually low dipole moments usually exhibit strong electrical interactions between the dipoles due to molecular association and/or from internal electric field compensation. Internal field compensation usually occurs when more than one dipole is present in a single molecule. Water associates strongly with itself by very strong polar forces or 'hydrogen bonding' which reduces the net dipole character of the associated molecules when determined from bulk external electrical measurements. Methanol also associates strongly with itself in a similar manner. When a molecule contains two dipoles, the electric field from one dipole can oppose that from the other, resulting in a reduction in the *net* field as *measured externally*. As a result, *bulk* properties measured on the material will not reflect the true value for the dipole moment of the individual dipoles. However another molecule approaching a

The Distribution Coefficient

strongly polar molecule such as water or methanol will experience the uncompensated field of the single dipole and interact accordingly.

As an example of internal compensation, consider the low dipole moment of dioxane *i.e.*, 0.45 Debye. Compared with diethyl ether, which has a dipole moment of 1.15 Debyes, one would expect the dipole moment to be about half that of dioxane. However, the strong internal compensation between the dipoles from each of the ether groups within the molecule of dioxane reduces the field as measured by external techniques and gives a value for dioxane of only about one-third of that of a diethyl ether molecule. Nevertheless, another molecule approaching one ether group of the dioxane molecule will be subject to the uncompensated field of a single dipole and interact accordingly. It follows that, although dioxane has a very low dipole moment of 0.45, it is still a very polar substance that is completely miscible with water.

It has been shown that the *polarizability* of a substance containing no dipoles will indicate the *strength of* any *dispersive* interactions that might take place with another molecule. In comparison, due to self-association or internal compensation that can take place with polar materials, the *dipole moment* determined from bulk dielectric constant measurements will often *not* give a true indication of the *strength of any polar interaction* that might take place with another molecule. An impression of a dipole-dipole interaction is depicted in Figure 11.

The dipoles are shown interacting directly as would be expected. Nevertheless, it must be emphasized that behind the dipole-dipole interactions will be dispersive interactions from the random charge fluctuations that continuously take place on both molecules. In the example given above, the net molecular interaction will be a combination of both dispersive interactions from the fluctuating random charges and polar interactions from forces between the two dipoles. Examples of substances that contain permanent dipoles and can exhibit polar interactions with other molecules are alcohols, esters, ethers, amines, amides, nitriles, etc.

Dipole-Induced Dipole Interactions

All compounds that can exhibit polar interactions need not contain permanent dipoles. Certain compounds, for example those that contain an aromatic nucleus (and thus π electrons), are termed polarizable. On close proximity of such compounds with a molecule having a permanent dipole, the electric field from the permanent dipole induces a counter-dipole in the polarizable molecule.

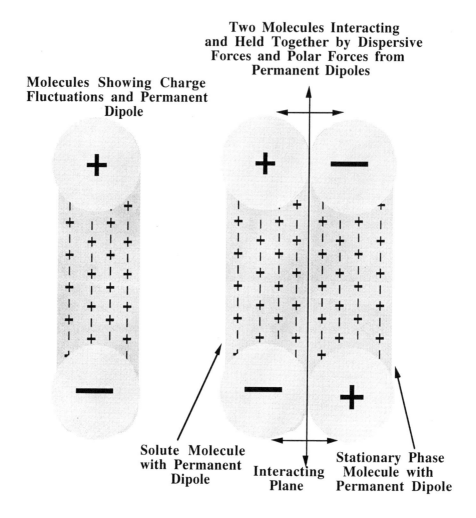

Figure 11. Polar Interactions: Dipole-Dipole Interactions

The induced counter-dipole can act in a similar manner to a permanent dipole and the electric forces between the two dipoles (permanent and induced) result in strong polar interactions. Typically, polarizable compounds are the aromatic hydrocarbons; examples of their separation using induced dipole interactions to affect retention and selectivity will be given later. Dipole-induced dipole interaction is depicted in Figure 12. Just as dipole-dipole interactions occur coincidentally with dispersive interactions, so are dipole-induced dipole interactions accompanied by dispersive interactions. It follows that using an n-alkane stationary phase, aromatic

The Distribution Coefficient

hydrocarbons can be retained and separated by purely dispersive interactions as in GC.

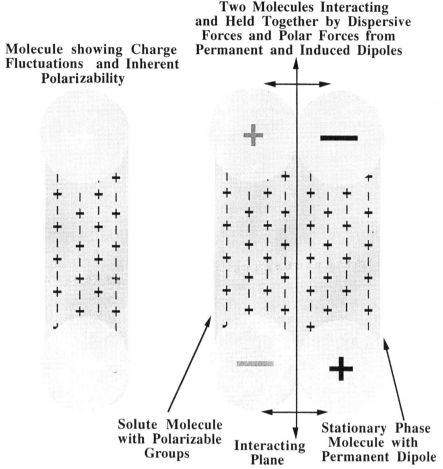

Figure 12. Polar Interactions: Dipole-Induced Dipole Interactions

Alternatively, using a polyethylene glycol stationary phase, aromatic hydrocarbons can also be retained and separated primarily by dipole-induced dipole interactions combined with some dispersive interactions. Molecules can exhibit multiple interactive properties. For example, phenyl ethanol possesses both a dipole as a result of the hydroxyl group and is polarizable due to the aromatic ring. Complex molecules such as biopolymers can contain many different interactive groups.

Ionic Forces

Compounds that possess dipoles have no net charge on the molecule. Ions, however, possess a net charge and interact strongly with ions having an opposite charge usually called *counter-ions*. Ion exchange chromatography separates ionic materials by ionic interactions which result from electrical forces between oppositely charged ions. In practice, the counter-ions to the ions being separated are chosen to be situated in the stationary phase. Ionic interactions are always accompanied by dispersive interactions and, in most cases, are also associated with polar interactions. Nevertheless, the dominant forces controlling retention in ion exchange chromatography result from ionic interactions. Ionic interaction is depicted diagrammatically in Figure 13.

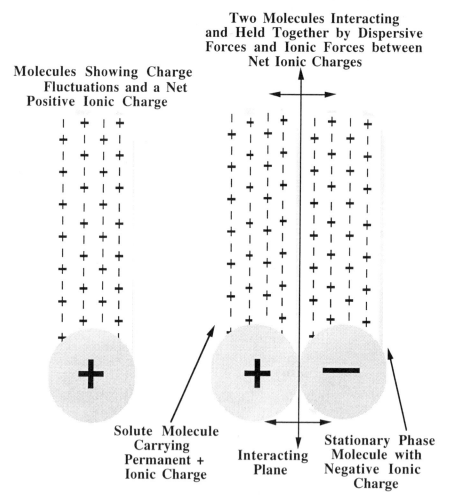

Figure 13. Ionic and Dispersive Interactions

The interactive character of a molecule can be very complex and a molecule can have many interactive sites. These sites will comprise the three basic types of interaction, *i.e.*, dispersive, polar and ionic. Some molecules (for example, large molecules such as biopolymers) can have many different interactive sites dispersed throughout the entire molecule. The interactive character of the molecule as a whole will be

The Distribution Coefficient

determined by the net effect of all the sites. For example, if the molecule has a preponderance of dispersive sites, the overall property of the molecule will be dispersive, and will be termed by the biotechnologists as "hydrophobic" or "lyophobic". Conversely, if the molecule is dominated by a preponderance of dipoles and polarizable sites, then the overall property of the molecule will be polar, which will be termed by the biotechnologists as "hydrophilic" or "lyophilic". These alternative terms are not based on physical chemical considerations, but have evolved, somewhat arbitrarily, from the discipline of biology. Their meanings are very significant to biologists and biochemists and must, therefore, be discussed in some detail

Hydrophobic and Hydrophilic Interactions

The term "hydrophobic interaction" unfortunately implies some form of molecular repulsion, which, outside the van der Waals radii of a molecule, is quite impossible. The term "hydrophobic force" literally means "fear of water" force. The term hydrophobic has been introduced as an alternative to *dispersive* but means the same. It is not clear from the literature how the word hydrophobic originated, but it may have been provoked by the immiscibility of a dispersive solvent such as *n*-heptane with a very polar solvent such as water.

n-Heptane and water are immiscible, not because water molecules *repel* heptane molecules; they are immiscible because the forces between heptane molecules and the forces between water molecules are much greater than the forces between a heptane molecule and a water molecule. Immiscibility occurs because water molecules and heptane molecules interact *very much more strongly* with *themselves* than with *each other*.

In fact, water has a small but finite solubility in *n*-heptane, and similarly *n*-heptane has a small but finite solubility in water. Despite *water-water* interactions and *hydrocarbon-hydrocarbon* interactions being very much stronger than *water-hydrocarbon* interactions, the latter do exist. In addition, *water-hydrocarbon* interactions are sufficiently strong to allow some solution to take place. Even taking into account the strong interactive forces that exist between hydrocarbons, at normal temperatures, a small fraction will randomly acquire sufficient kinetic energy to part. Thus, equilibrium occurs in a *saturated solution* of any hydrocarbon in water when the *probability* of a pair of hydrocarbon molecules gaining sufficient energy to *part* is equal to the statistical *probability* of a hydrocarbon molecule colliding with a water molecule and interacting. The probability of an interacting hydrocarbon pair gaining sufficient energy to *part* is very small, so the saturated concentration of heptane in water (which determines the probability of collision) must also be small. *Ipso facto*, the mutual solubility of water and hydrocarbons must be very low.

The term "hydrophilic force", literally meaning "love of water" force, was introduced as a complement to "hydrophobic force". Hydrophilic forces are equivalent to polar forces, and polar solvents that interact strongly with water are called hydrophilic solvents.

The origin of the terms "lyophobic" (meaning fear of lye) and "lyophilic" (meaning love of lye) to describe interactive forces is somewhat more obscure and, in fact, as they are essentially alternatives to the terms hydrophobic and hydrophilic, are somewhat irrelevant. The terms stemmed from the early days of the soap industry when soap was prepared by boiling a vegetable oil with an alkaline solution obtained from leaching "wood ash'" with water. The alkaline product from the wood ash was a crude solution of sodium and potassium carbonates called "lye". The result of boiling vegetable oil with the lye was the soap (sodium and potassium salts of long-chained fatty acids) which, due to the dispersive interactions between the fatty acid alkane chains, separated from the lye and, consequently, was called "lyophobic". It is clear that "lyophobic", from a physical chemical point of view, is the same as "hydrophobic", and hydrophobic and lyophobic interactions are dispersive. The other product from soap-making is glycerol, which, being strongly polar, remained in the lye and was consequently termed "lyophilic". Glycerol is very polar because of its many hydroxyl groups and is completely miscible with water and, hence, is a "hydrophilic" or "lyophilic" substance. These alternative terms describing molecular interactions are somewhat impertinent and confusing to the physical chemist and perhaps should be avoided by those not involved in the biological sciences. Nevertheless, they are extensively employed in biotechnology to describe the interactive character of a molecule and so their meaning must be understood. Their use becomes more understandable when one considers the need for a term that describes not merely a particular molecular interaction, but one that can summarize the overall character of a biopolymer, for example a polypeptide. A peptide contains a large number of different amino acids, each having quite different interactive groups. All the carbonyl and amide groups will exhibit polar interactions but each amino acid will also contribute its own unique interactive character to the peptide. The diagrams

The Distribution Coefficient 73

in Figure 14 endeavor to illustrate the interactive character of two different polypeptides.

(A) The Interactive Character of a "Hydrophobic" Macromolecule Depicted Diagrammatically

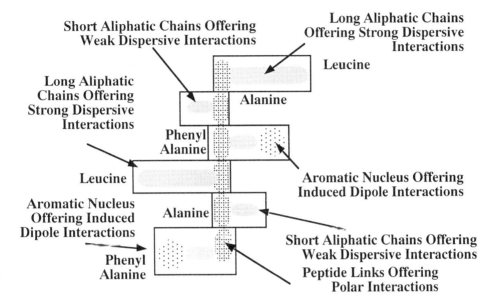

(B) The Interactive Character of a "Hydrophilic" Macromolecule Depicted Diagrammatically

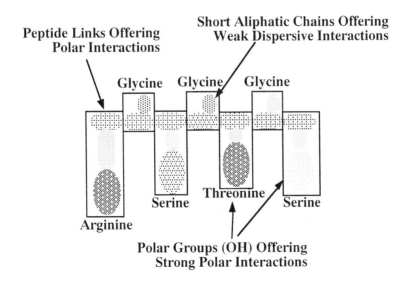

Figure 14. Examples of Hydrophobic and Hydrophilic Molecules

The spatial arrangements in Figure 14 are completely imaginary; the intent of the diagram is solely to show how the net interactive character of a large molecule is made up of the individual contributions from the different chemical groups. The overall character, that is, the resultant interactive effect from all the groups, can be dispersive (hydrophobic) or polar (hydrophilic). This will determine the nature of their overall interaction with the environment. The upper peptide (A) might be the hexapeptide

leucine-alanine-phenylalanine-leucine-alanine-phenylalanine

or of similar composition. It is seen that there is an excessive number of aliphatic chains in the molecule, long chains contained by the two leucine fragments and short chains contained by the alanine fragments. The peptide bonds are the sole contributors to the polar character to the molecule, augmented somewhat by the polarizability of the two aromatic rings from the phenylalanine fragments. It is clear that the net interactive character of the hexapeptide will be dispersive and the peptide would be classed as a *hydrophobic* or *lyophobic* substance.

In contrast, the lower peptide (B) might be considered to be the following heptapeptide, or of similar composition.

arginine-glycine-serine-glycine-threonine-glycine-serine

Now, it is seen that polar groups dominate the molecular structure, resulting from hydroxyl groups from the two serine and threonine fragments in addition to the peptide bonds themselves. Only weak dispersive interactions will be contributed by glycine fragments (CH_2 groups).

Potential ionic interactions may also be provided by the free amine groups of the arginine fragment. As a consequence, the net character of the heptapeptide will be polar, and the peptide would be classed as *hydrophilic* or *lyophilic* in nature.

The terms hydrophilic and hydrophobic are mostly used to describe the net interactive character of a large molecule and not so much to describe individual group interactions. Consequently, they are alternative terms commonly used to express the overall interactive properties of a molecule where either polar or dispersive interactions dominate.

The Distribution of Standard Free Energy Between Different Types of Molecular Interactions

In contrast to apportioning the standard free energy between different groups in the solute molecule, the standard free energy can also be dispensed between the different types of forces involved in the solute/phase-phase distribution. This approach has been elegantly developed by Martire *et al.* [13]. In a simplified form, the standard free energy can be divided into portions that result from the different types of interaction, *e.g.*,

$$\Delta G = \Delta G_{Dispersive} + \Delta G_{Polar} + \Delta G_{Ionic}$$

The polar interactions are often divided into weak, moderate and strong interactions that have been, somewhat arbitrarily, given terms such as π/π interactions, *dipole/dipole* interactions, and *hydrogen bonding*, *e.g.*,

$$\Delta G = \Delta G_D + \Delta G_{\pi/\pi\, Interact.} + \Delta G_{Dipole/Dipole\, Interact.} + \Delta G_{Hydrogen\, Bonding}$$

Nevertheless, it is important to realize that they are not different *types* of interaction but are all *polar interactions* but of *different strength.* In fact, many more terms have been introduced to describe subtly different enthalpic and entropic contributions to retention. Again one should remember Einstein's comment on *first order* effects being simple. However, in fairness, it must be said that, in many cases, these extra terms are often introduced to take into account many *second order* effects. One final point, the standard free energy is a bulk property, and in this approach where a portion is allotted to a particular type of interaction, *e.g.*, dispersive, it will include *all* dispersive interactions throughout the whole molecule. It will account for the *net* effect of many interactions involving all dispersive groups and not a *specific* interaction that arises from a particular dispersive group.

This approach has been used to determine the so-called 'binding' constants between solute and stationary phase and also to investigate 'complexation constants'. Binding constants and complexation constants have been introduced to distribution system theory, largely, to account for the wide range of strengths associated with polar interactions between molecules. Some schools of thought consider that polar interactions need to be treated in a significantly different manner to dispersive interactions. As a consequence, alternative terms have been introduced to more clearly describe the different polar interaction strengths that can occur. The portion of the molecule that is negatively charged due to an asymmetric accumulation of electrons has been given the term *electron donor site,* whereas the area of electron depletion in the molecules (that electrically balances the site of electron accumulation) has been given the term *electron acceptor site.*

More simply, these are basically negatively and positively charged sites on the molecule, as previously discussed. Again, because polar interactions have such a range of strengths, certain ranges of polarity have been assumed to be due to specific types of electrical interaction and have been given the terms [14] *nonbonding lone pair, bonding π orbital, vacant orbital, antibonding π orbital, antibonding σ orbital, etc.,* all of which are based on the accepted electron configuration of the molecule. In practice however, whatever the rationale, the strength of the interaction and, thus, its effect on solute retention, are directly related to the *intensity* of the interacting charges. From the standing of separation technology, whether the introduction of these alternative terms helps to explain the separation process is a moot point, as, even in theory, the individual effects can be extremely difficult to completely isolate from one another.

According to the theory of complexation in chromatography, when a polar interaction occurs between a solute molecule and a molecule of stationary phase, an associate (complex) is assumed to occur which virtually removes the solute from the migration process. This type of interaction has been termed complexation. In gas chromatography the complex is assumed to have no vapor pressure and, thus, remains stationary in the column until, as a result of the equilibrium kinetics, dissociation again generates the individual components. The uncomplexed solute molecule can now continue to migrate along the column until complexing again takes place. In the authors' opinion, the difference between complexing and strong interaction is by no means clear and, in any event, the net effect of both will be the same, *i.e.,* strong solute retention.

This is one approach to the explanation of retention by polar interactions, but the subject, at this time, remains controversial. Doubtless, complexation can take place, and probably does so in cases like olefin retention on silver nitrate doped stationary phases in GC. However, if dispersive interactions (electrical interactions between randomly generated dipoles) can cause solute retention without the need to invoke the

The Distribution Coefficient

concept of complexation, it is not clear why polar interactions (electrical interaction between permanent or induced dipoles) need to be considered differently.

The Concept of Complex Formation

In an attempt to explain the nature of polar interactions, Martire *et al.* [15] developed a theory assuming that such interactions could be explained by the formation of a complex between the solute and the stationary phase with its own equilibrium constant. Martire and Riedl adopted a procedure used by Langer *et al.* [16], and divided the solute activity coefficient into two components.

$$V_g^o = \frac{273R}{\gamma_1^\infty (1-c) p_1^o MW_L}$$

where (V_g^o) is the specific retention volume of the solute,

(γ_1^∞) is the fully corrected solute activity coefficients,

(p_1^o) is the solute bulk vapor pressure,

(MW_L) is the stationary phase molecular weight,

(R) is the gas constant,

and (c) is the fraction of the solute that is complexed.

The complex is considered to have zero vapor pressure and is stationary in the column. Thus, there are now two parts of the denominator, (γ_1^∞) the free solute activity coefficient and (1-c) the fraction of free solute.

The thermodynamic equilibrium constant (K_1) is contained in the function

$$K_{EQ} = \frac{\alpha_{DA}}{\alpha_D \alpha_A} = K_1 \frac{\gamma_{DA}}{\gamma_D \gamma_A}$$

where (K_{EQ}) is the thermodynamic equilibrium constant,

(α_{DA}) is the activity of the complex in the stationary phase,

(α_D) is the activity of the solute in the stationary phase,

(α_A) is the activity of the stationary phase,

(γ_{DA}) is the activity coefficient of the complex,

(γ_D) is the activity coefficient of the solute,

and (γ_A) is the activity coefficient of the stationary phase.

At infinite dilution, (γ_{DA}) and (γ_D) will tend to unity, (C_{DA}) and (C_D^o) will tend to zero and (α_{DA}) and (α_D) will tend to (C_{DA}) and (C_D) respectively.

Thus, $$K_{EQ} = \frac{C_{DA}}{C_D^o \alpha_A}$$

Defining a new constant (K'), where $K' = \frac{C_{DA}}{C_D^o} = \alpha_A K_{EQ}$,

The fraction (c) of the complexed solute is given as,

$$c = \frac{C_{DA}}{C_{DA} + C_D^o} = \frac{C_{DA}}{\frac{C_{DA}}{K'} + C_D^o} = \frac{K'}{1 + K'} \quad \text{or} \quad 1 - c = \frac{1}{1 + K'}$$

Substituting for (1-c) in the equation for (V_g^o),

$$V_g^o = \frac{273R}{\gamma_1^\infty p_1^o MW_L}(1 + K')$$

Unfortunately, it is impossible to separate the terms (γ_1^∞) and $(1 + K')$ and, thus, the procedure must be complicated further by the introduction of four additional specific retention volumes. Experimental values for these can be used (with certain assumptions) to obtain a value for (K') and consequently a value for (K_{eq}). They are as follows,

$$V_g^{o(A)} = \frac{273R}{\gamma_A^\infty p_N^o MW_N}(1 + K')$$

where $(V_g^{o(A)})$ is the specific retention volume of a noncomplexing solute with a vapor pressure (p_N^o) and a noncomplexing stationary phase having a molecular weight (MW_N);

$$V_g^{o(B)} = \frac{273R}{\gamma_B^\infty p_N^o MW_C}(1 + K')$$

where $(V_g^{o(B)})$ is the specific retention volume of the same noncomplexing solute with a pure complexing stationary phase having a molecular weight (MW_C);

$$V_g^{o(C)} = \frac{273R}{\gamma_C^\infty p_C^o MW_C}(1 + K')$$

where $(V_g^{o(C)})$ is the specific retention volume of a complexing solute with a vapor pressure (p_C^o) with a pure complexing stationary phase having a molecular weight (MW_C);

The Distribution Coefficient

$$V_g^{o(D)} = \frac{273R}{\gamma_D^\infty p_C^o MW_N}(1 + K')$$

where ($V_g^{o(D)}$) is the specific retention volume of a complexing solute with a vapor pressure (p_N^o) with a pure noncomplexing stationary phase having a molecular weight (MW_N).

If the two stationary phases have the same physical properties, except for one being polar (having a complexing capability), then a reasonable assumption would be that,

$$\frac{\gamma_D^\infty}{\gamma_C^\infty} = \frac{\gamma_A^\infty}{\gamma_B^\infty}$$

By rearranging,

$$K' = \frac{V_g^{o(A)} V_g^{o(C)}}{V_g^{o(B)} V_g^{o(D)}} - 1$$

To obtain a value for (K_{EQ}) then the pure stationary phase activity (a_A), or the activity coefficient (γ_A) must be determined. It was argued that (γ_A) could be defined as follows,

$$\gamma_A = \frac{\gamma_A^\infty}{\gamma_B^\infty} = \frac{V_g^{o(B)} MW_C}{V_g^{o(A)} MW_N}$$

This relationship depends on the assumption that two similar stationary phases, irrespective of their polarity, can be considered to differ by measuring the ratio of the activity coefficients of two noncomplexing solutes (this basically implies the solute is nonpolar and will only interact with the stationary phase by dispersion forces). If this were true then,

$$K_{EQ} = \frac{K'}{\alpha_A} = \frac{K' \overline{V}_A}{\gamma_A}$$

where ($\overline{V}_A$) is the molar volume of the complexing phase.

Thus,

$$K_{EQ} = \left(\frac{\overline{V}_A}{\gamma_A}\right)\left(\frac{V_g^{o(A)} V_g^{o(C)}}{V_g^{o(B)} V_g^{o(D)}} - 1\right)$$

Similar substances were chosen to meet the assumptions of the theory that would offer approximately the same dispersive interactions but quite different polar interactions, for example di-*n*-octylmethylamine and *n*-octadecane. Consequently, retention differences were then assumed to be solely due to polar interaction and, thus, the 'complexation constant' could be calculated from retention data. Data obtained only agreed moderately well with values calculated by other techniques. It is clear, however, that the assumptions made in the development of the theory are, at the least, open to debate and although complexation may occur under certain circumstances, it is not necessary to evoke the concept to explain all polar interactions. From a chromatographic perspective, if it is accepted that dispersive forces cause solute retention by simple and direct interaction between molecules, without the need to invoke the concept of complexation, it is not clear why retention by polar interactions need to be considered differently. In the next chapter, examples will be given of retention that can be explained simply by assuming a distribution system that is controlled solely by polar and dispersive forces modified by the probability of interaction.

Applied Thermodynamics

An interesting and practical example of the use of thermodynamic analysis is to explain and predict certain features that arise in the application of chromatography to chiral separations. The separation of enantiomers is achieved by making one or both phases chirally active so that different enantiomers will interact slightly differently with the one or both phases. In practice, it is usual to make the stationary phase comprise one specific isomer so that it offers specific selectivity to one enantiomer of the chiral solute pair. The basis of the selectivity is thought to be spatial, in that one enantiomer can approach the stationary phase closer than the other. If there is no chiral selectivity in the stationary phase, both enantiomers (being chemically identical) will coelute and will provide identical $\log(V_r')$ against $1/T$ curve. If, however, one enantiomer can come closer to the stationary phase than the other, then its entropy change will be decreased and, as it is closer to the stationary phase, interactions will be stronger and the *enthalpy* will also be *increased*. Now irrespective of the change in entropy, if the enthalpy increases, then the slope of the $\log(V_r')$ against $1/T$ curve must also increase. *This means that each enantiomer will have $\log(V_r')$ against $1/T$ curves of slightly different slopes and,* therefore, **they must intersect**. Consequently, there will be a specific temperature at which both enantiomers will still coelute, despite the chiral selectivity of the stationary phase. It follows that the use of any chiral stationary phase to separate a particular pair of enantiomers must be examined over a range of temperatures to identify that temperature where chiral

The Distribution Coefficient

selectivity is sufficient to achieve separation. This phenomenon was elegantly demonstrated by Heng Liang Jin [17] and his results are shown in Figure 15.

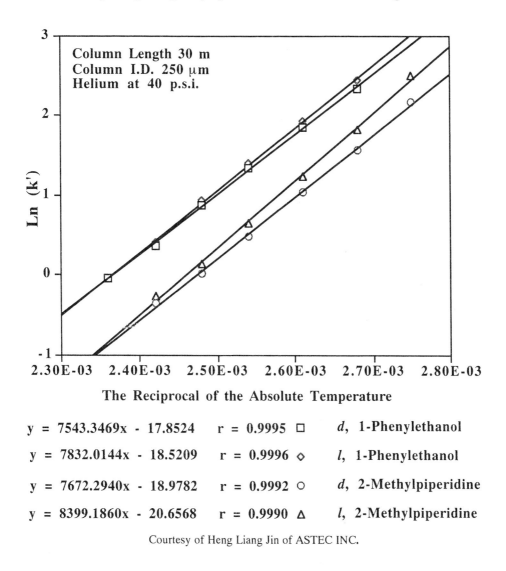

Courtesy of Heng Liang Jin of ASTEC INC.

Figure 15. Curves Relating Log(k') against 1/T for Two Pairs of Enantiomers

The curves relate log(k') against 1/T for two pairs of enantiomers on a chiral BBP column containing 1,6-dipentyl-buteryl cyclodextrin that was 20 m long, 250 μm I.D. The data has been curve fitted to a linear function and thus the enthalpy and entropy contributions are extracted as the slope and intercept of each curve.

Thus $\quad \text{Log}(k'_1) = \dfrac{\psi_1}{T} - \phi_1 \text{ and } \text{Log}(k'_2) = \dfrac{\psi_2}{T} - \phi_2$

where (ψ_1) and (ψ_2) are the slopes of the curves for enantiomer (1)

and enantiomer (2) respectively,

and (ϕ_1) and (ϕ_2) are the intercepts of the curves for enantiomer (1)

and enantiomer (2) respectively.

When $k'_1 = k'_2$ Then, $\dfrac{\psi_1}{T} - \phi_1 = \dfrac{\psi_2}{T} - \phi_2$ or $T = \dfrac{\psi_1 - \psi_2}{\phi_1 - \phi_2}$ (2)

It is seen that if the enthalpies and entropy's differ for two enantiomer pairs, there will always be a temperature where they elute coincidentally and cannot be separated. From the curves and intercepts given in Figure 15, the temperature for coincident retention of the two phenyl ethanol enantiomers is 432°K or 159°C and for the methylpiperidine enantiomers is 433°K or 160°C, which agrees excellently with the curves shown in Figure 15. Now, assume that, in order to separate a pair of enantiomers, a separation ratio of ($\alpha_{1/2}$) is required. Assuming, $\psi_1 \neq \psi_2$ and $\phi_1 \neq \phi_2$, it is now possible to calculate the temperatures at which a separation ratio of ($\alpha_{1/2}$) can be realized.

Now, $\alpha_{1/2} = \dfrac{k'_1}{k'_2}$

Thus, $\quad \ln(\alpha_{1/2}) = \ln(k'_1) - \ln(k'_2)$

$$= \left(\dfrac{\psi_1}{T} + \phi_1\right) - \left(\dfrac{\psi_2}{T} + \phi_2\right)$$

$$= \dfrac{\psi_1 - \psi_2}{T} + (\phi_1 - \phi_2) \quad\quad (3)$$

Therefore, $\quad T = \dfrac{\psi_1 - \psi_2}{\ln(\alpha) + (\phi_2 - \phi_1)}$

This equation allows the temperature to be calculated at which the separation ratio between the solutes would be (α). From the preceding equation, in addition,

$$\alpha = e^{\dfrac{\psi_1 - \psi_2}{T} + (\phi_1 - \phi_2)}$$

Thus, the separation ratio that will be obtained for the solute pair can be calculated for any temperature. The results of Heng Liang Jin not only confirm the existence of a

The Distribution Coefficient

temperature at which closely eluting enantiomers will coelute but, in addition, the results also show that for any closely eluting pair of solutes that have different enthalpies, there will always be a temperature of coelution. Moreover, in all forms of chromatography, GC, LC and TLC, temperature will be an important variable to consider when separation ratios are small and resolution is difficult.

Synopsis

The distribution of a solute between two phases can be explained thermodynamically and by molecular interactions. The natural logarithm of the distribution coefficient is equal to minus the standard free energy, which, in turn, is equal to a function of the standard enthalpy and the standard entropy. The standard enthalpy results from the forces involved in molecular interaction between the solute and the two phases and the standard entropy results from the change in randomness or freedom of the molecule when moving from one phase to the other. If the standard enthalpy is the major factor controlling retention, then the distribution is said to be 'energy driven'. If the standard entropy is the major factor controlling retention, then the distribution is said to be 'entropically driven'. If the logarithm of the corrected retention volume is plotted against the reciprocal of the absolute temperature, a straight line is produced and the slope will give a measure of the standard enthalpy and the intercept a measure of the standard entropy. If a straight line is not produced, then this indicates that the actual distribution system is changing with temperature. For example, in a water/methanol mobile phase the proportion of unassociated water and methanol water associate will also vary with temperature, which changes the distribution system. The standard free energy, the standard enthalpy and the standard entropy can be apportioned in two ways. It can be assigned to different parts of a molecule (*e.g.* methyl groups, methylene groups halogen atoms etc.) and thus explain how each group contributes to the distribution. Alternatively it can be assigned to different types of interactions (*e.g.* dispersive, polar etc.) and thus explain how each type of interaction contributes to the distribution. The standard entropy term tends to increase with the standard enthalpy term. An increase in enthalpy indicates that more energy is used up in the association of the solute molecule with the molecules of the stationary phase. This means that the intermolecular forces are *stronger* and thus the stationary phase molecules hold the solute molecules *more tightly* and the freedom of the molecule is reduced. Distribution can also be explained on the basis of molecular interactions which are all electrical in nature. There are three types of molecular interaction; dispersive, polar and ionic. Dispersion forces arise from random charge fluctuations throughout a molecule resulting from electron/nuclei vibrations. They are a statistical effect and occur in all molecular interactions. Polar interactions arise from electrical forces between localized charges residing on different parts of the molecule that result from permanent or induced dipoles. Polar interactions cannot occur in isolation, but must be always accompanied by statistically generated dispersive interactions. Polar interactions can vary widely in strength depending on the magnitude of the charges on each dipole and are sometimes given alternative terms depending on their strength, *e.g.*, π–π interactions, dipole-dipole interactions, hydrogen bonding etc. Polar compounds, although possessing dipoles, have no net charge on the molecule. In contrast, ions possess a net charge and, consequently, can interact strongly with ions having an opposite charge. Ionic interactions are always accompanied by dispersive interactions and often, also with polar interactions. Hydrophobic interaction is another term for dispersive interaction, but the term hydrophobic is usually employed to describe the overall property of a large molecule where dispersive groups are dominant. Similarly, hydrophilic interaction is another term for polar interactions but the term hydrophilic is, again, usually employed to describe the overall property of a large molecule where polar groups dominate. The standard free energy can also be distributed between the different types of forces involved in the solute/phase-phase distribution. For example, the standard free energy, standard enthalpy and standard entropy can be distributed between dispersive interactions, polar interactions and ionic interactions. In fact, the portion allotted to polar interactions has been further divided in to portions representing π-π interactions, dipole-dipole interactions, hydrogen bonding etc. The standard free energy is a bulk property and, in this approach, where a portion is allotted to a particular type of interaction, *e.g.*, dispersive, it will include *all* dispersive interactions throughout the whole molecule. It will account for the *net* effect of many interactions involving all dispersive groups and not a *specific* interaction that arises from a particular dispersive group. Polar interactions have also been interpreted as a complexing phenomenon with a complexing equilibrium constant. Theories have been developed based on complexation to explain polar contributions to retention in gas chromatography. However, it is difficult to separate the polar interactions from the dispersive interactions. One solution was to use very similar substances that would offer approximately the same dispersive interactions but quite different polar interactions, for example di-*n*-octylmethylamine and *n*-octadecane. Consequently, retention differences were then assumed to be solely due to polar interaction and thus the 'complexation constant' could be calculated from retention data. The results, compared with other methods of determination, were only partly convincing. From a chromatographic perspective, if it is accepted that dispersive interactions cause solute retention by simple direct interaction without the need to invoke the concept of complexation it is not clear why retention by polar interactions need to be considered differently. The thermodynamic description of retention implies that for closely eluting peaks such as those in chiral separations, irrespective of the change in entropy, if the enthalpy differs, then the slope of the $\ln(V_r')$ against $1/T$ curves will also differ. This means that each enantiomer will have $\text{Log}(V_r')$ against $1/T$ curves that will intersect. Consequently, there will be a specific temperature at which both enantiomers will still coelute, despite the chiral selectivity of the stationary phase. In addition, at temperatures either side of the coelution temperature, separation will be possible.

References

1. R. P. W. Scott and C. F. Simpson, *Faraday Discussions of the Royal Society of Chemistry, No 15,* Faraday Symposium 15, (1980)69.
2. R. P. W. Scott, "Silica gel and Bonded Phases", John Wiley and Sons, Chichester (1993)147.
3. R. P. W. Scott and J. G. Lawrence, *J. Chromatogr. Sci.*, 7(1969)65.
4. A. J. P. Martin, Private Communication, *Symposium on Vapor Phase Chromatography,* London (1956).
5. D. E. Martire and P. Reidl, *J. Phys. Chem.,* 72(1968)3478.
6. Y. B. Tewari, J. P. Sheriden, and D. E. Martire, *J. Phys. Chem.,* 74(1970)3263.
7. Y. B. Tewari, D. E. Martire and J. P. Sheriden, *J. Phys. Chem.,* 74(1970)2345.
8. J. P. Sheriden, D. E. Martire and Y. B. Tewari, *J. Am. Chem. Soc.,* 94(1972)3294.
9. H. L. Liao and D. E. Martire, *J. Am. Chem. Soc.,* 96(1974)2058.
10. H. L. Liao and D. E. Martire, *J. Phys. Chem.,* 96(1974)2058.
11. F. London, *Phys. Z,.* 60(1930)245.
12. S. Glasstone, "Textbook of Physical Chemistry," D. Van Nostrand Co, New York, (1946) 298 and 534.
13. D. E. Martire, *Unified Approach to the Theory of Chromatography Incompressible Binary Mobile Phase (Liquid Chromatography)* in *Theoretical Advancement in Chromatography and Related Separation Techniques* (Ed. F. Dondi, G. Guiochon, Kluwer, Academic Publishers, Dordrecht, The Netherlands,(1993)261.
14. R. J. Laub and R. L. Pecsok, *Physicochemical Applications of Gas Chromatography,* John Wiley and Sons, Chichester, New York, (1978)153.
15. D. E. Martire and P. Reidl, *J. Phys. Chem.,* 72(1968)3478.
16. S. H. Langer, C. Zahn and G. Pantazoplos, *J. Chromatogr.,* 3(1960)154.
17. T. E. Beesley and R. P. W. Scott, *Chiral Chromatography,* John Wiley and Sons, Chichester-New York, (1998), 46.

Chapter 4

The Theory of Mixed Phases in Chromatography

Introduction

Traditionally, the stationary and mobile phases have been chosen as single substances. They were selected to possess, in their molecular structure, the different groups that would provide the appropriate interactions, and, consequently, the necessary selectivity, to resolve the sample in question. It must again be pointed out that, in GC, the interactions of the solute molecules with the molecules of mobile phase are infrequent and weak and thus the choice of the gas was rarely used to influence retention or selectivity. However, it is also possible to obtain the required blend of molecular interactions to achieve the desired selectivity, by mixing two or more substances of different polarity in appropriate proportions. In a binary mixture that constitutes one of the phases, the second solvent, usually the more polar solvent, is often called the *moderating solvent* or the *moderator*. This terminology will be employed throughout this book. In the past, the best composition of the mobile or stationary phase was determined by experiment, but today the theory of mixed phases is fairly well understood. Consequently, the optimum composition of the phase, that will give the best selectivity, can usually be identified using the results from some preliminary experiments with each solvent. In GC the *stationary phase* can be made from a blend of different substances but in LC it is usually the *mobile phase* that is made up of a mixture of solvents. Most mobile phases are binary mixtures but sometimes a ternary mixture may be necessary. More than three components in the mobile phase is rarely needed in LC, as the required selectivity, even if very difficult to achieve, can usually be obtained by a subtle blending of three solvents only. In LC the effect of mobile composition on solvent retention can differ significantly with the level of the moderator. At low concentrations of moderator, a layer of the moderator is built up on the surface of the stationary phase according to the Langmuir adsorption isotherm. Thus, initially, increasing the moderator concentration changes both the interactive nature of the stationary phase as well as that of the mobile phase. When the layer of moderator adsorbed on the stationary phase surface is complete (which usually takes place at a relatively low level of moderator), then the character of the stationary phase remains constant and changes in moderator concentration only change the interactions in the mobile phase. Thus, the effect of solvent composition on retention will be considered in two parts, the first at *low* moderator concentrations and the second at *high* moderator concentrations.

Effect of Low Concentrations of Moderator

In LC, at very low concentrations of moderator in the mobile phase, the solvent distributes itself between the two phases in much the same way as the solute. However, as the dilution is not infinite, the adsorption isotherm is not linear and takes the form of the Langmuir isotherm.

Langmuir's Adsorption Isotherm for a Single Layer

Consider 1 cm^2 of surface carrying an adsorbed layer of a moderator at a concentration of (C_s) g.cm^{-2} in contact with a liquid containing (C_m) g/cm^{-3}. of the solvent depicted in Figure 1. Let the molecular weight of the moderator be (M) and the area covered by an adsorbed moderator molecule be (S_A).

Then, area of exposed surface = $1 - \dfrac{C_s}{M} N_A S$ where (N_A) is Avogadro's number.

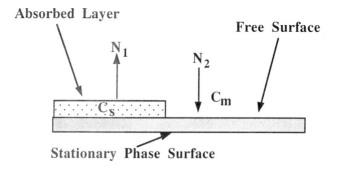

Figure 1. Monolayer Adsorption onto a Surface

The number of molecules (N_1) leaving the surface per unit time will be proportional to the concentration of the adsorbed molecules.

i.e., $N_1 = \beta C_s$ where (β) is a constant at a given temperature.

Mixed Phases

Similarly, the number of molecules (N_2) striking the surface per unit time and adhering will be proportional to the concentration of the moderator in the mobile phase *i.e.*,

$$i \quad N_2 = \alpha\left(1 - \frac{C_s N_A S}{M}\right) C_m \quad \text{where } (\alpha) \text{ is another constant at the same temperature.}$$

Under equilibrium conditions,

$$N_1 = N_2 \quad \text{or} \quad \beta C_s = \alpha\left(1 - \frac{C_s N_A S}{M}\right) C_m \qquad (1)$$

Thus,
$$\alpha C_m - \frac{\alpha C_s N_A S C_m}{M} = \beta C_s$$

and
$$C_s\left(\beta + \frac{\alpha N_A S C_m}{M}\right) = \alpha C_m \qquad (2)$$

or
$$C_s = \frac{\alpha C_m}{\beta + \frac{\alpha N_A S C_m}{M}}. \qquad (3)$$

Equation (3) is one form of the Langmuir isotherm and it should be noted that, when (C_m) tends to zero,

$$\beta >> \frac{\alpha N_A S C_m}{M} \quad \text{and} \quad C_s = \frac{\alpha C_m}{\beta}$$

i.e., the isotherm is linear. This is the condition under which chromatographic separations are carried out.

When (C_m) is very large,
$$\beta << \frac{\alpha N_A S C_m}{M}$$

Then,
$$C_s = \frac{M}{N_A S} \quad \text{i.e. } (C_s) \text{ tends to a constant value}$$

Thus, the surface is completely covered with solute and the chromatographic properties of the surface remains constant.

Thus,
$$\frac{C_s}{C_m} = \frac{\alpha}{\beta + \frac{\alpha N_A S C_m}{M}} = K \quad \text{or} \quad K = \frac{1}{g + \frac{N_A S C_m}{M}} \qquad (4)$$

where $g = \beta/\alpha$ the desorption/absorption coefficient of the moderator.

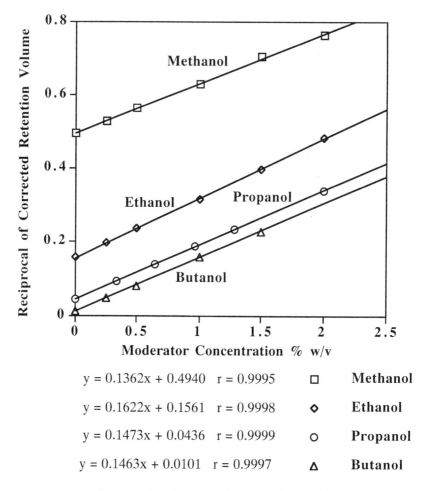

y = 0.1362x + 0.4940	r = 0.9995	□	Methanol
y = 0.1622x + 0.1561	r = 0.9998	◇	Ethanol
y = 0.1473x + 0.0436	r = 0.9999	○	Propanol
y = 0.1463x + 0.0101	r = 0.9997	△	Butanol

Courtesy of the Royal Society of Chemistry (ref.1)

Figure 2. Graph of the Reciprocal of the Corrected Retention Volume against the Moderator Concentration for a Series of Aliphatic Alcohols

Now taking (C_m) as the concentration of the moderator and (φ) the total chromatographically available surface area, then the corrected retention volume (V'_r) will be given by, $V'_r = K\varphi$.

Thus from (4),

$$V'_r = \frac{\varphi}{g + \frac{N_A S C_m}{M}}$$

or

$$\frac{1}{V'_r} = \frac{g}{\varphi} + \frac{N_A S}{M\varphi} C_m \tag{5}$$

Thus, if the slope and intercept of the curve relating the reciprocal of the corrected retention volume to the concentration of the moderator are (μ) and (ψ) respectively,

then
$$\frac{\mu}{\psi} = \frac{gM}{NS}, \; g = \frac{\mu NS}{\psi M}, \text{ and } \varphi = \frac{g}{\mu}$$

It is seen, from equation (5), that a graph relating the reciprocal of the corrected retention volume to the concentration of the moderator can provide values for the adsorption/desorption coefficient and the surface area of the stationary phase. Scott and Simpson [1] used this technique to measure the surface area of a reversed phase and the curves relating the reciprocal of the corrected retention volume to moderator concentration are those shown in Figure 2.

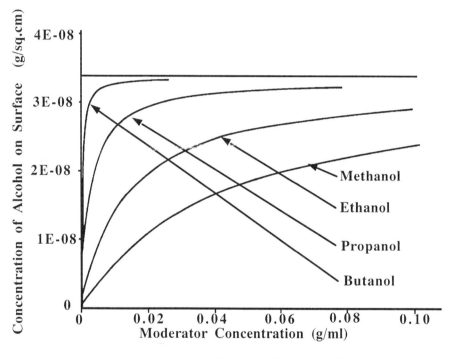

Courtesy of the Royal Society of Chemistry (ref.1)

Figure 3. Adsorption Isotherms for some Aliphatic Alcohols on a Reversed Phase

The mobile phase was water in which the moderator alcohols were dissolved. It is seen that the linear relationship is completely validated and the data can provide the adsorption isotherms in the manner discussed. The mean surface area was found to be 199 m^2/g with a standard deviation between the different alcohols of 11 m^2/g.

The adsorption isotherms calculated from their data are shown in Figure 3. It is seen that the propanol and butanol rapidly cover the surface at low moderator concentrations but the surface is hardly covered with methanol even at a concentration of 20%w/v. This means that only above 25%w/v of this moderator are the interactive properties of the stationary phase surface constant and the moderator is only effecting interactions in the mobile phase.

Effect of Bonding Characteristics

The above data were obtained on a polymeric bonded phase and not a 'brush' phase. The so-called 'brush' phases are made from *monochloro*-silanes (or other active group) and, thus, the derivative takes the form of chains attached to the silica surface [2]. The bulk phases are synthesized from *polyfunctional* silanes in the presence of water and, thus, are cross linked and form a rigid polymeric structure covering the silica surface. These two types of phases behave very differently at low concentrations of moderator.

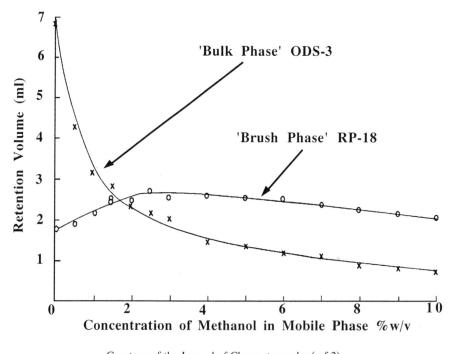

Courtesy of the Journal of Chromatography (ref.3)

Figure 4. Graph of Retention Volume of Ethanol against Concentration of Methanol for Two Different Types of Reversed Phase

Curves relating the corrected retention volume to the concentration of moderator (methanol) in the mobile phase [3] are shown in Figure 4. In pure water, the hydrocarbon chains of the brush phase interact with each other and collapse onto the surface in much the same way as drops of an hydrocarbon will coalesce on the surface of water. This means that the effective chromatographic surface is also reduced and, thus, retention is diminished. As the concentration of methanol is increased, more methanol is adsorbed onto the chains and greater interaction between the alkane chains and the mobile phase is possible. Consequently, the chains begin to dissemble, which results in an increase in the interactive surface area and an initial increase in solute retention. This process continues as more methanol is adsorbed until the methanol concentration is sufficient to allow all the alkane chains to be completely dispersed in the mobile phase, at which point the retention of the solute reaches a maximum. Subsequently, further increase in the methanol content causes methanol to be adsorbed according to the Langmuir function, and solute retention falls in the expected manner.

In contrast, the alkane chains on the polymeric phase cannot collapse in an environment of water as they are rigidly held in the polymer matrix. Thus, the retention of the solute now continuously falls as the methanol concentration increases as shown in Figure 4. It should be pointed out that if the nature of the solute-stationary phase interactions on the surface of a bonded phase is to be examined in a systematic manner with solvents having very high water contents, then a polymeric phase should be used and brush type reversed phases avoided if possible.

Langmuir's Adsorption Isotherm for a Double Layer

Silica gel, *per se*, is not so frequently used in LC as the reversed phases or the bonded phases, because silica separates substances largely by polar interactions with the silanol groups on the silica surface. In contrast, the reversed and bonded phases separate material largely by interactions with the dispersive components of the solute. As the dispersive character of substances, in general, vary more subtly than does their polar character, the reversed and bonded phases are usually preferred. In addition, silica has a significant solubility in many solvents, particularly aqueous solvents and, thus, silica columns can be less stable than those packed with bonded phases. The analytical procedure can be a little more complex and costly with silica gel columns as, in general, a wider variety of more expensive solvents are required. Reversed and bonded phases utilize blended solvents such as hexane/ethanol, methanol/water or acetonitrile/water mixtures as the mobile phase and, consequently, are considerably more economical. Nevertheless, silica gel has certain areas of application for which it is particularly useful and is very effective for separating *polarizable* substances such as the polynuclear aromatic hydrocarbons and substances of weak polarity such as phenols, esters and aliphatic ethers. It is also used frequently in exclusion chromatography after appropriate removal of high activity adsorption sites.

When the silica surface is in contact with a solvent, the surface is covered with a layer of the solvent molecules. If the mobile phase consists of a mixture of solvents, the solvents compete for the surface and it is partly covered by one solvent and partly by the other. Thus, any solute interacting with the stationary phase may well be presented with two, quite different types of surface with which to interact. The probability that a solute molecule will interact with one particular type of surface will be statistically controlled by the proportion of the total surface area that is covered by that particular solvent.

Mono-layer Adsorption

A solvent can be adsorbed from a solvent mixture on the surface of silica gel according to the Langmuir adsorption isotherm as previously discussed.

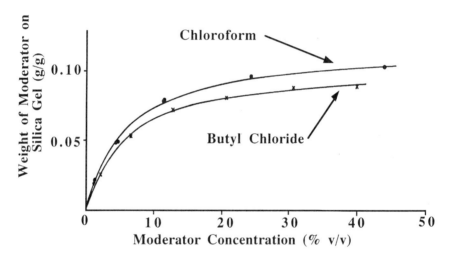

Courtesy of the Journal of Chromatography (ref.4)

Figure 5. Langmuir Adsorption Isotherms for Benzene, Butyl Chloride and Chloroform

Examples of mono-layer adsorption isotherms obtained for chloroform and butyl chloride are shown in Figure 5. The adsorption isotherms of the more polar solvents, ethyl acetate, isopropanol and tetrahydro-furan from *n*-heptane solutions on silica gel were examined by Scott and Kucera [4]. Somewhat surprisingly, it was found that the experimental results for the more polar solvents did *not* fit the simple mono-layer

Mixed Phases

adsorption equation. Consequently, the possibility of bi-layer adsorption on the silica gel surface was investigated. Bi-layer adsorption is not uncommon and the bi-layer adsorption isotherm equation can be derived by a simple extension of the procedure used to derive the Langmuir adsorption isotherm.

Bi-layer Adsorption

Consider the bi-layer adsorption of strongly polar solvent (B) (*e.g.*, ethyl acetate) from a solution in a dispersive solvent (A) (*e.g.*, *n*-heptane) onto a silica gel surface, as depicted in Figure 6.

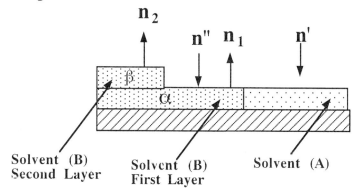

Figure 6. Bi-layer Adsorption on a Surface

Let the concentration of solvent (B) in equilibrium with the silica gel surface be (c) g/ml. Let a fraction (α) of the surface be covered with a *mono-layer* of the polar solvent (B) and, of that fraction (α), let a fraction (β) be covered by a *second* layer of the polar solvent (B). The number of molecules striking and adhering to the surface covered with a mono-layer of polar solvent (A) and that covered with a mono-layer of solvent (B) per unit time will be (n') and be (n") respectively. Furthermore, let the number of molecules of solvent (A) leaving the mono-layer surface and the bi-layer surface per unit time be n_1 and n_2 respectively. Now, under conditions of equilibrium,

$$n' = n_1 \quad \text{and} \quad n'' = n_2$$

Furthermore, at a given temperature,

$$n'' = ac\alpha(1-\beta) \quad \text{and} \quad n_2 = b\alpha\beta \quad \text{where, (a) and (b) are constants}$$

Thus, $$ac\alpha(1-\beta) = b\alpha\beta$$

Solving for (β),
$$\beta = \frac{ac}{b + ac} \qquad (6)$$

This equation is the mono-layer adsorption isotherm of solvent (B) and is exactly the same as the previously derived Langmuir adsorption equation but somewhat differently expressed.

Now, in a similar manner, $\quad n' = dc(1-\alpha)$ and $n_1 = e\alpha(1-\beta)$

$$dc(1-\alpha) = e\alpha(1-\beta)$$

and, solving for (α),
$$\alpha = \frac{bdc + adc^2}{eb + bdc + adc^2}$$

The total area covered (S), expressed as fraction of the total area, will be

$$S = \alpha + \alpha\beta$$

Substituting for (β) from equation (6) and simplifying,

$$S = \alpha + \alpha\beta = \alpha\left(1 + \frac{ac}{b+ac}\right) = \alpha\left(\frac{b+2ac}{b+ac}\right)$$

Substituting for (α),

$$S = \frac{bdc + adc^2}{eb + bdc + adc^2}\left(\frac{b+2ac}{b+ac}\right)$$

or
$$S = \frac{bdc + 2adc^2}{eb + bdc + adc^2}$$

Simplifying,
$$S = \frac{Ac + 2Bc^2}{D + Ac + Bc^2} \qquad (7)$$

where $A = bd$, $B = ad$, and $D = eb$

Equation (7) is the Langmuir function for bi-layer adsorption. The expression gives a value for the total amount of moderator on the surface ($C_{g/g}$), which is measured when determining the adsorption isotherm and is given by

$$C_{g/g} = \left(\frac{Ac + 2Bc^2}{D + Ac + Bc^2}\right)\phi$$

where (ϕ) is the mass of a single layer of the polar solvent covering the surface of one gram of silica gel.

A theoretical curve for bi-layer adsorption was calculated from experimental data [3] and is given in Figure 7. The actual values obtained are superimposed on the theoretical curve. An excellent fit is obtained and it is clear that, by comparing the curve shape given in Figure 7 with that given in Figure 5, the single layer adsorption function could not fit the experimental data.

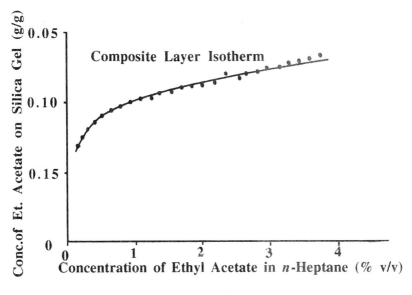

Courtesy of the Journal of Chromatography (ref.4)

Figure 7. The Bi-layer Adsorption Isotherm of Ethyl Acetate on Silica Gel

It should be noted that, due to the strong polarity of the hydroxyl groups on the silica, the initial adsorption of the ethyl acetate on the silica surface is extremely rapid. The individual isotherms for the two adsorbed layers of ethyl acetate are shown in Figure 8. The two curves, although similar in form, are quite different in magnitude. The first layer, which is very strongly held, is complete when the concentration of ethyl acetate is only about 1%w/w. At concentrations in excess of 1%w/w, the second layer is only just being formed. The formation of the second layer is much slower and the interactions between the solvent molecules with those already adsorbed on the surface are much weaker.

Assuming that the total area covered by the first layer will be the same as the area covered by the second layer, then only about onethird of the layer is complete at a concentration of about 4%w/v. This is in striking contrast to the formation of the first layer which is virtually complete at an ethyl acetate concentration of 1%w/v.

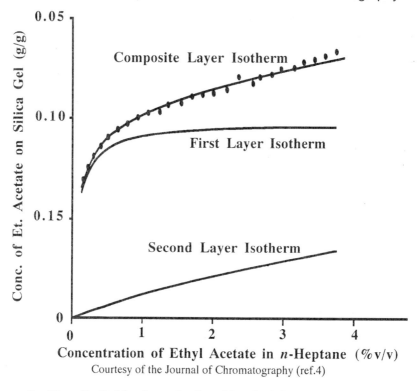

Figure 8. The Individual and Combined Adsorption Isotherms for Ethyl Acetate on Silica Gel

From the point of view of solute interaction with the structure of the surface, it is now very complex indeed. In contrast to the less polar or dispersive solvents, the character of the interactive surface will be modified dramatically as the concentration of the polar solvent ranges from 0 to 1%w/v. However, above 1%w/v, the surface will be modified more subtly, allowing a more controlled adjustment of the interactive nature of the surface It would appear that multi-layer adsorption would also be feasible. For example, the second layer of ethyl acetate might have an absorbed layer of the dispersive solvent *n*-heptane on it. However, any subsequent solvent layers that may be generated will be situated further and further from the silica surface and are likely to be very weakly held and sparse in nature. Under such circumstances their presence, if in fact real, may have little impact on solute retention.

Interactive Mechanisms with a Stationary Phase Surface in LC

Retention is controlled by solute interactions with both the mobile phase and the stationary phase and each will be discussed in this chapter. Interactions in the mobile phase are three dimensional and their effect on retention is controlled by the nature of the interaction and the probability of it occurring. The same is partly true for the stationary phase but, as in most cases the stationary phase takes the form of a surface, the interactions are two-dimensional. In addition, when multicomponent solvents are employed, if the surface is not completely covered with one component of the mobile phase, then the interacting surface constitutes a two-dimensional mixture in much the same way as a three-dimensional mixture is present in the mobile phase. It is clear that the interaction of the solute molecules with the stationary phase surface can be quite complex and also change with the composition of the mobile phase.

Interactions of a Solute with a Stationary Phase Surface

There are two ways a solute can interact with a stationary phase surface. The solute molecule can interact with the adsorbed solvent layer and rest on the top of it. This is called *sorption interaction* and occurs when the molecular forces between the solute and the stationary phase are relatively weak compared with the forces between the solvent molecules and the stationary phase. The second type is where the solute molecules displace the solvent molecules from the surface and interact directly with the stationary phase itself. This is called *displacement interaction* and occurs when the interactive forces between the *solute* molecules and the stationary phase surface are much stronger than those between the *solvent* molecules and the stationary phase surface. An example of sorption interaction is shown in Figure 9.

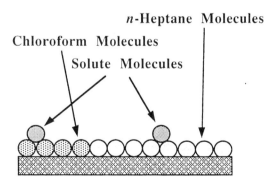

Figure 9. **Sorption Interaction**

It diagramatically represents a silica surface in contact with a low concentration of chloroform in *n*-heptane where the surface is partly covered with chloroform, the remainder covered with *n*-heptane. The solute molecules can either rest on the surface of the chloroform layer or on the surface of the layer of adsorbed *n*-heptane.

The second type of interaction, displacement interaction, is depicted in Figure 10. This type of interaction occurs when a strongly polar solute, such as an alcohol, can interact directly with the strongly polar silanol group and displaces the adsorbed solvent layer. Depending on the strength of the interaction between the solute molecules and the silica gel, it may displace the more weakly adsorbed solvent and interact directly with the silica gel but interact with the other solvent layer by sorption. Alternatively, if solute–stationary phase interactions are sufficiently strong, then the solute may displace both solvents and interact directly with the stationary phase surface.

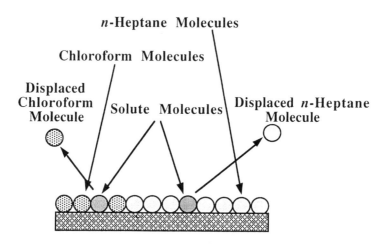

Figure 10. **Displacement Interaction**

If one solvent is very strongly interactive, as in the case of ethyl acetate and *n*-heptane, then, as already discussed, a very complex range of interactive possibilities exist which are depicted in Figure 11.

It is clear that such a surface offers a wide range of sorption and displacement processes that can take place between the solute and the stationary phase surface. Due to the bi-layer formation there are three different surfaces on which a molecule can interact by sorption and three different surfaces from which molecules of solvent can be displaced and allow the solute molecule to penetrate to the next layer. During a chromatographic separation under these circumstances, all the alternatives are possible. Nevertheless, depending on the magnitude of the forces between the solute molecule and the molecules in each layer, it is likely that one particular type of interaction will dominate. The various types of interaction are included in Figure 11.

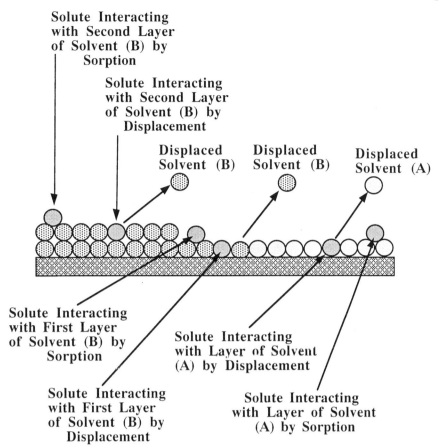

Figure 11. The Different Types of Solute Interaction that can Occur on a Silica Surface Containing a Solvent Bi-layer

Where there are multi-layers of solvent, the most polar is the solvent that interacts directly with the silica surface and, consequently, constitutes part of the first layer the second solvent covering the remainder of the surface. Depending on the concentration of the polar solvent, the next layer may be a second layer of the same polar solvent as in the case of ethyl acetate. If, however, the quantity of polar solvent is limited, then the second layer might consist of the less polar component of the solvent mixture. If the mobile phase consists of a ternary mixture of solvents, then the nature of the surface and the solute interactions with the surface can become very complex indeed. In general, the stronger the forces between the solute and the stationary phase itself, the more likely it is to interact by displacement even to the extent of displacing both layers of solvent (one of the alternative processes that is not depicted in Figure 11). Solutes that exhibit weaker forces with the stationary phase are more likely to interact with the surface by sorption.

Experimental Support for the Sorption and Displacement Processes

Scott and Kucera [4] carried out some experiments that were designed to confirm that the two types of solute/stationary phase interaction, sorption and displacement, did, in fact, occur in chromatographic systems. They dispersed about 10 g of silica gel in a solvent mixture made up of 0.35%w/v of ethyl acetate in *n*-heptane. It is seen from the adsorption isotherms shown in Figure 8 that at an ethyl acetate concentration of 0.35%w/v more than 95% of the first layer of ethyl acetate has been formed on the silica gel. In addition, at this solvent composition, very little of the second layer was formed. Consequently, this concentration was chosen to ensure that if significant amounts of ethyl acetate were displaced by the solute, it would be derived from the first layer on the silica and not the less strongly held second layer.

The container was sealed with a serum cap and thermostatted at 25°C. 100 mg aliquots of the solute were added sequentially to the mixture by means of a hypodermic syringe. After each addition, the container was shaken, thermal equilibrium allowed to become established over a period of about 30 minutes and then a 5 µl sample of the solvent taken for GC analysis. Corrections were made for the volume of the samples taken and the ethyl acetate and solute they contained. The amount of ethyl acetate in the solvent was assayed and, from the amount of ethyl acetate originally added, the amount of ethyl acetate on the silica gel calculated.

From the concentration of the solute in the solvent, and the total amount added, the quantity of solute adsorbed on the stationary phase was also calculated. The results obtained for the solutes anisole and nitrobenzene are shown as graphs in Figure 12. One pair of curves refers to the polar solvent and relates the concentration of ethyl acetate in the solvent (E_m) and the concentration of ethyl acetate in the stationary phase (E_s) to the total mass of solute added. The other pair of curves refers to the solute and relates the concentration of solute in the solvent (S_m) and the concentration of solute in the stationary phase (S_s) to the total mass of solute added.

It should be first noted that the curves relating the concentration of ethyl acetate in the solvent mixture and on the stationary phase are straight and horizontal. As the initial concentration of ethyl acetate in mobile phase was 0.35%w/v, the volume of mobile phase was 100 ml and the mass of silica was 10 g. It follows that, although a total of about 1.2 g of solute was added to the system, about a third of which resided on the silica surface, neither anisole nor nitrobenzene displaced any ethyl acetate from the silica gel.

Mixed Phases

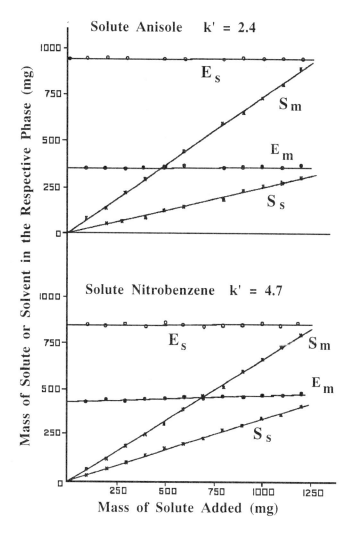

E_m is the Mass of Ethyl Acetate in Mobile Phase
S_m is the Mass of Solute in Mobile Phase
E_S is the Mass of Ethyl Acetate on Silica Gel
S_S is the Mass of Solute on the Silica Gel

Courtesy of the Journal of Chromatography (ref.4)

Figure 12. Graphs of Mass of Solute and Mass of Polar Solvent Contained in the Mobile Phase and on the Silica Gel, Respectively, against Mass of Solute Added

In addition, the concentration of solute in both phases increased almost linearly with the amount of solute added. This would indicate that the system was operating over that part of the adsorption isotherm that was linear. Thus, the solute was interacting with the stationary phase by a sorption process. The two solutes, anisole and nitrobenzene, are not well retained in a chromatographic column. In fact, on a silica gel column packed with the same silica and chromatographed with the same solvent mixture, anisole and nitrobenzene were eluted at (k') values of 2.4 and 4.7 respectively.

The authors repeated the experiment with two, more strongly retained, solutes *m*-dimethoxy benzene and benzyl acetate. These solutes were found to elute at (k') values of 10.5 and 27.0 respectively on a silica column operated with the same mobile phase. The results obtained are shown as similar curves in Figure 13. The *m*-dimethoxy benzene, which eluted at a (k') of 10.5, also failed to displace any ethyl acetate from the silica gel even when more than 0.5 g of solute resided on the silica surface. Consequently, the *m*-dimethoxy benzene must have also interacted with the surface by a sorption process.

However, the solute benzyl acetate, which was more polar and eluted at a (k') of 27, behaved in an entirely different manner. As soon as the solute was added to the system, ethyl acetate began to be displaced from the silica surface in a regular manner. The concentration of ethyl acetate in the mobile phase began to increase, accompanied by an equivalent decrease in the amount of ethyl acetate residing on the silica gel. It is also seen, that the curve relating the concentration of benzyl acetate on the silica gel with the amount added, is no longer linear but takes on the characteristic profile of the Langmuir adsorption isotherm. This demonstrated that the more polar solute was interacting with the silica surface by a displacement process in contrast to the other three solutes. The results of the experiments confirm that, in contrast to the less polar solutes, benzyl acetate interacted with the surface by displacing the ethyl acetate and interacted directly with the silica surface. Again, as the initial concentration of ethyl acetate in mobile phase was 0.35%w/v, the volume of mobile phase 100 ml and the mass of silica 10 g, the ethyl acetate must have been removed directly from the silica surface. The mode of interaction with the stationary phase surface will differ for diverse solutes. Other distribution models may exhibit sorption and displacement interactions with very different solute types and at widely different solvent concentrations. Solute retention does not depend only on the interactions of the solute with the stationary phase but also on interactions with the mobile phase. Changes in solvent composition will change the nature of the surface and consequently, the solute interactions with the surface.

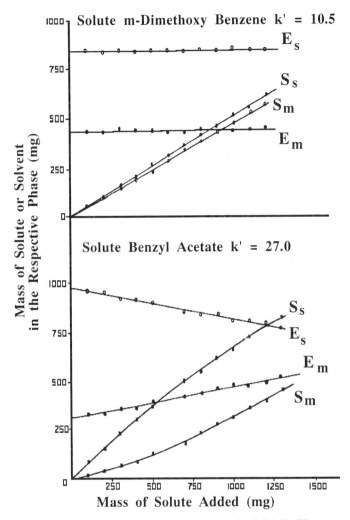

E_m = Mass of Ethyl Acetate in the Mobile Phase.
S_m = Mass of Solute in the Mobile Phase.
E_S = Mass of Ethyl Acetate on Silica Gel.
S_S = Mass of Solute on the Silica Gel.

Courtesy of the Journal of Chromatography (ref.4)

Figure 13. Graphs of Mass of Solute and Mass of Polar Solvent Contained in the Mobile Phase and on the Silica Gel, Respectively against Mass of Solute Added

However, the same changes in the mobile phase will also modify the interactions of the solute with the mobile phase, and the effect on solute retention can be as great or even greater than the modification of the stationary phase. Changes in the interactive character of the stationary phase usually occur at relatively low concentrations of moderator, and after the surface has been covered with the moderator, the effect of mobile phase composition on retention is solely due to changes in the interactive character of the mobile phase.

High Concentrations of Moderator

Concentrations of moderator at or above that which causes the surface of a stationary phase to be completely covered can only govern the interactions that take place in the mobile phase. It follows that retention can be modified by using different mixtures of solvents as the mobile phase, or in GC by using mixed stationary phases. The theory behind solute retention by mixed stationary phases was first examined by Purnell and, at the time, his discoveries were met with considerable criticism and disbelief. Purnell *et al.* [5], Laub and Purnell [6] and Laub [7], examined the effect of mixed phases on solute retention and concluded that, for a wide range of binary mixtures, the corrected retention volume of a solute was linearly related to the volume fraction of either one of the two phases. This was quite an unexpected relationship, as at that time it was tentatively (although not rationally) assumed that the retention volume would be some form of the exponent of the stationary phase composition. It was also found that certain mixtures did not obey this rule and these will be discussed later. In terms of an expression for solute retention, the results of Purnell and his co-workers can be given as follows,

$$V'_{r(AB)} = K_A \alpha V_S + K_B (1-\alpha) V_S \tag{7}$$

where ($V'_{r(AB)}$) is the retention volume of the solute on a mixture of stationary phases (A) and (B),

(K_A) is the distribution coefficient of the solute with respect to the pure stationary phase (A),

(K_B) is the distribution coefficient of the solute with respect to the pure stationary phase (B),

(V_S) is the total volume of stationary phase in the column

and (α) is the volume fraction of phase (A) in the stationary phase mixture

That is, $V'_{r(AB)} = \alpha V'_A + (1-\alpha) V'_B$

where (V'_A) is the retention volume of the solute on the same volume of pure phase (A)

and (V'_B) is the retention volume of the solute on the same volume of pure phase (B)

Mixed Phases

Rearranging,

$$V'_{AB} = \alpha(V'_A - V'_B) + V'_B \tag{8}$$

This remarkably simple relationship is depicted in Figure 14. It was apparent from his results that the volume fraction of the solvent determined the probability of interaction with the solute in much the same way that the partial pressure of a gas determines the probability of collision. It also indicated that the influence of each stationary phase component was independent and unaffected by presence of the other.

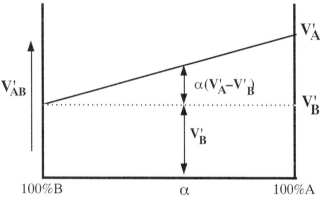

Figure 14. Graph of Corrected Retention Volume against Volume Fraction of Stationary Phase

Purnell proved the independent effect of each component of the stationary phase by the following three simple but very elegant experiments. In the first experiment, the two fractions of stationary phase were mixed, coated on some support, and packed into the column and the retention volume measured. In the second experiment, the same fractions of stationary phase were then independently coated on some support, the supports mixed and packed into a column. The retention volume of the same solute was again measured. In the third experiment, separate samples of supports were coated with the same fractions of stationary phase and each packed into separate columns and the columns joined in series. The retention volume of the solute was again measured. It should be emphasized that each column system contained exactly the same amount of each stationary phase. The three columns gave exactly the same corrected retention volume for any given solute. The experiments are fundamentally important and the experiment is depicted in Figure 15. The relationship described by equations (7) and (8), however, was found *not to be universal* and did break down when there appeared to be strong association between the two components of the stationary phase. Under such circumstances the stationary phase would no longer be a simple binary mixture but would also contain the associate of the two phases as a third component. It follows that with three components present, the simple linear relationship obtained for a binary mixture could not be expected to hold.

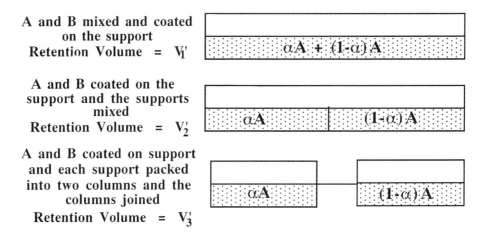

Purnell experimentally demonstrated that $V'_1 = V'_2 = V'_3$

Figure 15. Alternative Methods for Combining Volume Fractions of Stationary Phase in GC

There is another interesting outcome from the results of Purnell and his co-workers. For those phase systems that gave a linear relationship between retention volume and volume fraction of stationary phase, it was clear that the linear functions of the distribution coefficients *could be summed directly*, but their *logarithms could not*. This casts a doubt on the thermodynamic procedure for describing the effect of phase composition on solute retention, where the stationary phase composition is often taken into account by including an extra term in the expression of the standard free energy of distribution. The results of Purnell indicate that, providing strong association does not take place between the phase components, then solute retention and the distribution coefficient are *linearly,* **not** *exponentially,* related to the stationary phase composition.

It follows that stationary phases of intermediate polarities can be formed from binary mixtures of two phases, one strongly dispersive and one strongly polar. This procedure is not used extensively in commercial columns, although is the easiest and most economic method of fabricating columns having intermediate polarities.

Mixed Phases

Nevertheless, mixed phases are always worth considering as a flexible alternative to the use of a specific proprietary material.

Mixed Mobile Phases in LC

Although, for most moderators, the surface of a stationary phase in LC can be considered stable at moderator concentrations above about 5%v/v, the results from the same experiments as those carried out by Purnell and his group could still be considered invalid and, at best, would not lead to unambiguous conclusions. Katz *et al.* [9] avoided this problem by examining liquid/liquid distribution systems using water as one phase and a series of immiscible solvent mixtures as the other and by measuring absolute distribution coefficients as opposed to retention volumes.

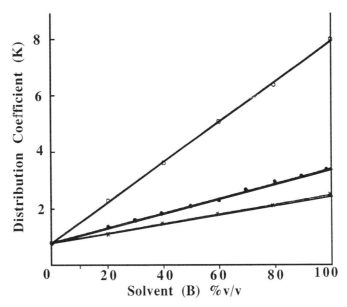

Solvent A	Solvent B	
1 *n*-Heptane	*n*-Heptyl Acetate	O
2 *n*-Heptane	Toluene	●
3 *n*-Heptane	*n*-Heptyl Chloride	X

Courtesy of the Journal of Chromatography (ref.10)

Figure 16. Graphs Showing the Distribution Coefficient of *n*-Pentanol between Water and Three Binary Solvent Mixtures Plotted against Solvent Composition

They measured the distribution coefficient of *n*-pentanol between water and mixtures of *n*-heptane and chloroheptane, *n*-heptane and toluene, and *n*-heptane and heptyl acetate. The two phase system was thermostatted at 25°C and, after equilibrium had been established, the concentration of solute in the two phases was determined by GC analysis. The results they obtained are shown in Figure 16. The same linear relationship between solvent composition and distribution coefficient was obtained for all three solvent mixtures, which closely matched the results that Purnell and Laub obtained in their gas chromatography experiments. The relationship between distribution coefficient and solvent composition will be described by exactly the same equation. Consequently, if the distribution coefficient of pentanol between water and any of the pure solvents is known, then the distribution coefficient can be calculated for any binary mixture of those solvents and water. Again the results confirm that the solute interacts with each component of the phase mixture independently. It also confirms that, in the distribution systems examined, the actual distribution coefficients can be summed directly to provide the overall distribution coefficient but not their logarithms. The simplest explanation of the effects observed is that the concentration of any solvent in the mixture determines the probability of interaction between the solute and that solvent. For example, in the pure solvent the probability of interaction is unity. In a 50%v/v mixture of two solvents, the probability of interaction between the solute and either solvent is 0.5.

Katz *et al.* tested the theory further and measured the distribution coefficient of *n*-pentanol between mixtures of carbon tetrachloride and toluene and pure water and mixtures of *n*-heptane and *n*-chloroheptane and pure water. The results they obtained are shown in Figure 17. The linear relationship between the distribution coefficient and the volume fraction of the respective solvent was again confirmed. It is seen that the distribution coefficient of *n*-pentanol between water and pure carbon tetrachloride is about 2.2 and that an equivalent value for the distribution coefficient of *n*-pentanol was obtained between water and a mixture containing 82%v/v chloroheptane and 18%v/v of *n*-heptane. The experiment with toluene was repeated using a mixture of 82 %v/v chloroheptane and 18% *n*-heptane mixture in place of carbon tetrachloride which was, in fact, a ternary mixture comprising of toluene, chloroheptane and *n*-heptane. The chloroheptane and *n*-heptane was always in the ratio of 82/18 by volume to simulate the interactive character of carbon tetrachloride.

The results confirmed that the chloroheptane/*n*-heptane mixture behaves in an identical manner to carbon tetrachloride and all the points were on the same straight line as that produced using a mixture of carbon tetrachloride and toluene. These experiments are similar to normal phase chromatography using pure water instead of

Mixed Phases

silica gel, except that the water phase is not modified by the solvents in the way a silica gel surface would be.

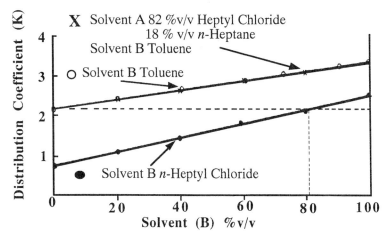

Courtesy of the Journal of Chromatography (ref.9)

Solvent A	Solvent B
Carbon Tetrachloride	Toluene
n-Heptane	*n*-Heptyl Chloride
82%v/v Heptyl Chloride, 18%v/v n-Heptane	Toluene

Figure 17. Distribution Coefficients of *n*-Pentanol between Water and Two Binary and One Ternary Solvent Mixtures Plotted against Solvent Composition

Reiterating the equation proposed by Purnell, for two solvents (A) and (B) in GC

$$V'_{AB} = \alpha(V'_A - V'_B) + V'_B \qquad (9)$$

The results of Katz *et al.* can be algebraically expressed in a similar form as in LC,

$$K_{AB} = \alpha(K_A - K_B) + K_B \qquad (10)$$

For chromatography purposes the product of the distribution coefficient and the volume of stationary phase, or stationary phase surface area, gives the corrected retention volume, *i.e.*,

$$V'_{AB} = K_{AB}\phi, \; V'_A = K_A\phi \text{ and } V'_B = K_B\phi$$

In the experiments of Katz *et al.*, that validated the relationship given in equation (10), the distribution coefficients (K) were referred to the *solvent phase* (mobile phase), whereas in LC it is the mobile phase composition that is changed and the distribution coefficients (K") are referred to the *stationary phase*.

Thus,
$$K"_{AB} = \frac{1}{K_{AB}}, \quad K"_A = \frac{1}{K_A} \text{ and } K"_B = \frac{1}{K_B}$$

or,
$$V"_{AB} = \frac{1}{V'_{AB}}, \quad V"_A = \frac{1}{V'_A} \text{ and } V"_B = \frac{1}{V'_B}$$

Substituting for the corrected retention volumes in equation (9) for inverse phase system,

$$V"_{AB} = \frac{1}{\alpha\left(\dfrac{1}{V"_A} - \dfrac{1}{V"_B}\right) + \dfrac{1}{V"_B}}$$

Simplifying,
$$V"_{AB} = \frac{V"_A V"_B}{\alpha(V"_B - V"_A) + V"_A} \tag{11}$$

Thus, when $\alpha = 0$, $V"_{AB} = V"_B$ and when $\alpha = 1$ $V"_{AB} = V"_A$

Practically a more convenient way of expressing solute retention in terms of solvent concentration for a binary solvent mixture as the mobile phase is to use the inverse of equation (11), *i.e.*,

$$\frac{1}{V"_{AB}} = \alpha\left(\frac{1}{V"_A} - \frac{1}{V"_B}\right) + \frac{1}{V"_B} \tag{12}$$

If the corrected retention volume in the pure strongly eluting solute is very small compared with the retention volume of the solute in the other pure solvent. *i.e.*, $V"_A \ll V"_B$, which is very often the case in practical LC, then equation (12) simplifies to the simple reciprocal relationship

$$V"_{AB} = \frac{V"_A}{\alpha} \tag{13}$$

Under these conditions the reciprocal relationship fits the data extremely well, particularly at volume fractions below 0.5 of the strongly eluting solute. In addition, under these conditions the inverse will also apply, *i.e.*,

$$\frac{1}{V"_{AB}} = \frac{\alpha}{V"_A} \tag{14}$$

Mixed Phases

Equation (14) has been validated by the results from a number of workers [10-12] and an example of the expected correlation is given by the curves in Figure 18 [13].

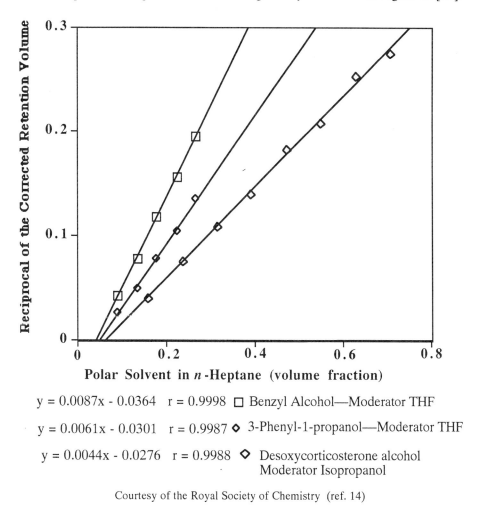

y = 0.0087x - 0.0364   r = 0.9998   □ Benzyl Alcohol—Moderator THF

y = 0.0061x - 0.0301   r = 0.9987   ◊ 3-Phenyl-1-propanol—Moderator THF

y = 0.0044x - 0.0276   r = 0.9988   ◊ Desoxycorticosterone alcohol
                                      Moderator Isopropanol

Courtesy of the Royal Society of Chemistry (ref. 14)

Figure 18. Graphs Relating the Reciprocal of the Corrected Retention Volume to the Volume Fraction of the Polar Solvent in n-Heptane

The column was 25 cm long, 4.6 mm I.D. and packed with Partisil 10. It is seen that linear curves were obtained for three different solutes and two different moderators in *n*-heptane. Scott and Beesley [14] obtained retention data for the two enantiomers, (S) and (R) 4-benzyl-2-oxazolidinone. The column chosen was 25 cm long, 4.6 mm I.D. packed with 5 mm silica particles bonded with the stationary phase Vancomycin (Chirobiotic V provided by Advanced Separations Technology Inc., Whippany, New Jersey). This stationary phase is a macrocyclic glycopeptide Vancomycin that has a molecular weight of 1449.22, and an elemental composition of 54.69% carbon,

5.22% hydrogen, 4.89% chlorine, 8.70% nitrogen and 26.50% oxygen. Vancomycin contains 18 chiral centers surrounding three 'pockets' or 'cavities' which are bridged by five aromatic rings and thus can readily offer unique enantiomeric selectivity to a wide range of chiral substances.

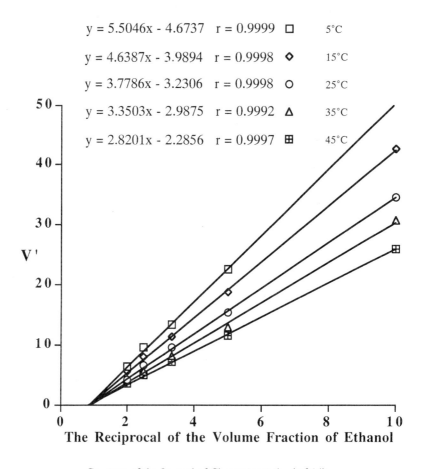

Courtesy of the Journal of Chromatography (ref.14)

Figure 19 Graph of Corrected Retention Volume of the (S) 4-Benzyl-2-oxazolidinone against the Reciprocal of the Volume Fraction of Ethanol

The column was operated in the normal phase mode using mixtures of *n*-hexane and ethanol as the mobile phase. Equation (13) is validated by the curves relating the corrected retention volume to the reciprocal of the volume fraction of ethanol in Figure 19. It is seen that an excellent linear relationship is obtained between the corrected retention volume and the reciprocal of the volume fraction of ethanol.

Ternary Mixtures

Extending the relationship given in equation (10) to a ternary mixture of solvents (A), (B) and (C)

$$V''_{AB} = \frac{1}{\alpha \frac{1}{V''_A} + \beta \frac{1}{V''_B} + \gamma \frac{1}{V''_C}} \quad (15)$$

Simplifying,

$$V''_{AB} = \frac{V''_A V''_B V''_C}{\alpha(V''_B V''_C) + \beta(V''_A V''_C) + \gamma(V''_A V''_B)} \quad (16)$$

where (α), (β) and (γ) are the volume fractions of solvents (A), (B) and (C), respectively, and $\alpha + \beta + \gamma = 1$

Thus, when $\alpha=1$, $\beta=0$ and $\gamma=0$, then $V''_{ABC} = V''_A$; when $\alpha=0$, $\beta=1$ and $\gamma=0$, then $V''_{ABC} = V''_B$; and finally when $\alpha=0$, $\beta=0$ and $\gamma=1$, then $V''_{ABC} = V''_C$.

Equation (16) was tested against some data obtained for (R) 4-phenyl-2-oxazolidinone using a range of mixtures of ethanol, acetonitrile and *n*-hexane as the mobile phase. The column chosen was similar to that previously used for the separation of the 4-phenyl-2-oxazolidinone which was 25 cm long, 4.6 mm I.D. packed with 5 mm silica particles bonded with the stationary phase Vancomycin. The results obtained are shown in Table 1 and this is the data used in subsequent computer calculations.

Table. 1 Retention Volume of (R) 4-Phenyl-2-oxazolidinone Determined at Different Solvent Compositions

Volume Fraction	Retention Volume (V_r) (ml)				
Acetonitrile →	0.1	0.2	0.3	0.4	0.5
Isopropanol ↓					
0.5	5.57	4.58	4.07	3.78	3.57
0.4	6.20	5.00	4.43	4.04	3.84
0.3	7.65	5.84	4.94	4.36	4.10

The range of concentrations that could be used was somewhat restricted due to the immiscibility of certain solvent mixtures. The data given in Table 1 was fitted to equation (16) and the various constants determined. Employing the constants derived from the curve fitting process, the theoretical values for the retention volumes, at the different solvent compositions, were calculated and are shown plotted against the experimental data in Figure 20.

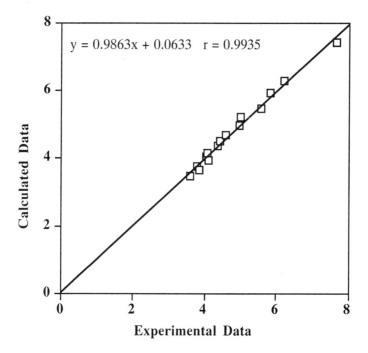

Figure 20. Calculated Retention Volume of 4-Phenyl-2-oxazolidinone Plotted against Those Experimentally Determined

It is seen that there is a good correlation between experimental and calculated values. The scatter that does exist may be due to the dead volume of the column not being precisely independent of the solvent composition. The dead volume will depend, to a small extent, on the relative proportion of the different solvents adsorbed on the stationary phase surface, which will differ as the solvent composition changes. A constant value for the dead volume was assumed in the computer program that derived the equation.

The overall effect of solvent composition on retention can be better understood by constructing a 3-D graph relating retention on the (z) axis to the volume fraction of ethanol and acetonitrile on the (x) and (y) axes. A set of such a curves is shown in Figure 21. As would be expected, the greatest retention is observed at the lowest concentrations of both acetonitrile and isopropanol. The contours showing the relationship between the retention volume and volume fraction of each component are smooth and can be accurately predicted. At high concentrations of acetonitrile, the effect of changes in volume fraction of isopropanol on retention is minimal. At low

Mixed Phases

concentrations of acetonitrile, however, the amount of isopropanol present can significantly effect solute retention. Overall, the more polar solvent, acetonitrile, has by far the greatest control over the elution rate of the solutes.

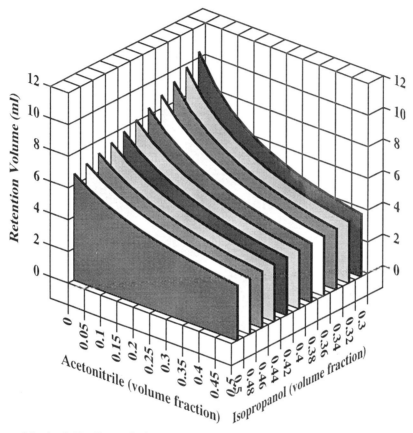

Figure 21 A 3-D Set of Curves Relating the Retention Volume of 4-Phenyl-2-oxazolidinone to the Composition of a Ternary Solvent Mixture

It is clear that if the same data was produced for the other enantiomer, then a three dimensional graph relating the separation ratio of the enantiomer pair against the solvent composition could be constructed. Repeating the computer iteration program for the other isomer, the separation ratio of the pair of enantiomers was calculated and the data presented as curves relating separation ratio to solvent composition (Figure 22). The relationship between separation ratio and solvent composition is somewhat unexpected. The largest separation ratio is obtained at the highest concentrations of the two polar solvents, where retention is the smallest for the solvent composition range examined. It is also clear that the more polar solvent, acetonitrile, has a greater effect on the separation ratio of this particular pair of enantiomers than does the less polar solvent isopropanol.

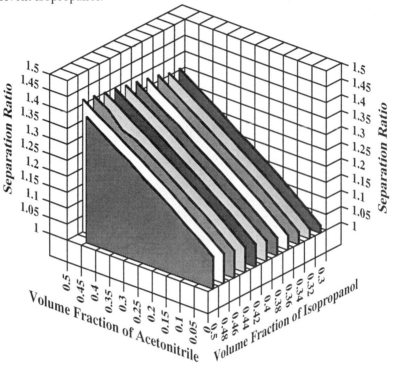

Figure 22. A 3-D Set of Curves Relating the Separation Ratio of the Two Enantiomers of 4-Phenyl-2-oxazolidinone to the Composition of a Ternary Solvent Mixture

Although it is often possible to predict the effect of the solvent on retention, due to the unique interactive character of both the solvents and the enantiomers, it is virtually impossible to predict the subtle differences that control the separation ratio from present knowledge. Nevertheless, some accurate retention data, taken at different solvent compositions, can allow the retention and separation ratios to be calculated over a wide range of concentrations using the procedure outlined above. From such data the phase system and the column can be optimized to provide the separation in the minimum time, a subject that will be discussed later in the treatment of chromatography theory.

The Combined Effect of Temperature and Solvent Composition on Solute Retention

If the relationship between retention volume and temperature is known, as well as its dependence on solvent composition, then the combined effect of both solvent

Mixed Phases

composition and temperature on retention can be calculated. Equations (17) and (18) were experimentally derived by Scott and Beesley [14] to express the retention volume of the two enantiomers of 4-benzyl-2-oxazolidinone in terms of column temperature and solvent composition.

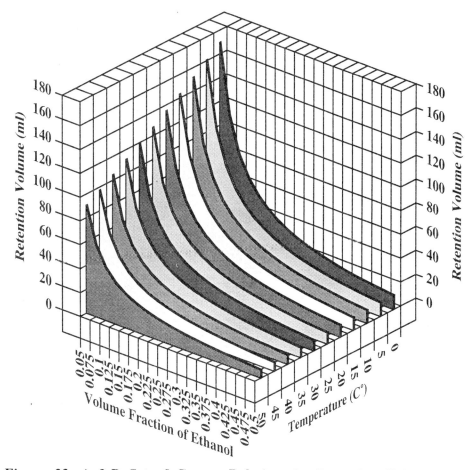

Figure 23. A 3-D Set of Curves Relating the Retention Volume of 4-Phenyl-2-oxazolidinone to the Column Temperature and the Volume Fraction of Ethanol in the Solvent Mixture

$$V'_{S(c,T)} = \frac{10^{\frac{726.93}{T}-1.782}}{c} - 10^{\frac{636.98}{T}-1.571} \qquad (17)$$

$$V'_{R(c,T)} = \frac{10^{\frac{641.48}{T}-1.563}}{c} - 10^{\frac{661.33}{T}-1.698} \qquad (18)$$

The numerical constants were obtained over the temperature range of 5°C to 45°C and a concentration range of 0 to 0.5 volume fraction of ethanol in *n*–hexane. The effect of temperature and solvent composition on solute retention can, again, be best displayed by the use of 3-D graphs, and curves relating both temperature and solvent composition to the retention volume of the (S) enantiomer of 4-benzyl-2-oxazolidinone are shown in Figure 23. Figure 23 shows that the volume fraction of ethanol in the solvent mixture has the major impact on solute retention.

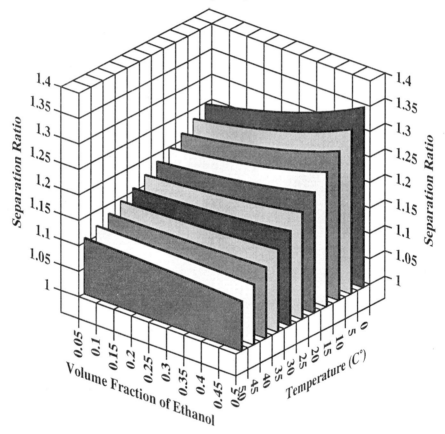

Figure 24. Curves Relating the Separation Ratio of the Two Enantiomers of 4-Phenyl-2-oxazolidinone to Temperature and Solvent Composition

The effect of temperature, although significant, is not nearly as great as that from the ethanol content and is greatest at low concentrations of the polar solvent. It is clear, that the solute retention is the least at high ethanol concentrations and high temperatures, which would provide shorter analysis times providing the selectivity of the phase system was not impaired. The combined effect of temperature and solvent composition on selectivity, however, is more complicated and to some extent

Mixed Phases

unexpected. The relationship between the separation ratio, of the two enantiomers, solvent composition and temperature is shown in Figure 24.

It is seen that at about 45°C the separation ratio appears to be independent of the solvent composition. This remarkable relationship can be examined theoretically.

Restating equation (10) but employing (β) as the volume fraction of solvent instead of (α)

$$K_{AB} = \beta(K_A - K_B) + K_B$$

It should be recalled that the distribution coefficients are referenced to the solvent mixture and not the stationary phase and are thus the inverse of the distribution coefficient employed in the chromatography elution equation.

Now the separation ratio (α) of a pair of solutes (1) and (2) is defined by

$$\alpha_{(1)(2)} = \frac{V'_{(1)}}{V'_{(2)}} = \frac{K_{(1)}V_S}{K_{(2)}V_S} = \frac{K_{(1)}}{K_{(2)}}$$

where ($\alpha_{(1)(2)}$) is the separation ratio of solute (1) to solute (2),
($V'_{(1)}$) is the corrected retention volume of solute (1),
($V'_{(2)}$) is the corrected retention volume of solute (2),
($K_{(1)}$) is the distribution coefficient of solute (1) with respect to the stationary phase,
($K_{(2)}$) is the distribution coefficient of solute (2) with respect to the stationary phase,
and (V_S) is the volume of stationary phase in the column.

Bearing in mind that the distribution coefficients given in the above equation are the reciprocal of the distribution coefficients given by Katz *et al.*, then

$$\alpha_{AB(1)(2)} = \frac{\beta(K_{A(2)} - K_{B(2)}) + K_{B(2)}}{\beta(K_{A(1)} - K_{B(1)}) + K_{B(1)}} \tag{19}$$

where the subscripts (1) and (2) refer to solutes (1) and (2) and the subscripts (A), (B) and (AB) have the meaning previously ascribed to them.

Simplifying,

$$\alpha_{AB(1)(2)} = \frac{\beta a + b}{\beta c + d} \tag{20}$$

where $a = (K_{A(2)} - K_{B(2)})$, $b = K_{B(2)}$,

and $c = (K_{A(1)} - K_{B(1)})$, $d = K_{B(1)}$.

Now if $\dfrac{a}{c} = \dfrac{b}{d} = \varepsilon$, where ($\varepsilon$) is a constant, then $a = c\varepsilon$ and $b = d\varepsilon$.

Thus, substituting for (a) and (b) in equation (4),

$$\alpha_{AB(1)(2)} = \dfrac{\beta c\varepsilon + d\varepsilon}{\beta c + d} = \dfrac{\varepsilon(\beta c + d)}{\beta c + d} = \varepsilon$$

Consequently, if or when $\alpha_{AB(1)(2)} = \dfrac{a}{c} = \dfrac{b}{d}$, then the separation ratio will be independent of the composition of the solvent mixture.

Thus, from equation (19) the condition for the independence of the separation ratio from the solvent composition will be when

$$\dfrac{K_{A(2)} - K_{B(2)}}{K_{A(1)} - K_{B(1)}} = \dfrac{K_{B(2)}}{K_{B(1)}}$$

or $\quad (K_{A(2)} - K_{B(2)})K_{B(1)} = K_{B(2)}(K_{A(1)} - K_{B(1)})$

or $\quad K_{A(2)}K_{B(1)} = K_{B(2)}K_{A(1)}$ (21)

Expressing equation (21) in thermodynamic terms using the conventional terms for the standard free energy,

$$e^{-\Delta G_{A(2)}} e^{-\Delta G_{B(1)}} = e^{-\Delta G_{A(1)}} e^{-\Delta G_{B(2)}}$$

Taking the logarithms, $\quad -\Delta G_{A(2)} - \Delta G_{B(1)} = -\Delta G_{A(1)} - \Delta G_{B(2)}$

Introducing the functions for standard free enthalpy and standard free entropy,

$$-\dfrac{\Delta H_{A(2)}}{RT} + \dfrac{\Delta S_{A(2)}}{R} - \dfrac{\Delta H_{B(1)}}{RT} + \dfrac{\Delta S_{B(1)}}{R} =$$
$$-\dfrac{\Delta H_{A(1)}}{RT} + \dfrac{\Delta S_{A(1)}}{R} - \dfrac{\Delta H_{B(2)}}{RT} + \dfrac{\Delta S_{B(2)}}{R}$$

Rearranging,

$$T = \dfrac{\Delta H_{A(1)} - \Delta H_{A(2)} + \Delta H_{B(2)} - \Delta H_{B(1)}}{\Delta S_{A(1)} - \Delta S_{A(2)} + \Delta S_{B(2)} - \Delta S_{B(1)}} \quad (22)$$

Mixed Phases

It is seen from equation (22) that there will, indeed, be a temperature at which the separation ratio of the two solutes will be independent of the solvent composition. The temperature is determined by the relative values of the standard free enthalpies of the two solutes between each solvent and the stationary phase, together with their standard free entropies. If the separation ratio is very large, there will be a considerable difference between the respective standard enthalpies and entropies of the two solutes. As a consequence, the temperature at which the separation ratio becomes independent of solvent composition may well be outside the practical chromatography range. However, if the solutes are similar in nature and are eluted with relatively small separation ratios (for example in the separation of enantiomers) then the standard enthalpies and entropies will be comparable, and the temperature/solvent-composition independence is likely be in a range that can be experimentally observed.

Validation of Temperature/Solvent-Composition Independence

The data from the separation of the enantiomers of 4-phenyl-2-oxazolidinone [14] gave an expression for the retention volume of the two enantiomers, which are reiterated as follows,

$$V'_{S(c,T)} = \frac{10^{\frac{726.93}{T} - 1.782}}{c} - 10^{\frac{636.98}{T} - 1.571}$$

$$V'_{R(cT)} = \frac{10^{\frac{641.48}{T} - 1.563}}{c} - 10^{\frac{661.33}{T} - 1.698}$$

Thus,

$$\alpha_{(S/R)} = \frac{\dfrac{10^{\frac{726.93}{T} - 1.782}}{c} - 10^{\frac{636.98}{T} - 1.571}}{\dfrac{10^{\frac{641.48}{T} - 1.563}}{c} - 10^{\frac{661.33}{T} - 1.698}}$$

or

$$\alpha_{(S/R)} = \frac{10^{\frac{726.93}{T} - 1.782} - \beta 10^{\frac{636.98}{T} - 1.571}}{10^{\frac{641.48}{T} - 1.563} - \beta 10^{\frac{661.33}{T} - 1.698}} \tag{23}$$

Developing equation (23) in the same way as equation (20), and applying the same arguments, the condition for separation ratio independence to solvent composition will be when

$$\frac{10^{\frac{726.93}{T}-1.782}}{10^{\frac{641.48}{T}-1.563}} = \frac{10^{\frac{636.98}{T}-1.571}}{10^{\frac{661.33}{T}-1.698}}$$

Simplifying and taking logarithms,

$$\frac{726.93}{T} - 1.782 - \frac{641.48}{T} + 1.563 = \frac{636.98}{T} - 1.571 - \frac{661.33}{T} + 1.698$$

Thus, $T = \frac{109.8}{0.346} = 317.3°A \equiv 43.9°C$

It is seen that the curves in Figure (24) become horizontal between 40°C and 45°C as predicted by the theory. It is also clear that there is likely source of error when exploring the effect of solvent composition on retention and selectivity. It would be important when evaluating the effect of solvent composition on selectivity to do so over a range of temperatures. This would ensure that the true effect of solvent composition on selectivity was accurately disclosed. If the evaluation were carried out at or close to the temperature where the separation ratio remains constant and independent of solvent composition, the potential advantages that could be gained from an optimized solvent mixture would never be realized.

Aqueous Solvent Mixtures

When the relationship between the distribution coefficient of a solute and solvent composition, or the corrected retention volume and solvent composition, was evaluated for aqueous solvent mixtures, it was found that the simple relationship identified by Purnell and Laub and Katz et al. no longer applied. The suspected cause for the failure was the strong association between the solvent and water. As a consequence, the mixture was not binary in nature but, in fact, a ternary system. An aqueous solution of methanol, for example, contained *methanol, water* and *methanol associated with water*. It follows that the prediction of the net distribution coefficient or net retention volume for a ternary system would require the use of three distribution coefficients: one representing the distribution of the solute between the stationary phase and water, one representing that between the stationary phase and methanol and one between the stationary phase and the methanol/water associate. Unfortunately, as the relative amount of association varies with the initial

Mixed Phases

composition of the methanol water mixture, the simple procedure used for the ternary mixture of unassociating solvents can not be used. The association of methanol and water was examined by Katz, Lochmüller and Scott [15] using data for the volume change on mixing and refractive index data and they established that the methanol/water solvent system was indeed a complex ternary system. Their theoretical development was as follows.

The relationship between water, methanol and water associated with methanol can be assumed to take the form,

$$\frac{[W][M]}{[WM]} = k \quad (24)$$

where (M) is the molar concentration of water,
(W) is the molar concentration of methanol,
(MW) is the molar concentration of methanol/water associate,
and (k) is the "association" constant

The molar concentration, in this context, is defined as the moles per milliliter *before* mixing. To be precise, as there is a volume change on mixing, equation (24) should be expressed in moles per milliliter after mixing. However, there is, at the maximum, only a 3% change in volume on mixing and so the value of (k) will only be modified by about 3%. Consequently, to simplify the algebra and still maintain a precision of 3% or better, the molar concentrations before mixing were used.

If the volume fraction of methanol in the original mixture was (α), then the volume fraction of water would be (1-α). The molar volume of a substance is the ratio of the molecular weight to the density and thus the molar concentration of methanol and water will be $(\frac{\alpha}{V_M})$ and $(\frac{(1-\alpha)}{V_W})$ respectively where (V_M) and (V_W) are the molar volume of methanol and water respectively.

Thus,
$$[M] + [MW] = \frac{\alpha}{V_M} \quad (25)$$

and
$$[W] + [MW] = \frac{(1-\alpha)}{V_W} \quad (26)$$

By the simple algebraic manipulation of equations (17), (18) and (19), it can be shown that

$$[W] = \frac{\left(-b \pm \left(b^2 - 4c\right)^{0.5}\right)}{2} \tag{27}$$

where
$$b = k + \frac{\alpha}{V_M} + \frac{\alpha}{V_W} - \frac{1}{V_W}$$

and
$$c = k\left(\frac{1}{V_W} - \frac{\alpha}{V_W}\right)$$

In addition,
$$[MW] = \frac{(1-\alpha)}{V_W} - [W] \tag{28}$$

and
$$[M] = \frac{\alpha}{V_M} + [MW] \tag{29}$$

As [M], [W] and [MW] were defined in moles per milliliter then the volume (v_i) after mixing will be

$$v_i = [W]V_W + [M]V_M + [MW]V_{MW} \tag{30}$$

where (V_{MW}) is the molar volume of the methanol/water associate.

The weight of the water contained in the solvent mixture (w_W) will be,

$$w_W = [W]M_W \tag{31}$$

The weight of the methanol contained in the solvent mixture (w_M) will be

$$w_M = [M]M_M \tag{32}$$

and the weight of the methanol/water associate contained in the solvent mixture (w_{MW}) will be

$$w_{MW} = [MW]M_{MW} \tag{33}$$

where (M_W), (M_M) and (M_{MW}) are the molecular weights of water, methanol and the methanol/water associate respectively.

Thus the weight (m_i) of the mixture originally containing a volume fraction (a_i) of methanol and a volume fraction ($1-\alpha_i$) of water will be

$$m_i = w_W + w_M + w_{MW}$$

and the density (d_i) of the mixture will be
$$d_i = \frac{m_i}{v_i} \tag{34}$$

Mixed Phases

Employing an iterative computer program, in conjunction with the above equations, Katz *et al.* examined a wide range of values for (k) and (V_{MW}) and calculated the volume change on mixing of a series of methanol/water mixtures having assumed volume fractions of methanol. The results for each selected values of (k) and (V_{MW}) and each volume fraction of methanol were compared with experimentally determined values of (v_i) and the specific values of (k) and (V_{MW}) that gave the minimum error between calculated and experimental results taken as the true values of (k) and (V_{MW}). The experimental values (represented as points) and the calculated values (represented as the curve) are shown in Figure 25.

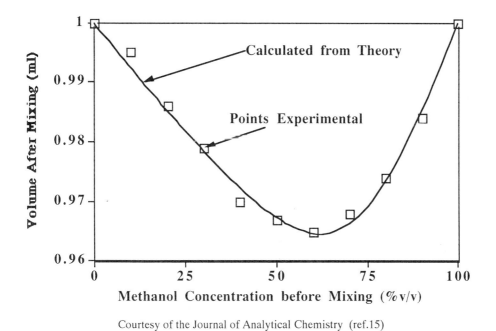

Courtesy of the Journal of Analytical Chemistry (ref.15)

Figure 25. Graph of Volume Change on Mixing against Solvent Composition for Methanol/Water Mixtures

Excellent agreement was obtained between experimental and calculated values of the volume change on mixing. Exactly the same approach was used to calculate values for (k) and (V_{MW}) using density data for the iteration program and the results obtained are shown in Figure 26. It is again seen that the selected values for (k) and (V_{MW}) that provide the minimum error between the experimental and calculated density values also provide a reliable estimate of the magnitude of these two parameters. Refractive index measurements can also provide an independent method of evaluating (k).

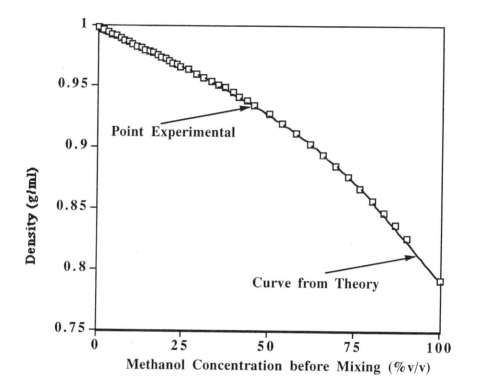

Courtesy of the Journal of Analytical Chemistry (ref.15)

Figure 26. Graph of Density against Solvent Composition for Methanol/Water Mixtures

The additive and constitutive properties of the molar refraction of a substance has been known for nearly a century and the molar refraction is defined by the following equation,

$$R = \frac{(n_r^2 - 1)V}{(n_r^2 + 2)} \quad (35)$$

where (R) is the molar refractivity of the substance,
(n_r) is the refractive index of the substance,
and (V) is the molar volume of the substance.

As the molar refractivity is additive and constitutive the molar refractivity of any mixture of (n) solvents is given by

Mixed Phases

$$R_{1,n} = X_1R_1 + X_2R_2 + X_3R_3 + \ldots + X_nR_n \tag{36}$$

where $(R_{1,n})$ is the molar refractivity of the mixture,
(R_1) is the molar refractivity of solvent (1),
(R_2) is the molar refractivity of solvent (2),
(R_3) is the molar refractivity of solvent (3),
(R_n) is the molar refractivity of solvent (n),
(X_1) is the molar fraction of solvent (1),
(X_2) is the molar fraction of solvent (2),
(X_3) is the molar fraction of solvent (3),
(X_n) is the molar fraction of solvent (n),
and (n) is the number of solvents in the mixture,

Taking the molar fractions of water, methanol and water/methanol associate to be $(X_M), (X_W)$ and (X_{MW}), respectively,

$$X_M = \frac{[M]}{[M] + [W] + [MW]} \tag{37}$$

$$X_W = \frac{[W]}{[M] + [W] + [MW]} \tag{38}$$

$$X_{MW} = \frac{[MW]}{[M] + [W] + [MW]} \tag{39}$$

For a mixture of methanol and water

$$R_i = X_M R_M + X_W R_W + X_{MW} R_{MW} \tag{40}$$

where (R_i) is the molar refractivity of the solvent mixture,
(R_M) is the molar refractivity of methanol,
(R_W) is the molar refractivity of water,
and (R_{MW}) is the molar refractivity of the methanol/water associate.

Thus, if (n_i) is the refractive index of the mixture and (V_i) is the equivalent molar volume of the mixture,

$$\frac{(n_i^2 - 1)V_i}{(n_i^2 + 2)} = X_M R_M + X_W R_W + X_{MW} R_{MW}$$

Now,
$$V_i = \frac{[M]M_M + [W]M_W + [MW]M_{MW}}{([M] + [W] + [MW])d_i} \tag{41}$$

Substituting for (V_i) from (41), for (X_M), (X_W) and (X_{MW}) from (37), (38) and (39); and simplifying,

$$\frac{(n_i^2 - 1)V_i}{(n_i^2 + 2)} = \frac{([M]R_M + [W]R_W + [MW]R_{MW})d_i}{[M]M_M + [W]M_W + [MW]M_{MW}} = Z \quad (42)$$

Thus,
$$n_i = \left(\frac{2Z+1}{1-Z}\right)^{0.5} \quad (43)$$

Taking known values for the molar refractivities of water and methanol, and again assuming a range of values for the equilibrium constant (k) and the refractive index (n_i) of the methanol/water associate, the actual values that fit the equation for these two parameters can be determined, using a similar iterative program as that used for the volume change on mixing and the density.

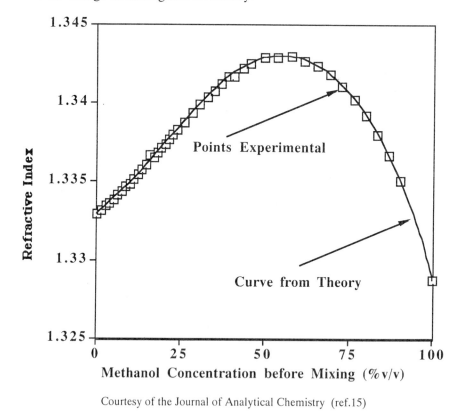

Courtesy of the Journal of Analytical Chemistry (ref.15)

Figure 27. Graph of Refractive Index against Solvent Composition for Methanol/Water Mixtures

The program identifies those values of (k) and (n_i) that provide the minimum error between the calculated values and the experiential values for the refractive index of the mixtures. The results obtained are shown in Figure (27). Excellent agreement between experimental and calculated values of the refractive indices was obtained for the different solvent mixtures. The results of all three estimates are summarized in Table 2.

Table 2. Summary of Equilibrium Data

Data Source	Equilibrium Constant (k)	Mol.Vol. of Associate	Density of Associate	Molar Refractivity of Associate	Refractive Index of Associate
Volume Change on Mixing	0.00443	55.46	0.9024		
Density	0.00565	54.90	0.9118		
Refractive Index	0.00504			11.88	1.3502
Mean	0.00504	55.18	0.9071	11.88	1.3502

It is seen that the three values for the equilibrium constant (k) range from 0.00443 to 0.00565 with an average value of 0.00504. The two values for the densities of the methanol/water associate are in reasonable agreement and have a magnitude that would be expected for the hydrogen bonded associate.

Using the average value for the equilibrium constant, the distribution concentration of the different components of a methanol water mixture were calculated for initial methanol concentrations ranging from zero to 100%v/v. The curves they obtained are shown in Figure 28. The molar refractivities of 11.88 is also in accordance with that expected since the molar refractivity's of water and methanol are 3.72 and 8.28 respectively. The refractive index of the associate of 1.3502 is, as would be expected, higher than that of either water or methanol.

It is seen from that there are three distinct ranges of methanol concentration where the solvent will behave very differently. At concentrations ranging from zero to 40%v/v of methanol, the solvent will behave as though it were a binary mixture of water and methanol associated with water. At concentrations ranging from 40%v/v to 80%v/v of methanol, the solvent will behave as though it were a ternary mixture of water, methanol and water associated with methanol. Finally, at concentrations ranging from 80%v/v to 100%v/v of methanol the solvent will behave as though it were again a binary mixture but this time a mixture of methanol and water associated with methanol.

132   Chromatography Theory

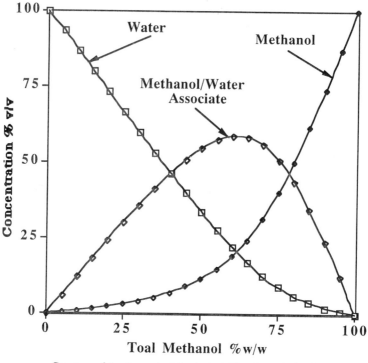

Courtesy of the Journal of Analytical Chemistry (ref.15)

Figure 28 Diagram of the Ternary Solvent System for Methanol/Water Mixtures

The curves shown in Figure 28 explain some of the unique characteristics of mobile phases consisting of methanol/water mixtures when used in reversed phase LC. It is seen that when the original methanol mixture contains 50%v/v of methanol, there is little free methanol available in the mobile phase to elute the solutes, as it is mostly associated with water. At higher methanol concentrations, however, the amount of methanol unassociated with water increases rapidly in the solvent mixture with an accompanying increase in the eluting strength of the mobile phase. If gradient elution is employed, this rapid increase eluting strength must be accommodated by the use of a convex gradient profile. The <u>convex</u> gradient profile will compensate for the strongly <u>concave</u> concentration profile of the unassociated methanol as shown in Figure 25. The strong association of methanol with water could also account for the fact that proteins can tolerate a significant amount of methanol in the mobile phase without them becoming denatured. This is because there is virtually no unassociated methanol present in the mixture which could cause protein denaturation since all the methanol is in a deactivated state by association with water.

Mixed Phases

Katz, Lochmüller and Scott also examined acetonitrile/water, and tetrahydrofuran (THF)/water mixtures in the same way and showed that there was significant association between the water and both solvents but not nearly to the same extent as methanol/water. At the point of maximum association for methanol, the solvent mixture contained nearly 60% of the methanol/water associate. In contrast the maximum amount of THF associate that was formed amounted to only about 17%, and for acetonitrile the maximum amount of associate that was formed was as little as 8%. It follows that acetonitrile/water mixtures would be expected to behave more nearly as binary mixtures than methanol/water or THF/water mixtures.

The Thermodynamic Properties of Water/Methanol Association

It is of interest to obtain values for the excess free enthalpy and excess free entropy of the association of water and methanol in order to assess the strength of the association and to disclose the effect of temperature on the formation of the methanol/water associate. In order to do this, the equilibrium constant needs to be determined over a series of temperatures to obtain the necessary van't Hoff curves. Gutierrez [16] as part of a Ph.D. Thesis at Georgetown University, Washington, DC, reported the density of a range of methanol/water mixtures at 25°C, 35°C, 45°C and 55°C. From this data, using an iterative program similar to that previously described, for volume change on mixing and refractive index, Scott [17] calculated the equilibrium constant at each temperature and the values he obtained are shown plotted against the reciprocal of the absolute temperature in Figure 29.

Taking the data from the linear curve fit to the curve in Figure 29,

$$\log(K) = \frac{-\Delta H}{RT} + \frac{\Delta S}{R} = \frac{-964.2}{T} + 1.03 \qquad (44)$$

Thus, $\quad \dfrac{-\Delta H}{R} = -964.2 \quad$ or $\quad \Delta H = 964.2R$

and $\quad \dfrac{\Delta S}{R} = 1.03 \quad$ or $\quad \Delta S = 1.03R$

Taking a value of R of 1.987 Kcal/°/mol, the standard free enthalpy and standard free entropy of distribution is 1916 cal/mol and 2.05 cal/°/mol. respectively. Bearing in mind that, in the pure state, water is strongly associated with water and methanol strongly associated with methanol, it would appear that the energy involved in the association of water with methanol is much greater than either.

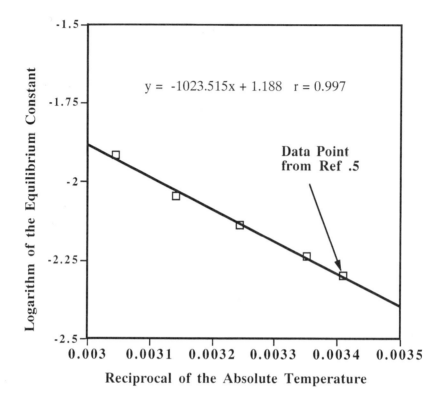

Figure 29. Graph of the Logarithm of the Water/Methanol Association Equilibrium Constant against the Reciprocal of the Absolute Temperature

This would mean, from the point of view of chromatography, that the interactive potential of both the methanol and the water with any dissolved solute is diluted by removing a significant proportion of both in the formation of the strongly bound and consequently less interactive associate. The net effect is to reduce the overall eluting capacity of the solvent. It is also interesting to note that the standard free entropy of association is relatively small indicating that the molecules are just as restrained when associated with themselves as when they are associated with each other, *i.e.,* their random character changes little.

Using the results calculated from the density data, the concentration of the individual components of the methanol/water mixture at different temperatures can be computed thus disclosing the effect of temperature on the elution properties of methanol/water mixtures. The results are shown in Figure 30.

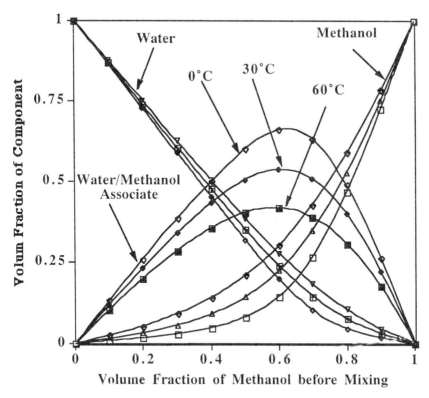

Figure 30. Curves Relating Volume Fractions of Methanol, Water and Methanol/Water Associate to Volume Fraction of Methanol before Mixing for Three Different Temperatures

It is clear from Figure 30 that raising the temperature reduces the amount of association and thus, at any given methanol content, there is more methanol free to interact with the solute at the higher temperatures. As methanol is, in most cases, the stronger eluting solvent, raising the temperature increases the elution rate in two ways. It increases the interaction between the solutes and the methanol and secondly, reduces the standard free enthalpy of distribution of the solute which also reduces retention and elution time. However, even at a temperature of 200°C the equilibrium constant (k) is still about 0.1 and so it is not practical to raise the operating temperature to a level where the methanol mixture behaves as a binary mixture and the simple relationship between retention and volume fraction of solvent applies.

Solute Interactions with Associated Solvents

Testing the applicability of equation (10) to liquids where the solvent components associate with themselves is experimentally difficult. Katz *et al.* attempted to do this by measuring the distribution coefficients of some solutes between hydrocarbon and methanol water mixtures; the solvents they used were *n*-pentanol, cyclohexyl acetate, vinyl acetate and benzene. The distribution system chosen was hexadecane (to closely emulate a reverse phase) and methanol/water mixtures having compositions ranging from 0 to 80%v/v of methanol in the original mixture made up. Obviously, due to the strong association of water and methanol the actual quantity of unassociated methanol in the solvent mixture will be very much less than the actual methanol added.

Considering the hexadecane/water-methanol system the same arguments and treatment can be afforded to the methanol/water mixture on the assumption that it is a ternary mixture containing methanol, water and methanol associated with water. Thus, the equation used for the system of Katz *et al.* reduces to

$$K = K_M \alpha_M + K_{MW} \alpha_{MW} + K_W \alpha_W$$

where (K_M) is the distribution coefficient of the solute between methanol unassociated with water and *n*-hexadecane,

(K_{MW}) is the distribution coefficient of the solute between methanol associated with water and *n*-hexadecane

(K_W) is the distribution coefficient of the solute between water unassociated with methanol and *n*-hexadecane,

(α_M) is the volume fraction of methanol in the original mixture,

(α_{MW}) is the volume fraction of the methanol/water associate in the original mixture,

and (α_W) is the volume fraction of water in the original mixture.

The results obtained by Katz *et al.* [15] are shown as experimental points on the curves relating the distribution coefficient of the solute against volume fraction of methanol added to the original mixture in Figure 31. Due to the difficulty of measuring the distribution coefficient of each solute between pure water and hexadecane (because of their extremely high retention), the values were obtained from a polynomial curve fit to the data which gave a value for (K) at $\alpha = 0$.

From the data obtained for each of the solutes and a knowledge of each respective value of α_1, α_2, and α_3 (obtained from the work of Katz *et al.* as previously discussed), the values of K_M and K_{MW} were calculated with the aid of a simple iterative computer program and the results obtained are included in each graph.

Mixed Phases 137

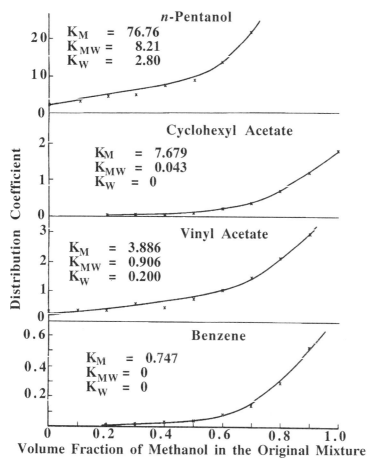

Figure 31. Graph of the Distribution Coefficient of the Solutes between Methanol/Water Mixtures and Hexadecane against Volume Fraction of Methanol in the Original Mixture

Using the values of K_M and K_{MW} for each solute, the theoretical relationship between volume fraction of methanol and solute distribution coefficient was calculated and the values obtained are included as the curves in Figure 31. The agreement between the calculated curves and the experimental points is quite convincing. The values of the individual distribution coefficient shown in Figure 31 give some insight into the nature of the solutes. Obviously, pentanol, the most polar has significant interactions with both the methanol associated with water and water itself. In contrast benzene apparently has little or no interaction with methanol associated with water or water as the values of K_{MW} and K_W are zero. Interaction of benzene with the aqueous system appears to occur solely with the methanol unassociated with water. Cyclohexyl acetate associates weakly with methanol associated with water but not with water itself whereas vinyl acetate, like pentanol, interacts with all three components of the aqueous mixture albeit only very weakly with water.

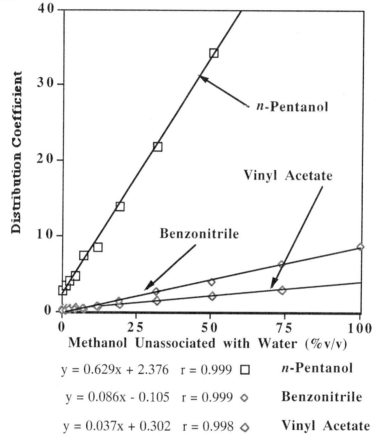

Figure 32. Graph of Distribution Coefficient of *n*-Pentanol, Benzonitrile and Vinyl Acetate against Concentration of Unassociated Methanol

Katz *et al.* also plotted the distribution coefficient of *n*-pentanol, benzonitrile and vinyl acetate against the concentration of unassociated methanol in the solvent mixture and the results are shown in Figure 32. It is seen that the distribution coefficient of all three solutes is predominantly controlled by the amount of unassociated methanol in the aqueous solvent mixture. In addition, the distribution coefficient increases linearly with the concentration of unassociated methanol for all three solutes over the entire concentration range. The same type of curves for anisole and benzene, shown in Figure 33, however, differ considerably. Although the relationship between distribution coefficient and unassociated methanol concentration is approximately linear up to about 50%v/v of unassociated methanol, over the entire range the

Mixed Phases

relationship appears more accurately described by a second order polynomial expression.

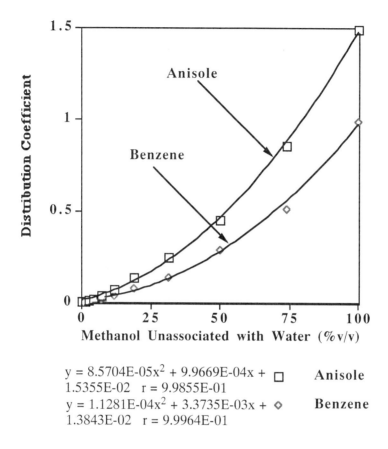

$y = 8.5704\text{E-}05x^2 + 9.9669\text{E-}04x + 1.5355\text{E-}02 \quad r = 9.9855\text{E-}01$ □ Anisole

$y = 1.1281\text{E-}04x^2 + 3.3735\text{E-}03x + 1.3843\text{E-}02 \quad r = 9.9964\text{E-}01$ ◇ Benzene

Figure 33. Graph of Distribution Coefficient of Anisole and Benzene against Concentration of Unassociated Methanol

The reason for this relationship is not clear, but as there is poor interaction between the solutes and methanol due to their highly dispersive character, they may need to interact with two methanol molecules in order to become sufficiently solvated to disperse in the aqueous mixtures.

The interaction of one solute molecule with two solvent molecules would lead directly to a second order equation in terms of concentration but more experimental evidence is required before this can be confirmed.

General Comments

The full explanation of solute retention on silica, bonded phases or for that matter in liquid/liquid systems is still elusive and controversial. The thermodynamic approach can explain the different mechanisms that are effective and unambiguously identify distribution systems that are energy driven or entropically driven. It cannot predict a specific change in excess free energy for a particular solute distribution system and, thus, cannot predict retention.

In contrast molecular interaction kinetic studies can explain and predict changes that are brought about by modifying the composition of either or both phases and, thus, could be used to optimize separations from basic retention data. Interaction kinetics can also take into account molecular association, either between components or with themselves, and contained in one or both the phases. Nevertheless, to use volume fraction data to predict retention, values for the distribution coefficients of each solute between the pure phases themselves are required. At this time, the interaction kinetic theory is as useless as thermodynamics for predicting specific distribution coefficients and absolute values for retention. Nevertheless, it does provide a rational basis on which to explain the effect of mixed solvents on solute retention.

More work is necessary before solute distribution between immiscible phases can be *quantitatively* described by classical physical chemistry theory. In the mean time, we must content ourselves with largely empirical equations based on experimentally confirmed relationships in the hope that they will provide an approximate estimate of the optimum phase system that is required for a particular separation.

Synopsis

Phase selectivity can be accomplished by choosing substances that have specific groups to offer appropriate interactions with the solute or, alternatively, by using a mixture of different substances with unique interacting properties that can achieve the same effect. The result of mixed phases differs somewhat with the level of the moderating solvent. In LC, low concentrations of moderator in the mobile phase largely modifies retention by changing the surface characteristics of the stationary phase. Mono- or bi-layers of moderator can be built up on the surface according to the pertinent Langmuir adsorption isotherms, which can change the nature of solute interactions with the surface and thus modify retention and selectivity. If one component of the mobile phase is water and it is in extreme excess, the surface of the stationary phase can also be modified in a different way. If the surface contains hydrocarbon chains that are free to move ('brush' phases), then they can interact with one another and collapse on the surface thus reducing their availability for interaction with any solute. With 'bulk' phases (polymeric phases) where the hydrocarbon

Mixed Phases

chains are held rigidly in the polymer matrix, the chains can not interact with one another and thus their availability for interaction with the solutes is not affected by high water concentrations in the mobile phase. Solute molecules can interact with the stationary phase surface in two ways, by resting on top of the adsorbed layer of solvent molecules (by sorption interaction) or by interacting directly with the stationary phase surface and displacing the adsorbed solvent layer (by displacement interaction). The two types of interaction have been confirmed experimentally. If bilayers or multi-layers exist on the stationary phase surface, then there will be a number of different sorption and displacement interactions that can take place, although, depending on the nature of the solute, one type of interaction often dominates. The stationary phase surface is usually stabilized at fairly low moderator concentrations, and at higher concentrations the effect of the moderator is to increase specific types of interaction in the *mobile phase* and thus decrease retention. Work on mixed stationary phases in GC, where the only significant molecular interactions present are in the stationary phase, demonstrated that contribution to retention was linearly related to the volume fraction of the respective stationary phase component. This can be explained on the basis that the volume fraction of the component determines the probability of interaction with the solute. The control of interaction probability by the volume fraction of component has also been confirmed in LC for binary mixtures of solvents. As the interactions were in the mobile phase, the retention volume of a solute varied *inversely* as the volume fraction of the stronger eluting solute. However, this simple relationship was found to be only valid if the components of the mobile phase did not associate strongly with each other. The relationship is not valid for aqueous solvent mixtures, *cf.* methanol/water, as, due to the strong association, the mixture is ternary in nature, not binary. If there is no strong association between the components of the mobile phase, the simple inverse relationship also appears to hold for ternary mixtures. The combined effect of temperature and solvent composition on selectivity is unexpected. There can be a temperature where the separation ratio of a pair of solutes is independent of the solvent composition. The pertinent temperature can be calculated form the enthalpies and entropies of distribution. This phenomenon can be easily observed for closely eluting solutes such as enantiomers, but the temperature at which it occurs may be out of normal operating temperature range if retention differences are large and the separation ratio is high. By using density, volume change on mixing and refractive index data, it is possible to determine the equilibrium constant between an associating solvent and water. From the equilibrium constant, the relative concentrations of water, solvent associate and solvent can be calculated and their individual effect on solute retention identified. Raising the temperature reduces the methanol/water association. Thus, as the temperature rises the system behavior moves towards that of a binary mixture. However, even at 200°C (well above the boiling points of the components) there is still significant association between water and methanol. It is found that for many solutes, it is the proportion of the solvent that is unassociated with water which determines the eluting properties of the mixture. Although solute interactions with the water associated solvent can also increase the elution rate, the contribution from water unassociated with the solvent usually appears to have a relatively small effect on solute retention

References

1. R. P. W. Scott and C. F. Simpson, *Faraday Discussions of the Royal Society of Chemistry, No 15,* Faraday Symposium 15, (1980)69.
2. R. P. W. Scott, "Silica gel and Bonded Phases", John Wiley and Sons, Chichester (1993)147.
3. R. P. W. Scott and C. F. Simpson, *J. Chromatogr.,* **197**(1980)11.
4. R. P. W. Scott and P. Kucera, *J. Chromatogr.,* **149**(1978)93
5. M. McCann, J. H. Purnell and C. A. Wellington, *Proceedings of the Faraday Symposium,* Chemical Society, (1980)83.
6. R. J. Laub and J. H. Purnell, *J. Chromatogr.* **112**(1975)71.
7. R. J. Laub, "*Physical Methods in Modern Chromatographic Analysis*" (Ed. P. Kuwana) Academic Press, New York (1983) Chapter 4.
9. E. D. Katz, K. Ogan and R. P. W. Scott, *J. Chromatogr.* **352**(1986)67.
10. W. K. Robbins and S. C. McElroy, *Liquid Fuel Technol.,* **2**(1984)113.
11. R. J. Hurtubise, A. Hussain and H. F. Silver, *Anal. Chem.,* **53**(1981)1993.
12. M. McCann, S. Madden, J. H. Purnell and C. A. Wellington, *J. Chromatogr.,* **294**(1984)349.
13. R. P. W. Scott, *J. Chromatogr.,* **122**(1976)35.
14. R. P. W. Scott and T. E. Beesley, *Analyst,* **124**(1999)713.
15. E. D. Katz, C. H. Lochmüller and R. P. W. Scott, *Anal Chem.* **61**(1981)344.
16. J. E. N. Gutierrez, Ph.D. Dissertation, Georgetown University, Washington, DC (1998).
17, R. P. W. Scott, *The Analyst,* **125**(2000)1543.

Chapter 5

Programming Techniques

There are three methods that can be used to accelerate the elution of strongly retained peaks during chromatographic development. The first technique is known as flow programming. In this procedure the flow of mobile phase is continuously increased during the development of the separation. Thus, the more retained peaks are eluted at higher flow rates with consequent reduction in elution time. The second procedure is called temperature programming and is the programming technique most commonly used in gas chromatography. The distribution coefficient of a solute has been shown to be exponentially related to the reciprocal of the absolute temperature and, thus, the peak is accelerated through the column as the temperature is increased. In a similar way to flow programming, late eluted peaks are accelerated more rapidly through the column at the higher temperatures. However, as the distribution coefficient is exponentially related to the reciprocal of the absolute temperature, the effect of temperature on late-peak acceleration is much greater than that of flow programming. The third programming technique, which is a procedure employed almost exclusively in liquid chromatography, is gradient elution where the composition of the mobile phase is progressively changed during chromatographic development. This procedure is not practical in gas chromatography because the interactions of the solute molecules with those of the carrier gas are extremely weak; consequently, changing the nature of the gas has little impact on the elution rate. Changing the nature of the mobile phase in LC can change the balance of the different types of molecular interactions that are taking place in the distribution system. For example, increasing the ethanol content of a *n*-hexane/ethanol mixture will increase the probability of polar interactions taking place in the mobile phase and, at the same time, decrease the probability of any dispersive interactions that occur. Consequently, if the solutes contain strong polar groups, then their migration rate will rise as a result of an increased number of polar interactions in the mobile phase.

The theory of all three programming methods will now be discussed and the simplest form, flow programming, will be considered first. However, the procedures that are used to theoretically calculate the effect of the different programming techniques needs some prior discussion. It is difficult to derive explicit equations to simulate the change in retention time with programming conditions and, even if such equations are derived, they are, with few exceptions, exceedingly complex and difficult to relate to the actual physical processes that are taking place. Today, with the extremely rapid calculating capabilities of the modern computer, the development of explicit equations to describe programming effects is unnecessary. The actual process can be imitated in small steps and the ultimate effect obtained by summing the effect of all the steps. This method of treatment will be applied to three forms of programming.

Flow Programming

In flow programming, the mobile phase flowrate is increased progressively during chromatographic development and the complexity of the theoretical treatment depends on whether the mobile phase is compressible or not. In gas chromatography, the mobile phase is compressible, but the compressibility of liquids is very small and, thus, in liquid chromatography, the density of the mobile phase can be considered constant along the column. In the first theoretical treatment, the simplest situation will be examined, and the effect of flow programming on solute retention in LC will be considered.

Flow Programming with a Non-Compressible Mobile Phase

If the mobile phase is not compressible, then the contribution (ΔV) to the retention volume (V_r) after time (t) for a period (Δt) will be given by

$$\Delta V = (Q_o + \alpha_Q t)\Delta t$$

where (Q_o) is the initial flow rate at (t=0),
and (α_Q) is the rate of change of flow rate per unit time.

Thus, if (t_r) is the elution time (in seconds) of a solute having a retention volume (V_r), (α_Q) is expressed as the change in flow rate per second, (Δt) is taken as 1 second and (n_t) the number of seconds.

Programming Techniques

Then, $\sum_{p=0}^{p=n_t}(Q_o + \alpha_Q p) = V_r$ and $t_r = n_t$ (1)

Using equation (1) with a simple computer program that identifies (n) when

$$\sum_{p=0}^{p=n_t}(Q_o + \alpha_Q p) = V_r$$

it is possible to calculate the retention time of a series of solutes having different retention volumes (V_r), chromatographed on a given column at a constant temperature, but different program rates (α). The results obtained from such calculations are shown in Figure 1.

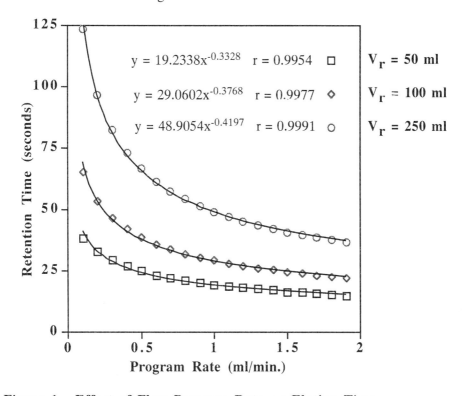

Figure 1. Effect of Flow Program Rate on Elution Time

It is seen that the greatest effect of flow programming is on the more retained solutes which, of course, is the *raison d'être* for the technique. The more retained solutes spend longer in the column and, thus, experience the higher flow rates that occur in the latter part of the flow program. The relationship between retention time and program rate is shown to be approximately a power function, but the magnitudes of the indices vary with the retention volume. Nevertheless, the relationship might be used to calculate approximate retention times for program rates outside the range experimentally examined. It is also seen that the effect of increased program rate becomes progressively less for the earlier eluted solutes as the program rate increases. This is due to the more rapid migrating solutes being eluted before the faster flow rates are reached in the program.

Flow Programming with a Compressible Mobile Phase

Flow programming with a compressible mobile phase is a far more complicated process to examine theoretically. We shall assume that, under flow programming conditions, the *mass* flow rate will be increased linearly with time, *i.e.* $(Q_o(t)) = (Q'_o + \alpha_Q t)$ where (Q'_o) is the *initial exit flow rate*, $(Q_o(t))$ is the exit flow rate after time (t) and (α_Q) is the flow program rate.

These conditions are usual for modern gas flow programming devices that utilize mass flow controllers which are often computer operated. Now, if $(\Delta V_{(o)})$ is an increment of exit flow measured at atmospheric pressure, then from chapter 2, the corrected gas flow (ΔV_r) will be

$$\Delta V_r = \frac{3}{2} \Delta V_{r(o)} \left(\frac{\gamma^2 - 1}{\gamma^3 - 1} \right)$$

where (γ) is the inlet/outlet pressure ratio of the column

Then, under the programming conditions defined above,

$$\Delta V_{r(t)} = \frac{3}{2} \left(\frac{\gamma_t^2 - 1}{\gamma_t^3 - 1} \right) (Q_o + \alpha_Q t) \Delta t \qquad (2)$$

where (γ_t) is the inlet/outlet pressure ratio at time (t)
and $(\Delta V_{r(t)})$ is the increment of volume flow at time (t).

Now as the flow is increased, the inlet pressure will also increase and, thus, the inlet/outlet pressure ratio (γ) will change progressively during the program. Consequently, as the inlet pressure increases, the *mean* flow rate will be reduced according to the pressure correction function and the expected decrease in elution rate will not be realized. Consider an open tubular column,

Programming Techniques

from Poiseuille's equation,
$$P_o Q_{(o)t} = \frac{(P_t^2 - P_o^2)\pi r^4}{16\eta l}$$

or
$$P_o(Q_o + \alpha_Q t) = \frac{(P_t^2 - P_o^2)\pi r^4}{16\eta l}$$

where (P_t) is the inlet pressure at time (t),
(P_o) is the outlet pressure (atmospheric),
(η) is the viscosity of the gas at the column temperature,
(l) is the length of the open tubular column,
and (r) is the radius of the open tubular column.

A similar equation would be used for a packed column except the constant ($\pi/16$) would be replaced by the D'Arcy constant for a packed bed.

Rearranging,
$$\frac{P_o(Q_o + \alpha_Q t)16\eta l}{\pi r^4} + P_o^2 = P_t^2$$

Thus
$$\gamma_t = \frac{P_t}{P_o} = \left[\frac{(Q_o + \alpha_Q t)16\eta l}{P_o \pi r^4} + 1\right]^{0.5} \qquad (3)$$

Assuming the column dimensions are 320 μm I.D. (radius r=0.0160 cm), 30 m long, and it is operated at 120°C using nitrogen as the carrier gas which, at that temperature, has a viscosity of 129 x 10^{-6} Poises, then by using equation (3), the change in (γ) can be calculated for different flow rates. The relationship between flow rate (as measured at the column exit and at atmospheric pressure) and the column inlet/outlet pressure ratio is shown in Figure 2. It is seen that the inlet/outlet pressure ratio will change significantly during a mass flow rate program and this will have a significant attenuating effect on the elution rate, as shown by the pressure correction factor. It is also seen that the curve is a close fit to a second order polynomial function and although this relationship is fortuitous, it could be used empirically to predict inlet/outlet pressure ratios. It is now possible to use the values for (γ_t) from equation (3) in equation (2) to calculate when the solute is eluted at the retention time (t_r) for different flow program rates. It is seen that, on summing equation (2), when the sum of all the increments of (ΔV_r) is equal to the retention volume (V_r), then (t) will be the retention time (t_r).

$$V_r = \sum_{t=0}^{t=t_r} \frac{3}{2}\left(\frac{\gamma_t^2 - 1}{\gamma_t^3 - 1}\right)(Q_o + \alpha_Q t)\Delta t \qquad (4)$$

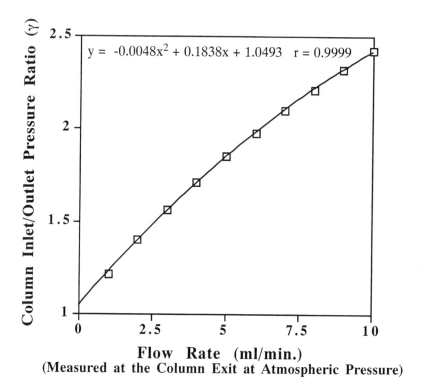

Figure 2. The Relationship between the Inlet/Outlet Pressure Ratio and Exit Flow Rate for an Open Tubular Column

In the following calculations, the time unit (Δt) is taken as 1 second and equation (3) is used to calculate (γ_t). Then, with a simple computer program in conjunction with equation (4), the retention time (t_r) can be calculated for a range of different program rates (α_Q) and for solutes having retention times of 10, 50 and 250 ml, respectively. The column was again assumed to be 320 mm I.D., 30 m long, and operated at 120°C using nitrogen as the carrier gas which, at that temperature, has a viscosity of 129×10^{-6} Poises. The results are shown in Figure 3.

It is seen that the effect of program rate on retention time is much as would be expected. It would appear that, for any individual solute, the retention time is related to some power of the solute retention volume but the indices vary significantly with the retention volume of the solute. This relationship does not have a theoretical explanation at this time, but might be useful for predicting retention times from experimental data. Despite the attenuating effect of the pressure correction factor, the use of flow programming is effective in reducing the retention time of strongly retained solutes. However, unless the diffusivity of the solutes in the mobile phase is

Programming Techniques

high (*i.e.*, mobile phases such as hydrogen or helium are employed) there will be significant peak dispersion at the higher velocities, the column efficiency will be reduced and resolution may be lost. The effect of mobile phase velocity on peak dispersion will be discussed in later chapters.

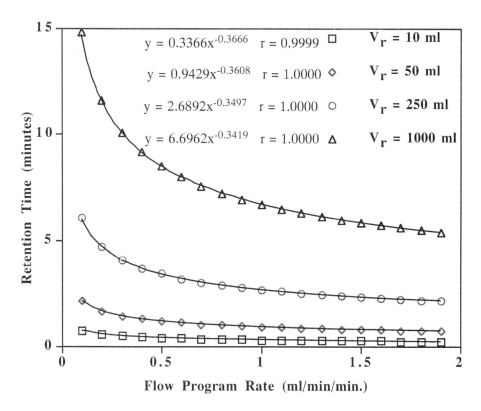

Figure 3. Curves Relating Elution Time to Flow Program Rate for Solutes having Different Retention Volumes

Temperature Programming

Temperature programming was introduced in the early days of GC and is now a commonly practiced elution technique. It follows that the temperature programmer is an essential accessory to all contemporary gas chromatographs and also to many liquid chromatographs. The technique is used for the same reasons as flow programming, that is, to accelerate the elution rate of the late peaks that would otherwise take an inordinately long time to elute. The distribution coefficient of a solute is exponentially related to the reciprocal of the absolute temperature, and as the retention volume is directly related to the distribution coefficient, temperature will govern the elution rate of the solute.

Now, the retention volume (V'_r) is given by $V'_r = KV_S$ and, from chapter 3,

$$\log K = -\frac{\Delta H_o}{RT} + \frac{\Delta S_o}{R}$$

Thus,

$$V'_{r(T_o)} = V_S e^{-\frac{\Delta H_o}{RT_o} + \frac{\Delta S_o}{R}}$$

Hence, for a period (Δt) at (T_t), the effective retention volume can be considered to be constant at

$$V'_{r(T_t)} = V_S e^{-\frac{\Delta H_o}{RT_t} + \frac{\Delta S_o}{R}}$$

Furthermore, after time (t_p) between (t_p) and ($t_p+\Delta t$) during a temperature program at a rate of $\alpha°C$ per unit time and an initial temperature of (T_o), the effective retention volume, $V'_{r(T_o+\alpha_T p)}$ can be considered to be

$$V'_{r(T_o+\alpha_T p)} = V_S e^{-\frac{\Delta H_o}{R(T_o+\alpha_T p)} + \frac{\Delta S_o}{R}}$$

It follows that if the retention time in seconds is (n_T) and (Δt) is conveniently taken as one second, then the mean value of the retention volume throughout the program will be

$$\overline{V}'_{r(T_o)to(T_o+\alpha_T p)} = \frac{\sum_{p=0}^{p=n_t} V_S e^{-\frac{\Delta H_o}{R(T_o+\alpha_T p)} + \frac{\Delta S_o}{R}}}{\sum_{p=0}^{p=n_t} \Delta n}$$

$$= \frac{1}{n} \sum_{p=0}^{p=n_t} V_S e^{-\frac{\Delta H_o}{R(T_o+\alpha_T p)} + \frac{\Delta S_o}{R}}$$

Thus, if the mean flow rate is ($\overline{Q}$), then

$$\overline{V}'_{r(T_o)to(T_o+\alpha_T n_t)} = \frac{1}{n} \sum_{p=0}^{p=n_t} V_S e^{-\frac{\Delta H_o}{R(T_o+\alpha_T p)} + \frac{\Delta S_o}{R}} = \overline{Q} n_t \quad (5)$$

Unfortunately, the mean flow rate ($\overline{Q}$) will not be simply obtained by applying the standard pressure correction for the initial conditions of the program. As a result of the effect of temperature on the viscosity of the gas, the inlet/outlet pressure (γ) will change continuously during the program and, consequently, so will the magnitude of the pressure correction.

For example, consider an open tubular column with the dimensions previously defined, operated at constant mass flow rate of helium (which is normal for temperature programming purposes), then, from Poiseuille's equation, after time (t),

$$P_o Q_{(o)} = \frac{(P_t^2 - P_o^2)\pi r^4}{16 \eta_{T_t} l}$$

where (η_{T_t}) is the viscosity of the gas at temperature (T_t) after the temperature program has progressed for time (t).

Thus,
$$\eta_{T_t} = \frac{P_o(\gamma_t^2 - 1)\pi r^4}{16 l Q_{(o)}}$$

and, as the exit pressure (P_o) (atmospheric) and the exit flow rate under conditions of constant mass flow rate (Q_o) are constant,

$$\frac{\eta_{T_t}}{\eta_{T_o}} = \frac{(\gamma_t^2 - 1)}{(\gamma_o^2 - 1)} \quad \text{or} \quad \gamma_t = \left(\frac{\eta_{T_t}(\gamma_o^2 - 1)}{\eta_{T_o}} + 1 \right)^{0.5} \tag{6}$$

From equation (3),

$$\gamma_o = \frac{P_t}{P_o} = \left[\frac{(Q_o) 16 \eta_o l}{P_o \pi r^4} + 1 \right]^{0.5} \tag{7}$$

Thus, using equation (6) to calculate (γ_o), equation (7) to calculate (γ_t), the mean flow rate ($\overline{Q_t}$) over a time period (t) can be calculated from

$$\overline{Q} = \frac{\sum_{t=0}^{t=n_t} \frac{3}{2}\left(\frac{\gamma_t^2 - 1}{\gamma_o^3 - 1}\right) Q_o}{\sum_{t=0}^{t=n_t} \Delta t} = \frac{1}{n_t} \sum_{t=0}^{t=n_t} \frac{3}{2}\left(\frac{\gamma_t^2 - 1}{\gamma_o^3 - 1}\right) Q_o \tag{8}$$

Thus, rearranging equation (5), the retention time at a particular program rate (α) can be calculated as that time (t_r) (*nota bena* $t_r = n_t$) that allows the following equation to be satisfied.

$$\frac{V_s}{n_t} \sum_{t=0}^{t=n_t} e^{-\frac{\Delta H_o}{R(T_o + \alpha_o t)} + \frac{\Delta S_o}{R}} = \frac{3}{2} \sum_{t=0}^{t=n_t} \left(\frac{\gamma_t^2 - 1}{\gamma_o^3 - 1}\right) Q_o \tag{9}$$

Note, if the units in which (n) is measured are taken as seconds, the flow rate (Q_o) must also be measured in ml/sec.

It is clear that to develop an explicit algebraic expression for (t_r) or (n) would be exceedingly cumbersome and, as already stated, in this modern day of the digital computer, is unnecessary. A simple program can be written, using equations (6), (7), (8) and (9), that searches for the time (t_r) that allows the equivalence defined in equation (9) to be identified; this can be carried out for a range of different program rates (*i.e.*, different values of (α)).

It is first necessary, however, to assign values to the various constants that are used in equations (6), (7), (8) and (9). These are summarized in the following table. This data would normally be taken from the pertinent chromatographic properties of the distribution system and those of the column.

Table 1. Constants Used in Programming Calculations

Column Length	30 m
Column Diameter	320 μm
Stationary Phase in Column	0.1 ml (arbitrarily chosen)
Viscosity of He at 59°C	0.0002118 Poises [1]
Flow Rate (constant mass flow rate)	5 ml/min. (0.08333ml/sec) (arbitrarily chosen)

Values for the viscosity of helium were calculated from the following equation given in Ref.1:

$$\eta_T = \eta_o \left(\frac{T}{273.4} \right)^{0.647}$$

Values for (η_T) for a range of different temperatures obtained using the above equation are shown plotted against temperature in Figure 4.

It is seen that the viscosity of the gas will change significantly during a temperature program and, thus, at a constant gas *mass flow rate*, the inlet pressure will rise proportionally. This increase in inlet pressure will result in an increase in the inlet/outlet pressure ratio and, as a consequence, will extend the retention time and oppose the effect of any increase in temperature. It also follows that the effect of viscosity change on inlet pressure can be very significant and must, therefore, be taken into account when examining the effect of temperature program rate on retention time.

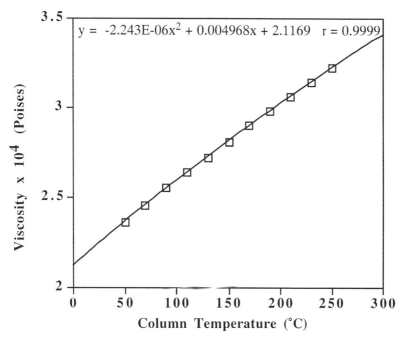

Figure 4. Graph of Helium Viscosity against Temperature

Finally, it is necessary to select values for the thermodynamic constants that are to be used in equation (9). The data selected were that published by Beesley and Scott [2], for the two enantiomers, (S) and (R) 4-benzyl-2-oxazolidinone. The values for the standard free enthalpy and standard free entropy for the (R) isomer were $\left(\frac{\Delta H^\circ}{R}\right) = 729.34$ and $\left(\frac{\Delta S^\circ}{R}\right) = 0.6255$, respectively. When calculating the separation ratio for the two isomers, under differing program conditions, the values taken for the (S) isomer were $\left(\frac{\Delta H^\circ}{R}\right) = 635.38$ and $\left(\frac{\Delta S^\circ}{R}\right) = 0.579$, respectively.

Using the defined data, the retention time of the (R) enantiomer eluted at different program rates and at initial temperatures of 25°C, 35°C and 45°C was calculated by an iterative procedure using equation (9). The results obtained are shown as retention time plotted against program rate for the three different initial temperatures in Figure 5. The results are much as would be expected for those experienced in the practical use of temperature programming in GC. However, by employing the pertinent chromatographic and physical chemical data, the retention times can now be accurately calculated for any program rate and, if necessary, the optimum program can be optimized. Although the advantage of this procedure is not apparent for the elution of a single solute, it can be a very useful method development procedure for more complex mixtures.

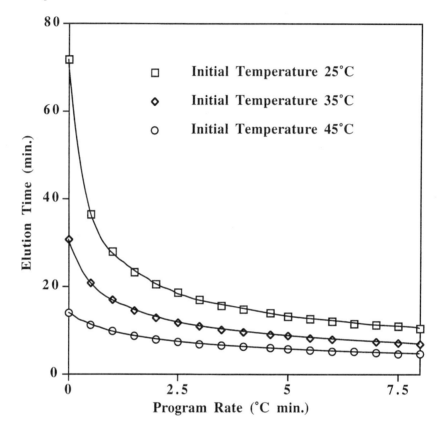

Figure 5. Graphs of Retention Time of (R) 4-Benzyl-2-oxazolidinone against Temperature Program Rate for Three Different Initial Temperatures

The curves in Figure 5 show some characteristic features that need discussing. It is clear that, for the solute examined, there is little advantage in using program rates much above 3°C per min. because the consequent reduction in retention time is minimal. This is because the solute is eluted so rapidly in the early part of the development that the higher temperatures in the program have little time to become effective. Solutes with high retention, separated at relatively low initial temperatures, benefit most from the programming procedure. Conversely, the more rapidly eluted solutes will show little advantage from temperature programming and, as will be seen later, are likely to lose resolution. For this reason, if there are many *early* eluting peaks in the mixture, it is best to operate isothermally for an initial period to elute the less retained solutes before the program is initiated. This will ensure adequate retention and resolution of the less retained solutes. Under such circumstances the temperature program needs to be initiated at a (k') value of about 5. In any event, the separation can be simulated in the same manner as that described, providing the effect of an isothermal period is inserted into equation (9).

Unless there are solutes that are very strongly retained, and the maximum operating temperature of the system is reached without all the solutes being eluted, there is usually no need for a final isothermal period of any significant length. There is, however, one exception where a final isothermal period is helpful, and that is for mixtures that contain *thermally labile* materials.

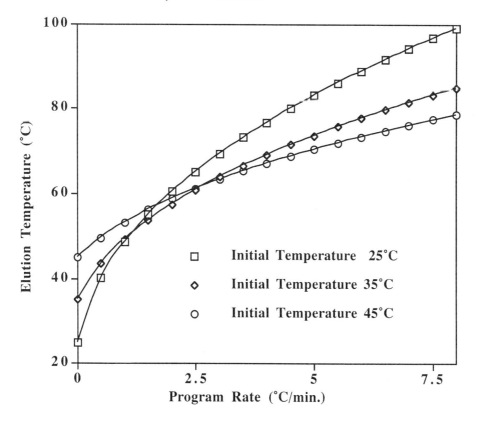

Figure 6. Graphs of Elution Temperature of (R) 4-Benzyl-2-oxazolidinone against Program Rate for Three Different Initial Temperatures

Under such circumstances the temperature can be programmed up to the maximum *stable* temperature for the solutes and any remaining solutes in the column must then be eluted isothermally. This situation can often occur in the separation of substances of biological origin, *e.g.*, the separation of essential oils containing labile materials such as terpenes.

Knowing the retention time and the program rate, the *elution temperature* can be easily calculated, the elution temperature is shown plotted against program rate in Figure 6. It is seen that the highest elution temperature, occurs when the initial temperature is the lowest. This is because the solute stays in the column longer and, thus, the program can reach a higher temperature before the solute is eluted. Conversely, if the elution is started at a higher temperature the solute retention time is much less and thus the program is effective for a much shorter time and the solute is eluted at a lower temperature. Employing the thermodynamic data for both isomers, the retention time of both isomers can be calculated and, thus, the effect of temperature program on resolution can be demonstrated. Curves relating resolution to program rate are shown in Figure 7.

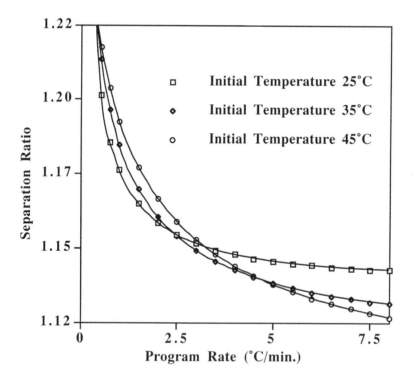

Figure 7. Graphs of Resolution of the Isomers of 4-Benzyl-2-oxazolidinone against Program Rate for Three Different Initial Temperatures

It is seen that the resolution rapidly falls with increased program rate. However, the manner in which the resolution changes with temperature is complicated by the fact that the standard free enthalpies of the two isomers differ and, thus, the effect of temperature on the retention of the two isomers will also differ. It is seen that at low program rates, the greater resolution is realized at the higher initial program temperature, whereas at high program rates, the higher resolution is obtained at the lowest initial temperature. Due to the different interacting effects of the unequal standard free enthalpies, it is difficult to interpret precisely the shapes and magnitude of the curves. It is interesting to note that, for this particular pair of isomers, although the resolution falls dramatically with program rate, the resolution never falls below unity and, thus, there is no program rate that will cause co-elution.

Gradient Elution or Solvent Programming

The technique of gradient elution, where the composition of the mobile phase is changed continuously during development in LC, is complementary to the technique of temperature programming used in GC. It allows a solute mixture containing components that span a wide range of polarity to be eluted in a reasonable time. There are three basic phase systems used in LC, of which all others are modified forms. In the first, the solutes are retained on the stationary phase by predominantly dispersive interactions and the mobile phase is made predominantly polar. The reversed phase system employing methanol-water or acetonitrile-water mixtures as the mobile phase is typical of this type. During solvent programming, the mobile phase is continuously made more dispersive by increasing the methanol or acetonitrile content. The increased modifier content provides more frequently competing dispersive interactions with the solutes and consequently reduces their retention. In the second phase system, the solutes are retained on the stationary phase by predominantly polar interactions and the mobile phase is made predominantly dispersive. A typical system would be the use of a polar bonded stationary phase with *n*-hexane/ethanol mixtures as the mobile phase. In this case, during solvent programming, the mobile phase is continuously made more polar by increasing the ethanol content. The increased ethanol content provides more frequently competing polar interactions with the solutes and, consequently, reduces their retention. It should be emphasized that increasing the ethanol content does not increase the *strength* of the polar interactions in the mobile phase, but increases the *probability* of their occurring, *i.e.,* increases the probability of an ethanol molecule interacting with a solute molecule and, thus, increases its affinity for the mobile phase. The third phase system consists of an ionic stationary phase that provides the counter-ion to that of the solute(s) and an aqueous mobile phase. During the program, the concentration of a competing ionic material in the mobile phase is gradually increased until the interaction between these competing ions become sufficiently strong to oppose the ionic interactions between the solute and the stationary phase and the solute(s) is eluted.

The effect of solvent programming can be simulated by means of a computer in much the same way as temperature programming can be simulated. Reiterating equation (13) from chapter 4

$$\frac{1}{V''_{AB}} = \alpha\left(\frac{1}{V''_A} - \frac{1}{V''_B}\right) + \frac{1}{V''_B} \qquad (10)$$

where (V''_{AB}) is the corrected retention volume of the solute with respect to the mixture of solvents (A) and (B),

(V''_A) is the corrected retention volume of the solute with respect to the pure solvent (A),

(V''_B) is the corrected retention volume of the solute with respect to the pure solvent (B),

and (α) is the volume fraction of solvent (A).

If the corrected retention volume in the pure strongly eluting solute is very small compared with the retention volume of the solute in the other pure solvent, *i.e.*, $V''_A \ll V''_B$ which is very often the case in practical LC, then equation (10) simplifies to the simple reciprocal relationship,

$$V''_{AB} \propto \frac{V''_A}{\alpha} \qquad (11)$$

In addition, under these conditions, the inverse will also apply, *i.e.*,

$$\frac{1}{V''_{AB}} \propto \frac{\alpha}{V''_A} \qquad (12)$$

Scott and Beesley [2] measured the corrected retention volumes of the enantiomers of 4-benzyl-2-oxazolidinone employing hexane/ethanol mixtures as the mobile phase and correlated the corrected retention volume of each isomer to the reciprocal of the volume fraction of ethanol. The results they obtained at 25°C are shown in Figure 8. It is seen that the correlation is excellent and was equally so for four other temperatures that were examined. From the same experiments carried out at different absolute temperatures (T) and at different volume fractions of ethanol (c), the effect of temperature *and* mobile composition was identified using the equation for the free energy of distribution and the reciprocal relationship between the solvent composition and retention.

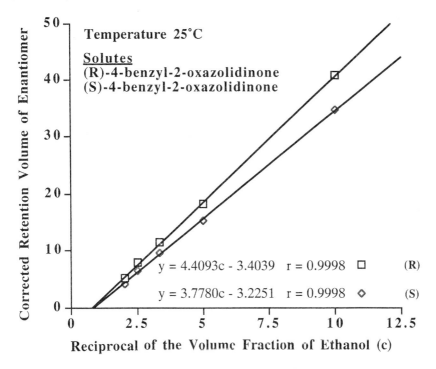

Figure 8. Graph of the Corrected Retention Volumes of the Two Enantiomers of 4-Benzyl-2-oxazolidinone against the Reciprocal of the Volume Fraction of Ethanol

The authors arrived at the following specific equations for the retention volume $V'_{S(c,T)}$ and $V'_{R(cT)}$ of the two isomers. These are the same equations used in chapter 4 when examining the combined effect of temperature and solvent composition on the separation ratio.

$$V'_{S(c,T)} = \frac{10^{\frac{726.93}{T}-1.782}}{\alpha} - 10^{\frac{636.98}{T}-1.571} \qquad (13)$$

$$V'_{R(cT)} = \frac{10^{\frac{641.48}{T}-1.563}}{\alpha} - 10^{\frac{661.33}{T}-1.698} \qquad (14)$$

At a temperature of 25°C equations (13) and (14) reduce to

$$V'_{S(c,25°C)} = \frac{4.509}{\alpha} - 3.661 \tag{15}$$

$$V'_{R(c,25°C)} = \frac{3.861}{\alpha} - 3.298 \tag{16}$$

Adopting a similar approach to other programming procedures, if volume fraction of ethanol (c) is increased linearly with time (t) according to the equation

$$c = e + \beta_\alpha t$$

where (e) is the initial solvent concentration
and (β_α) is the rate of change of volume fraction per second.

Then, after (t') seconds the mean effective corrected retention volumes of the two isomers will be given by

$$V'_{S(c,25°C)} = \frac{\sum_{p=1}^{p=t} \frac{4.509}{e + \beta_\alpha p} - 3.661}{t}$$

$$V'_{R(c,25°C)} = \frac{\sum_{p=1}^{p=t} \frac{3.861}{e + \beta_\alpha p} - 3.298}{t}$$

Now, if the column is operated at a flow rate (Q), then the solute will be eluted when the mean retention volume is equal to the product of the program time and the flow rate. Under these circumstances the program time will be the retention time. Thus,

$$V'_{S(c,25°C)} = \frac{\sum_{p=1}^{p=t_r} \frac{4.509}{e + \beta_t p} - 3.661}{t_r} = Q t_{r(S)} \tag{17}$$

$$V'_{R(c,25°C)} = \frac{\sum_{p=1}^{p=t_r} \frac{3.861}{e + \beta_t p} - 3.298}{t_r} = Q t_{r(R)} \tag{18}$$

When the above equations are satisfied, then $t = t_r$ for each respective solute. It is seen with the aid of a computer the value of (t) that satisfies equations (17) and (18) can be identified; for a specific program rate, the retention time (t_r) can be calculated. In addition, if the program is run for the second enantiomer, then the ratio of the retention times of the two isomers can be calculated (the separation ratio) for a range of different solvent program rates.

Programming Techniques

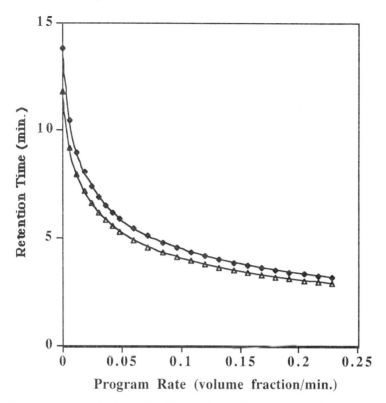

Figure 9. Curves Relating the Retention Time of the Two Enantiomers against Program Rate

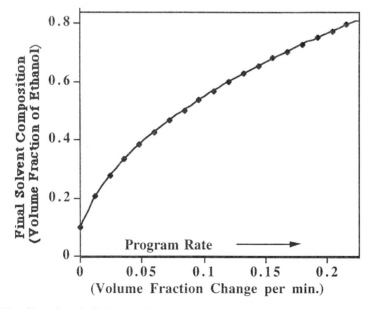

Figure 10. Graph of Solvent Concentration at Elution against Program Rate

Employing a flow rate of 3.0 ml/min. (0.05 ml/sec.) and a range of solvent program rates starting with a 0.1 volume fraction of pure ethanol ($e = 0.1$), the respective retention times were calculated for the two isomers and the relationship between program rate and elution time is shown in Figure 9. It is seen that the profiles of the curves are similar to temperature programming in GC. It should also be noted that the relationship between the logarithm of the retention time and program rate is far from linear. The computer program can also be used to calculate the solvent composition at the elution time of the solute.

The curve relating 'solvent composition on elution' to program rate is shown in Figure 10. It is seen that, even when the elution time of the enantiomers has been reduced to little more than three minutes by employing the most rapid program rate, the concentration of ethanol on elution is still little more than a volume fraction of 0.8. By taking the ratio of the two retention times, the separation ratio of the two enantiomers can be calculated and a graph relating separation ratio to program rate is shown in Figure 11.

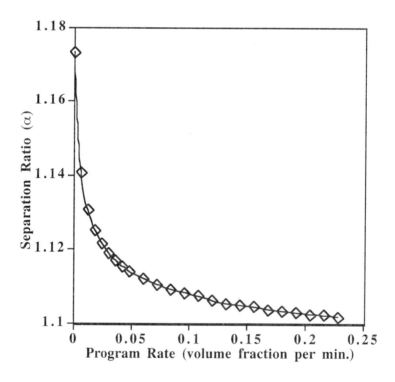

Figure 11. Graph of the Separation Ratio of the Two Enantiomers against Program Rate

Programming Techniques

The separation ratio rapidly decreases with increase in program rate, but at high program rates, the curve flattens out. However, the separation ratio has already been reduced to about 1.11 at a program rate of only 0.08 volume fraction per min. As a consequence, high program rates should be avoided if possible, but this must be achieved without extending the analysis times too extensively. It is clear that if an explicit equation is obtained for the retention time of a solute as a function of the volume fraction of strongly eluting solute, the effect of program rate on most of the important parameters of a separation can easily be calculated.

Synopsis

The three techniques generally used to accelerate the elution of well-retained solutes are flow programming, temperature programming and solvent programming or gradient elution. Flow programming used with an incompressible mobile phase, such as a liquid in LC, increases the elution rate of the solutes linearly with time. If the mobile phase is compressible as in GC, however, any increase in flow rate will also be accompanied by a rise in inlet pressure. Thus, the inlet/outlet pressure ratio will also be greater so, as a result of the pressure correction factor discussed in chapter 2, the effect of increasing the flow rate will be somewhat attenuated. An equation can be derived that will give the retention time of a solute that is eluted by flow programming that takes into account these inherent pressure changes. As the relationship between retention volume and temperature is defined thermodynamically, the effect of temperature programming on solute retention and selectivity can also be treated theoretically. However, in GC, as the temperature is increased during the program, the viscosity of the gas also rises, which in turn results in an increase in inlet pressure. Nevertheless, the effect of change in viscosity can also be included in the theoretical treatment and it is possible to derive an expression that will allow the elution time, elution temperature and separation ratio to be calculated for different program rates. Gradient elution in LC is easier to treat theoretically, as the mobile phase can be considered incompressible. However, appropriate equations can only be derived if the mobile phase contains solvents that do not associate; this, of course, excludes most aqueous solvent mixtures due to strong association of most miscible solvents with water. The relationship between solvent composition and retention, originally identified in GC by Purnell and confirmed for LC by Katz *et al.*, can be used to develop an equation that allows the retention time and separation ratio to be calculated for different programming rates. In general, with the aid of a computer, all the programming techniques can be simulated and many chromatographic properties of the separation can be predicted.

References

1. J. O. Hirschfelder, C. F. Curtis and R. B. Bird, *Molecular Theory of Gases and Liquids*, John Wiley and Sons, New York. (1954)631.
2. R. P. W. Scott and T. E. Beesley, *Analyst*, **124**(1999)713.

Chapter 6

Extensions of the Plate Theory

In this chapter, the elution curve equation and the plate theory will be used to explain some specific features of a chromatogram, certain chromatographic operating procedures, and some specific column properties. Some of the subjects treated will be second-order effects and, therefore, the mathematics will be more complex and some of the physical systems more involved. Firstly, it will be necessary to express certain mathematical concepts, such as the elution curve equation, in an alternative form. For example, the Poisson equation for the elution curve will be put into the simpler Gaussian or Error function form.

The Gaussian Form of the Elution Equation

In order to change the Poisson form of the elution equation into the Gaussian form, it is necessary to change the origin. Consider the elution curve as shown in Figure 1. The origin of the Poisson equation is at the point of injection, whereas the origin of the Gaussian equation will occur at the peak maximum, which will be (n) plate volumes from the injection point. Thus, a point X, (v) plate volumes from the point of injection will be $(v-n) = w$ plate volumes from the peak maximum. Consequently, $v = (n+w)$. This change of origin is depicted in Figure 1.

Now, the Poisson form of the elution equation is as follows,

$$X_{m(n)} = \frac{X_0 e^{-v} v^n}{n!}$$

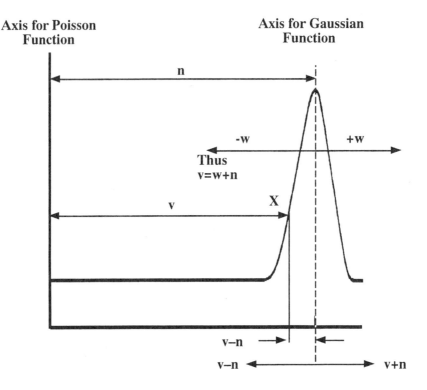

Figure 1. The Difference in Axes between the Poisson Function and the Gaussian Function

Thus, substituting (w+n) for (v),

$$X_{m(n)} = \frac{X_o e^{-(w+n)}(w+n)^n}{n!}$$

Employing Stirling's theorem, $\quad n! = e^{-n} n^n \sqrt{2\pi n}$

Consequently,
$$X_{m(n)} = \frac{X_o e^{-(w+n)}(w+n)^n}{e^{-n} n^n \sqrt{2\pi n}}$$

and
$$X_{m(n)} = \frac{X_o}{\sqrt{2\pi n}} e^{-w}\left(1 + \frac{w}{n}\right)^n$$

Now if x<1, $Log_e (1+x)$ can be expressed by the series

$$Log_e (1+x) = x - \frac{x^2}{2} + \frac{x^3}{3} - \frac{x^4}{4} + ...$$

Extensions of the Plate Theory

Expanding $(X_{m(n)})$ as a series,

$$\text{Log}_e X_{m(n)} = \log_e\left(\frac{X_o}{\sqrt{2\pi n}}\right) - w + n\left(\frac{w}{n} - \frac{w^2}{2n^2} + \frac{w^3}{3n^3} - \ldots\right)$$

$$= \log_e\left(\frac{X_o}{\sqrt{2\pi n}}\right) - w + w - \frac{w^2}{2n} + \frac{w^3}{3n^2} - \ldots$$

As (n) is large and virtually the whole elution curve for any solute is contained between $w = -2\sqrt{n}$ and $w = 2\sqrt{n}$ (*i.e.*, contained within four standard deviations of the Gaussian curve) then $\frac{w^2}{2n}$ will always be very much greater than $\frac{w^3}{3n^2}$ and thus, $\frac{w^3}{3n^2}$ and all higher terms can be ignored.

$$\text{Log}_e X_{m(n)} = \log_e\left(\frac{X_o}{\sqrt{2\pi n}}\right) + \frac{w^2}{2n}$$

Therefore,
$$X_{m(n)} = \frac{X_o e^{\frac{w^2}{2n}}}{\sqrt{2\pi n}} \qquad (1)$$

Equation (1) is the well-known Gaussian form of the elution curve equation and can be used as an alternative to the Poisson form in all applications of the Plate Theory.

Retention Measurements on Closely Eluting Peaks

The retention time, or retention volume, is one of the most important measurements to be made in any chromatographic analysis. In addition to providing data from which each eluted substances can be identified (either by the capacity ratio or the separation ratio), the retention times are also important in column design. It will be shown later in this chapter that with a knowledge of the capacity ratio and the separation ratio of a pair of peaks the column efficiency needed to ensure their complete resolution can be calculated. However, there is a serious source of error that can arise when the solutes of interest are eluted close together. Unfortunately, this error is particularly significant under those conditions where the accurate calculation of the required efficiency is particularly essential. The proximity of one peak to another can distort the positions of the peak maxima in the combined envelope, so that the apparent positions of the peak maxima are significantly different from their true positions when eluted individually. This effect is shown in Figure 2. The peaks are simulated employing the normal Error function (Gaussian) and are then added to provide the composite envelope.

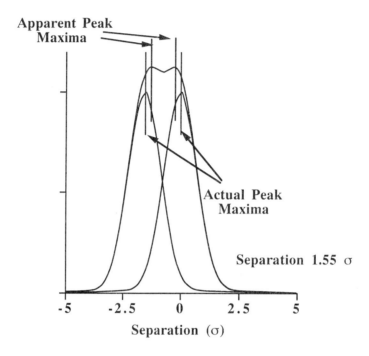

Figure 2. Errors Involved in Measuring Retention Data From Merging Peak Envelopes

The composite envelope is then plotted over the envelope of each individual peak. It is seen that the actual retention difference, if taken from the maxima of the envelope, will give a value of less than 80% of the true retention difference. Furthermore as the peaks become closer this error increases rapidly. Unfortunately, this type of error is not normally taken into account by most data processing software. It follows that, if such data was used for solute identification, or column design, the results can be grossly in error.

Another serious error can occur if it is known that there are two peaks which are unresolved, and the retention time of the maximum of the envelope is taken as the mean retention time of the two isomers. This measurement can only be true if the peaks are *absolutely symmetrical* and the two peaks are of *equal height*. The effect of different proportions of each isomer on the retention time of the composite envelope is shown in Figure 3. It is seen that the position of the peak maximum of the composite envelope is significantly different from the mean of the retention times of the individual peaks.

Extensions of the Plate Theory

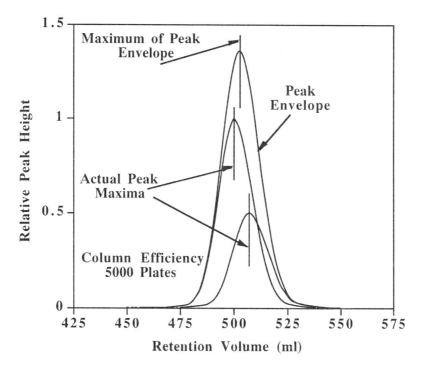

Figure 3. A Composite Peak Formed by Two Closely Eluting Peaks of Different Size

Furthermore, in the example given, the peaks were considered to be truly Gaussian in shape. Asymmetric peaks can distort the position or the peak maximum of the envelope to an even greater extent. In general, the retention time of a composite peak should never be assumed to have a specific relationship with those of the unresolved pair.

Another error can arise when two partially resolved peaks are asymmetrical, *e.g.*, the rear half of the peak is broader the front half. In such a situation, it is clear that there can be two sources of error, which are depicted in Figure 4. Firstly, the retention times, as measured from the peak envelope, will not be accurate. Secondly, because the peaks are asymmetrical (and most LC peaks tend to be asymmetrical to the extent shown in the Figure 4), the second peak appears higher. This can incorrectly imply that the second solute is present at a higher concentration in the mixture than the first. It follows that it is important to know the value of the specific separation ratio above which accurate measurements can still be made on the peak maxima of the individual peaks. The apparent peak separation ratio, relative to the actual peak separation ratio for columns of different efficiency, are shown in Figure 5. The data has been obtained from theoretical equations.

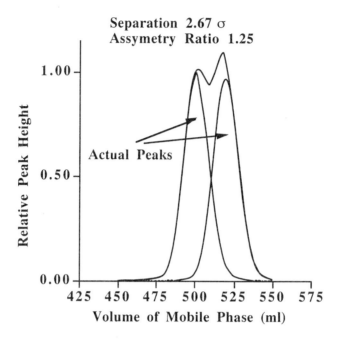

Figure 4. The Effect of Peak Asymmetry on the Apparent Composition of Closely Eluting Solutes

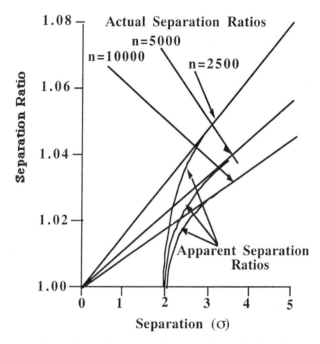

Figure 5. Curves Relating Apparent Separation Ratio Relative to Actual Separation Ratio for Two Closely Eluting Peaks

Extensions of the Plate Theory

It is seen that the separation ratio must be greater than about 1.055 for a low efficiency column (2500 theoretical plates) before accurate retention measurements can be made on the composite curve. On the high efficiency columns (10,000 theoretical plates), the separation ratio need only be in excess of about 1.035 before accurate retention measurements can be made on the composite curve. It will be seen later in this chapter that to optimize a column for a difficult separation, accurate retention data must be obtained over a range of temperatures and solvent compositions. It follows that:

Considerable care must be taken when accessing closely eluting peaks. If the resolution is inadequate, measurements must be taken on the individual solutes, chromatographed separately on the column.

Quantitative Analysis from Retention Measurements

There is an interesting consequence to the above discussion on composite peak envelopes. If the actual retention times of a pair of solutes are accurately known, then the *measured retention time* of the composite peak will be related to the *relative quantities* of each solute present. Consequently, an assay of the two components could be obtained from accurate retention measurements only. This method of analysis was shown to be feasible and practical by Scott and Reese [1]. Consider two solutes that are eluted so close together that a single composite peak is produced. From the Plate Theory, using the Gaussian form of the elution curve, the concentration profile of such a peak can be described by the following equation:

$$X_{AB} = X_A \frac{e^{\frac{-(v_A - n_{(A)})^2}{2 n_{(A)}}}}{\sqrt{2\pi n_{(A)}}} + X_B \frac{e^{\frac{-(v_B - n_{(B)})^2}{2 n_{(B)}}}}{\sqrt{2\pi n_{(B)}}} \qquad (2)$$

where (X_{AB}) is the concentration of solutes (A) and (B) in the composite peak,
(X_A) is the initial concentration of solute (A),
(X_B) is the initial concentration of solute (B),
($n_{(A)}$) is the column efficiency for solute (A),
($n_{(B)}$) is the column efficiency for solute (B),
(v_A) is the volume of mobile phase passed through the column in plate volume of solute (A),
and (v_B) is the volume of mobile phase passed through the column in plate volumes of solute (B).

Rearranging (multiplying the top and bottom of the exponent by (n_A) and taking (n_A^2) inside the brackets),

$$X_{AB} = X_A \frac{e^{-\frac{n_A}{2}(\frac{v_A}{n_{(A)}} - 1)^2}}{\sqrt{2\pi n_{(A)}}} + X_B \frac{e^{-\frac{n_B}{2}(\frac{v_B}{n_{(B)}} - 1)^2}}{\sqrt{2\pi n_{(B)}}} \qquad (3)$$

Equation (3) merely sums the two peaks to produce a single envelope. Providing retention times can be measured precisely, the data can be used to determine the composition of a mixture of two substances that, although having finite retention differences, are eluted as a single peak. This can be achieved, providing the standard deviation of the measured retention time is small compared with the difference in retention times of the two solutes. Now, there is a direct relationship between retention volume measured in plate volumes and the equivalent times, which is depicted in Figure 6.

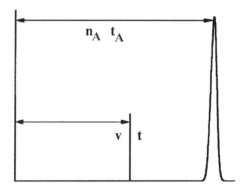

Figure 6. Relationship between Plate Volumes and Time

If (t_A) and (t_B) are the retention times of solutes (A) and (B) respectively from Figure 6 by simple proportion, $\frac{v}{n_{(A)}} = \frac{t}{t_A}$.

Thus equation (3) can be put in the form

$$X_{AB} = X_A \frac{e^{-\left(\frac{n_{(A)}}{2}\right)\left(\frac{t}{t_A} - 1\right)^2}}{\sqrt{2\pi n_{(A)}}} + X_B \frac{e^{-\left(\frac{n_{(B)}}{2}\right)\left(\frac{t}{t_B} - 1\right)^2}}{\sqrt{2\pi n_{(B)}}} \qquad (4)$$

Extensions of the Plate Theory

The variable (v) is replaced by the variable (t), the elapsed time, which is the parameter that can be accurately measured experimentally.

It is seen from equation (4) that when *only* (A) is present, the function will exhibit a maximum at $t = t_A$ and if *only* solute (B) is present, it will exhibit a maximum at $t = t_B$. It follows that the composite curve will give a range of maxima between (t_A) and (t_B) for different proportions of solutes (A) and (B). Thus, from the value for the retention time of the composite peak, the composition of the original mixture can be determined.

For closely eluted peaks $n_A = n_B$ and $\left(\sqrt{2\pi n_{(A)}}\right)$ and $\left(\sqrt{2\pi n_{(B)}}\right)$ are, in effect, average dilution factors resulting from the peak dispersion and can be replaced by a constant. The efficiencies $(n_{(A)})$ and $(n_{(B)})$ in the exponent function, however, can only be considered equal if the peaks are *symmetrical*. This is because in that part of the composite peak that determines its maximum, the rear part of the first peak merges with the front part of the second peak. In LC, the concentration profiles of eluted peaks are rarely completely symmetrical and, thus, $(n_{(A)})$ must represent the efficiency of the rear half of the first peak (solute A) and $(n_{(B)})$ the efficiency of the front half of the second peak (solute B). In practice, the efficiencies are calculated in the normal way except, when determining the efficiency of the front half of the peak, twice the front half peak width is used instead of the total peak width. Similarly, to determine the efficiency of the rear half of the peak, twice the rear half width is employed instead of the total peak width. The extent to which the two efficiency values will differ will depend on the quality of the column system, the injection device, and the phase system. In general well-packed columns give very similar efficiencies for the two halves of the peak but the front half of the peak is most often slightly more efficient than the rear half of the peak. In addition, the response of the detector to the specific solutes must be taken into account. Thus, if (D) is the voltage output from the detector, equation (28) can be put in the form

$$D = C \left(\alpha\, X_A\, e^{-\left(\frac{n_{(A)}}{2}\right)\left(\frac{t}{t_A}-1\right)^2} + \beta\, X_B\, e^{-\left(\frac{n_{(B)}}{2}\right)\left(\frac{t}{t_B}-1\right)^2} \right) \qquad (5)$$

where (C) is a constant,
   (α) is the response of the detector to solute (A),
and (β) is the response of the detector to solute (B).

Equation (5) was examined by Scott and Reese [1] employing mixtures of nitrobenzene and fully deuterated nitrobenzene as the test sample. Their retention times were 8.927 min. and 9.061 min., respectively, giving a difference of 8.04 seconds. The separation ratio of the two solutes was 1.023 and the efficiencies of the front and rear portions of the peaks were 5908 and 3670 theoretical plates, respectively. The detector was, not surprisingly, found to have the same response to both solutes, *i.e.*, $\alpha = \beta$. Thus, inserting these values in equation (5),

$$D = C\left(X_A e^{-\left(\frac{3670}{2}\right)\left(\frac{t}{8.927}-1\right)^2} + X_B e^{-\left(\frac{5908}{2}\right)\left(\frac{t}{9.061}-1\right)^2} \right) \quad (6)$$

A range of concentrations of the two substances were inserted in equation (6) and a curve constructed relating retention time of the composite peak (calculated by means of a computer) to mixture composition. The results are shown in Figure 7.

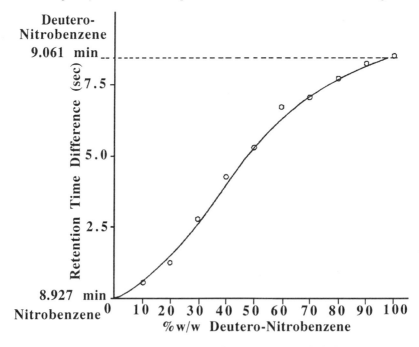

Courtesy of the Journal of Chromatography (ref. 1)

Figure 7. Graph of Retention Time of a Composite Peak against Composition of Mixture

The retention times of a series of mixtures of nitrobenzene and deuterated nitrobenzene were then actually measured and are included as points in Figure 7. It is seen that close agreement is obtained between the experimental points and the

Extensions of the Plate Theory

theoretical curve. The procedure described is an interesting alternative for the analysis of mixtures of closely eluted solutes to the difficult and often tedious construction of columns of extremely high efficiencies. With great care and the use of modern sophisticated computer programs, the required accuracy can be easily obtained, and it is surprising that this approach is not used more often.

Peak Asymmetry

There are a number of causes of peak asymmetry in both gas and liquid chromatography, including heat of adsorption, high activity sites on the support or absorbent, and nonlinear adsorption isotherms. Assuming that good quality supports and adsorbents are used, and the column is well thermostatted, the major factor causing peak asymmetry appears to result from nonlinear adsorption isotherms.

The equation for the retention volume of a solute, that was derived by differentiating the elution curve equation, can be used to obtain an equation for the retention time of a solute (t_r) by dividing by the flowrate (Q), thus,

$$\frac{V_r}{Q} = t_r = \frac{V_m + KV_s}{Q}$$

Now, the velocity of a solute band along the column (Z) is obtained by dividing the column length (L) by the retention time, (t_r); consequently,

$$Z = \frac{L}{t_r} = \frac{LQ}{V_m + KV_s}$$

Thus the band velocity (Z) is inversely proportional to ($V_m + KV_s$), and for a significantly retained solute, $V_m \ll KV_s$, consequently,

$$Z \propto \frac{1}{V_r'} \propto \frac{1}{KV_s} \propto \frac{1}{K} \qquad (7)$$

It is seen from the above equation that the band velocity is inversely proportional to the distribution coefficient with respect to the stationary phase. It follows that any changes in the distribution coefficient (K) will result directly in changes in the band velocity (Z). Consequently, if the isotherm is linear, then all concentrations will travel at the same velocity and the peak will be symmetrical.

Now, in gas/liquid chromatography, very small concentrations of solute are employed and linear absorption isotherms are to be expected. However, in LC the detectors have much lower sensitivities and as a result, significantly larger charges are necessary. Consequently, the solute concentration range can penetrate into that part of the Langmuir adsorption isotherm which is nonlinear. The adsorption isotherm for chloroform adsorbed on Partisil silica gel from a solution in *n*-heptane, obtained by Scott and Kucera [2], is shown in Figure 8. It is seen that the silica gel surface will not become completely covered with chloroform until the concentration of chloroform in the mobile phase is in excess of 40%w/w. The early part of the adsorption curve can be used to demonstrate the effect of the nonlinear portion of the isotherm on peak shape.

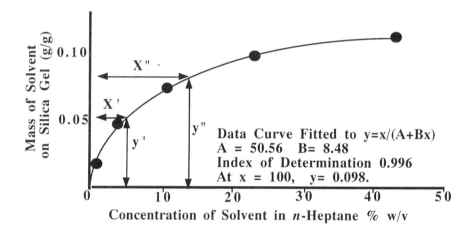

Courtesy of the Journal of Chromatography (ref. 2)

Figure 8. Langmuir Isotherm for Chloroform on Silica Gel

As already stated, while the isotherm is linear, then all concentrations in the peak will travel at the same speed and the resulting peak is symmetrical. Now, when the points y', x' and y", x" are reached, the situation changes. Taking the distribution coefficient as proportional to y'/x' and y"/x", it is clear that this ratio dectreases as the concentration of solute in the mobile phase increases.

It follows that as the peak velocity is inversely proportional to the distribution coefficient, then the higher concentrations in the peak will migrate faster than the lower concentrations. As a consequence, the peak is distorted with sharp front and a sloping back. This distortion is shown in Figure 9.

The major cause of peak asymmetry in GC is sample overload and this occurs mostly in preparative and semi-preparative separations. There are two forms of sample overload, *volume overload* and *mass overload*.

Extensions of the Plate Theory 177

Higher concentrations of solute move through the system more rapidly, thus reducing the retention time of the peak maximum which produces a peak with a sharp front and a sloping tail.

Figure 9. Peak Distortion by a Langmuir Type Isotherm

Volume overload results from too large a volume of sample being placed on the column, and this effect will be discussed later. It will be seen that volume overload does not, in itself, produce asymmetric peaks unless accompanied by mass overload, but it does *broaden* the peak. Mass overload, however, frequently results in a nonlinear adsorption isotherm. However, the isotherm is quite different from the Langmuir isotherm and is caused by an entirely different phenomenon.

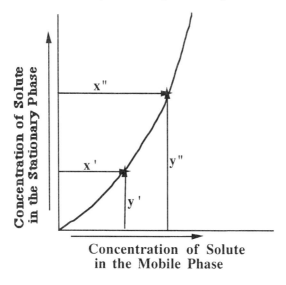

Figure 10. The Freundlich Isotherm

Consider the situation where a large charge is placed on a column and it is carrying a liquid stationary phase (*e.g.,* in GLC). The solute molecules, now in a relatively high concentration in the stationary phase, are no longer surrounded solely by solvent molecules, but by solvent and *solute* molecules.

Now, if the solutes interact with themselves more strongly than they do with the stationary phase, then their presence will increase the interaction of further solute with the stationary phase mixture. This gives an isotherm having the shape shown in Figure 10. This type of isotherm is called a Freundlich isotherm, the expression for which is given by

$$\varpi = \kappa c^{\frac{1}{n}}$$

where (ϖ) is the mass adsorbed per mass of adsorbent,

(k) is a constant,

and (n) is an empirical constant.

The isotherm will be close to linear at very low concentrations of solute in the mobile phase and thus, again, all concentrations in the peak will travel at the same speed and the resulting peak is symmetrical. However, when the points y', x' and y", x" are reached, the concentration of solute in the stationary phase is increasing more rapidly with the solute concentration in the mobile phase and thus the distribution coefficient at the higher concentrations is larger. It follows that, as the peak velocity is inversely proportional to the distribution coefficient, then the higher concentrations in the peak will migrate *more slowly* than the lower concentrations.

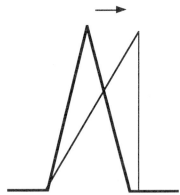

Higher concentrations of solute move through the system more slowly displacing the peak maximum to a later retention time and, thus, a peak is produced with a sloping front and a sharp back.

Figure 11. Peak Distortion Caused by a Freundlich Type Isotherm

As a consequence, the peak is distorted with sloping front and a sharp back as shown in Figure 11. There are a number of other effects that can cause peak distortion and

Extensions of the Plate Theory

some of these will be discussed later in the book. However, the major cause of peak asymmetry (the source of which is in the column and not other parts of the equipment) results from high solute concentrations in the distribution system. These high concentrations can cause the distribution to move into the curved parts of the Langmuir isotherm and produce peaks with a sharp front and sloping back. Conversely, the large charge can cause strong solute/solute interactions in the stationary phase (particularly when the stationary phase is a liquid) and the distribution then follows the Freundlich isotherm which produces peaks having a sloping front and a sharp back.

Column Efficiency

Recalling that a separation is achieved by moving the solute bands apart in the column and, at the same time, constraining their dispersion so that they are eluted discretely, it follows that the resolution of a pair of solutes is not successfully accomplished by merely selective retention. In addition, the column must be carefully designed to minimize solute band dispersion. Selective retention will be determined by the interactive nature of the two phases, but band dispersion is determined by the physical properties of the column and the manner in which it is constructed. It is, therefore, necessary to identify those properties that influence peak width and how they are related to other properties of the chromatographic system. This aspect of chromatography theory will be discussed in detail in Part 2 of this book. At this time, the theoretical development will be limited to obtaining a measure of the peak width, so that eventually the width can then be related both theoretically and experimentally to the pertinent column parameters.

Starting with the Poisson form of the elution equation, the peak width at the points of inflexion of the curve (which corresponds to twice the standard deviation of the normal elution curve) can be found by equating the second differential to zero and solving in the usual manner. Thus, at the points of inflexion,

$$\frac{d_2\left(X_o \frac{e^{-v} v^n}{n!}\right)}{dv^2} = 0$$

Now, $\dfrac{d_2\left(X_o \dfrac{e^{-v} v^n}{n!}\right)}{dv^2} = X_o \dfrac{e^{-v} v^n - e^{-v} n v^{(n-1)} - e^{-v} n v^{(n-1)} + e^{-v} n (n-1) v^{(n-2)}}{n!}$

Thus,
$$\frac{d_2\left(X_o \frac{e^{-v} v^n}{n!}\right)}{dv^2} = X_o \frac{e^{-v} v^{(n-2)}(v^2 - 2nv + n(n-1))}{n!}$$

At the points of inflexion,
$$\frac{d_2\left(X_o \frac{e^{-v} v^n}{n!}\right)}{dv^2} = 0$$

Hence
$$v^2 - 2nv + n(n-1) = 0$$

$$v = \frac{2n \pm \sqrt{(4n^2 - 4n(n-1))}}{2}$$

$$= \frac{2n \pm \sqrt{4n}}{2}$$

$$= n \pm \sqrt{n}$$

Thus, the points of inflexion occur after $n - \sqrt{n}$ and $n + \sqrt{n}$ plate volumes of mobile phase has passed through the column. It follows, the volume of mobile phase that has passed through the column *between* the inflexion points will be

$$n + \sqrt{n} - n + \sqrt{n} = 2\sqrt{n} \qquad (8)$$

Consequently, the peak width at the points of inflexion of the elution curve will be $2\sqrt{n}$ plate volumes. Converting this to *milliliters* of mobile phase by multiplying by the *plate volume*

$$\text{Peak Width} = 2\sqrt{n}(v_m + K v_s) \qquad (9)$$

The peak width at the points of inflexion of the elution curve is *twice the standard deviation* of the Poisson or Gaussian curve and thus, from equation (8), the variance (the square of the standard deviation) will be equal to (n), the total number of plates in the column.

The variance of the band (σ^2) in milliliters^2 of mobile phase will be given by

$$\sigma^2 = n(v_m + Kv_s)^2$$

Now,
$$V_r = n(v_m + Kv_s)$$

Thus,
$$\sigma^2 = \frac{V_r^2}{n}$$

Extensions of the Plate Theory

Thus, the variance of the peak is inversely proportional to the number of theoretical plates in the column. Consequently, the greater the value of (n), the more narrow the peak, and the more efficiently has the column constrained peak dispersion. As a result, the number of theoretical plates in a column has been given the term *Column Efficiency*. From the above equations, a fairly simple procedure for measuring the efficiency of any column can be derived.

Let the distance between the injection point and the peak maximum (the retention distance on the chromatogram) be (y) cm and the peak width at the points of inflexion be (x) cm. If a computer data acquisition and processing system is employed, then the equivalent retention times can be used.

Now, it has already been shown that the retention volume of a solute is given by $n(v_m + Kv_s)$, and twice the standard deviation of the peak at the inflexion points is given by $2\sqrt{n}(v_m + Kv_s)$

Thus, by simple proportion,

$$\frac{\text{Ret. Distance}}{\text{Peak Width}} = \frac{y}{x} = \frac{n(v_m + Kv_s)}{2\sqrt{n}(v_m + Kv_s)} = \frac{\sqrt{n}}{2}.$$

Then
$$n = 4\left(\frac{y}{x}\right)^2 \quad (10)$$

Using equation (10), the efficiency of any solute peak can be calculated for any column from measurements taken directly from the chromatogram (or, if a computer system is used, from the respective retention times stored on disk). The computer will need to have special software available to identify the peak width and calculate the column efficiency and this software will be in addition to that used for quantitative measurements. Most contemporary computer data acquisition and processing systems contain such software in addition to other chromatography programs. The measurement of column efficiency is a common method for monitoring the quality of the column during use.

Equation (10) also allows the peak width (2σ) and the variance (σ^2) to be measured as a simple function of the retention volume of the solute but, unfortunately, does not help to identify those factors that cause the solute band to spread, nor how to control it. This problem has already been discussed and is the basic limitation of the plate theory. In fact, it was this limitation that originally invoked the development of the

Rate Theory to describe the mechanism of band dispersion and which will be the major topic of Part 2 of this book.

The Position of the Points of Inflection

The measurement of efficiency is important, as it is used to monitor the quality of the column during use and to detect any deterioration that might take place. However, to measure the column efficiency, it is necessary to identify the position of the points of inflection which will be where the width is to be measured. The inflection points are not easily located on a peak, so it is necessary to know at what fraction of the peak height they occur, and the peak width can then be measured at that height.

As the peak represents the concentration profile of the eluting solute, the fraction of the peak height at which the points of inflexion are located will be the same as the ratio of the solute concentration after $(n - \sqrt{n})$ plate volumes of mobile phase has passed through the column to the solute concentration after (n) plate volumes of mobile phase have passed through the column.

Therefore, if (f) is the fraction of the height (h) at which the points of inflection occur, then

$$\frac{f}{h} = \frac{X_{m(n-\sqrt{n})}}{X_{m(n)}}$$

$$= \frac{\dfrac{X_o e^{-(n-\sqrt{n})}(n-\sqrt{n})^n}{n!}}{\dfrac{X_o e^{-n} n^n}{n!}}$$

$$= \frac{e^{-(n-\sqrt{n})}(n-\sqrt{n})^n}{e^{-n} n^n}$$

$$= e^{\sqrt{n}} \left(\frac{n-\sqrt{n}}{n}\right)^n$$

$$= e^{\sqrt{n}} \left(1-\frac{1}{\sqrt{n}}\right)^n$$

Thus, $$\log_e \frac{f}{h} = \sqrt{n} + n \log_e\left(1-\frac{1}{\sqrt{n}}\right) \qquad (11)$$

Extensions of the Plate Theory

Now, if $x \ll 1$, $\quad \log_e(1-x) = -x - \dfrac{x^2}{2} - \dfrac{x^3}{3} - \dfrac{x^4}{4}$

Consequently, since (n) will always be greater than 100,

$$n \log_e\left(1 - \dfrac{1}{\sqrt{n}}\right) = n\left[-\dfrac{1}{\sqrt{n}} - \dfrac{1}{2n} - \dfrac{1}{3n\sqrt{n}} - \dfrac{1}{4n^2} - \ldots\right]$$

$$= -\sqrt{n} - \dfrac{1}{2} - \dfrac{1}{3\sqrt{n}} - \dfrac{1}{4n} - \ldots \qquad (12)$$

Substituting for $n \log_e\left(1 - \dfrac{1}{\sqrt{n}}\right)$ from equation (12) in equation (11),

$$\log_e \dfrac{f}{h} = \sqrt{n} - \sqrt{n} - \dfrac{1}{2} - \dfrac{1}{3\sqrt{n}} - \dfrac{1}{4n} - \ldots$$

Now, $\dfrac{1}{3\sqrt{n}} \ll \dfrac{1}{2}$ Consequently, the term $\left(\dfrac{1}{3\sqrt{n}}\right)$ and those above can be ignored with respect to $\left(\dfrac{1}{2}\right)$. Then,

$$\log_e \dfrac{f}{h} = -\dfrac{1}{2} \quad \text{and} \quad \dfrac{f}{h} = e^{-\dfrac{1}{2}} = 0.6065 \qquad (13)$$

Equation (13) shows that the points of inflection occur at 0.6065 of the peak height. It follows that the peak width, at that height, will be equivalent to two standard deviations (2σ) of the Gaussian curve.

Resolving Power of a Column

Reiterating the conditions for a chromatographic separation once again, for two solutes to be resolved their peaks must be moved apart in the column and maintained sufficiently narrow for them to be eluted as discrete peaks. However, the criterion for two peaks to be resolved (usually defined as the *resolution*) is somewhat arbitrary and is usually defined as the ratio of the distance between the peak maxima to half the peak width (σ) at the points of inflection. To illustrate the various degrees of resolution that can be obtained, the separation of a pair of solutes 2σ, 3σ, 4σ, 5σ and 6σ apart are shown in Figure 12. Although, for *baseline resolution*, it is clear that the peak maxima should be separated by at least 6σ for most quantitative analyses, particularly where peak height measurements are employed, a separation of 4σ is usually satisfactory.

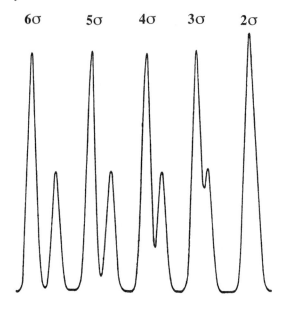

Figure 12. Separation of Solutes Pairs by 2σ, 3σ, 4σ, 5σ and 6σ

In addition, even if peak area measurements are used, a separation of 4σ will usually provide adequate accuracy and this is particularly so if computer data acquisition and processing is used with peak deconvolution software. In general, a resolution of 4σ will be assumed whenever dealing with resolution or column design. In most column optimization procedures the effect of increasing the criteria for resolution can easily be accommodated if so desired.

Although not apparent at this time, it will become clear that two adjacent peaks from solutes of different chemical type, or significantly different molecular weight, are not likely to have precisely the same peak widths (*i.e.*, exhibit the same efficiency). Nevertheless, in most cases, the difference will be relatively small and, in fact, likely to be negligible. As a consequence, the widths of closely adjacent peaks will, at this time, be assumed to be the same.

Consider the two peaks depicted in Figure 13. The difference between the peaks, for solutes (A) and (B), measured in volume flow of mobile phase will be

$$n(v_m + K_B v_S) - n(v_m + K_A v_S) = n(K_B + K_A)v_S \qquad (14)$$

Extensions of the Plate Theory

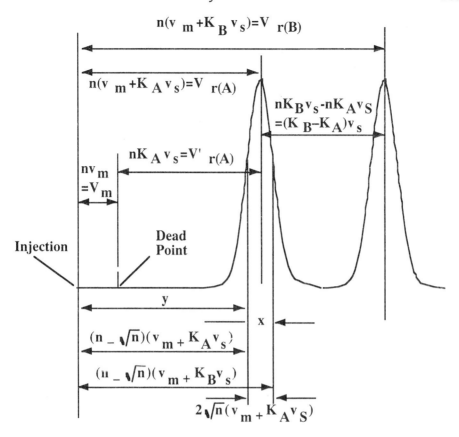

Figure 13. Chromatogram Showing Two Resolved Peaks

If the widths of the two peaks are the same, then the peak width in volume flow of mobile phase will be

$$2\sigma = 2\sqrt{n}\,(v_m + K_A v_S) \qquad (15)$$

where (K_A) is the distribution coefficient of the first eluted solute between the two phases. Assuming the criterion that resolution is achieved when the peak maxima of the pair of solutes are 4σ apart, then

$$4\sqrt{n}\,(v_m + K_A v_s) = n(K_B + K_A)v_s$$

Rearranging,
$$\sqrt{n} = \frac{4(v_m + K_A v_S)}{(K_B - K_A)v_S}$$

Dividing through by (v_m),
$$\sqrt{n} = \frac{4(1 + k_A)}{(k'_B - k_A)v_S}$$

Now (α), the separation ratio between the two solutes, has been defined as $\alpha_{B/A} = \dfrac{k_B}{k_A}$ Then, $\sqrt{n} = \dfrac{4(1+k_A)}{k_A(\alpha_{B/A} - 1)}$,

and
$$n = \left(\dfrac{4(1+k_A)}{k_A(\alpha_{B/A} - 1)}\right)^2 = 16 \dfrac{(1+k_A)^2}{k_A^2(\alpha_{B/A} - 1)^2} \qquad (16)$$

Equation (16) was first developed by Purnell [3] in 1959 and is extremely important. It can be used to calculate the efficiency required to separate a given pair of solutes from the capacity factor of the first eluted peak and their separation ratio. It is particularly important in the theory and practice of column design. In the particular derivation given here, the resolution is referenced to (k_A) the capacity ratio of the *first* eluted peak. A similar equation to that of equation (16), but in a slightly different form, can be derived employing (k_B), the capacity ratio of the second peak as the reference solute. Both equations will give the same numerical result.

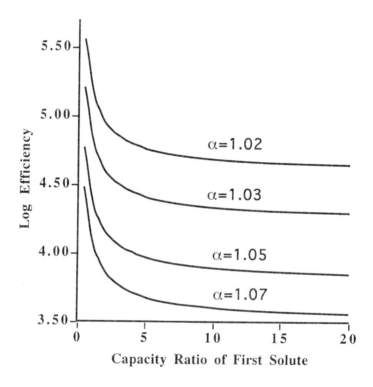

Figure 14. Graph of Log. Efficiency against Capacity factor for Solute Pairs having Different Separation Ratios

Extensions of the Plate Theory

Figure 14, shows curves relating (n) and (k'_A) for a range of solute pairs having separation ratios of 1.02, 1.03, 1.05 and 1.07 calculated using equation (16). It is seen that as the separation becomes more difficult (*i.e.*, the separation ratio ($\alpha_{A/B}$) becomes smaller), the efficiency (n) required to achieve the separation increases rapidly. This is to be expected, as when ($\alpha_{A/B}$) is small, and the peaks become closer, then the peak dispersion must be more constrained to reduce their width, which means the column must be made more efficient. However, the reason for the dramatic increase in (n) when the value of the capacity factor becomes small is not so obvious. This is a serious obstacle to achieving short analysis times, as the smaller the value of (k'), the faster the solutes are eluted. However, if a greater efficiency is required to realize the separation, this may mean that longer columns are necessary which will *extend* the analysis times. It is seen that a separation ratio of 1.02 would require an efficiency of 360,000 theoretical plates to resolve a given pair of solutes if the capacity ratio of the first eluted peak was only 0.5. Capillary columns in GC can provide such efficiencies but, in LC, although possible, would be extremely difficult and costly to produce. It follows that, as a general rule, the phase system should be chosen such that the solutes that are closest in the chromatogram are not eluted at very low (k') values. This will reduce the efficiency needed and, thus, lessen the necessary *column length* and, consequently, the *analysis time*. At (k') values that exceed 10, the efficiency required does not change very much as the capacity ratio increases. It should be emphasized again that, for fast analyses, the phase system should be selected not simply to provide a large separation ratio, but also to ensure that the first peak is eluted at a capacity ratio of 10 or greater. This indicates that the phase system should be chosen to have high selectivity and retentive capacity so that the column can be as short as possible. Column optimization is a complex procedure that will be dealt with later in this book.

Effective Plate Number

The idea of the *effective plate number* was introduced and employed by Purnell [4], Desty [5] and others in the late 1950s. Its conception was evoked as a direct result of the introduction of the capillary column or open tubular column. Even in 1960, the open tubular column could be constructed to produce efficiencies of up to a million theoretical plates [6]. However, it became immediately apparent that these high efficiencies were only obtained for solutes eluted at very low (k') values and, consequently, very close to the column dead volume. More importantly, on the basis of the performance realized from packed columns, the high efficiencies did not correspond to the greatly increased resolution that would be expected. The poor performance of columns having high efficiencies at low (k') values is precisely the effect just described.

In an attempt to rationally provide a measure of column performance that would be commensurate with the resolution that was obtained, the *effective plate number* was introduced. The effective plate number is calculated in a similar manner to the column efficiency in theoretical plates, but employs the *corrected retention distance*, as opposed to the total retention distance in conjunction with the width of the peak. As a result, the effective plate number is significantly smaller than the number of theoretical plates in the column, particularly for solutes eluted at low (k') values. The column efficiency and the effective plate number converge to the same value at high (k') values. As a consequence, the effective plate number more nearly corresponds to the actual resolving power of the column. The effective plate number might also be considered as an indirect way of numerically defining resolution. Although the theoretical plate, as defined by the plate theory, has a practical significance and is commonly used in column design, the concept of the *effective plate* is not theoretically unsound and is related directly to the theoretical plate.

The efficiency of a column (n), in number of theoretical plates, has been shown to be given by the following equation,

$$n = 4 \frac{y^2}{x^2}$$

where (y) is the retention distance
and (x) is the peak width.

Now, the number of *effective plates* (N_E), by definition, is given by

$$N_E = 4 \frac{(y - y_0)^2}{x^2} \qquad (17)$$

where (y_0) is the retention distance of an unretained solute (the position of the dead point).

Now from the plate theory,
$$\frac{y}{x} = \frac{n(v_m + Kv_s)}{2\sqrt{n}(v_m + Kv_s)}$$

Thus
$$\frac{y - y_0}{x} = \frac{n(v_m + Kv_s) - n v_m}{2\sqrt{n}(v_m + Kv_s)}$$

Extension of the Plate Theory

By dividing through by (v_m), and noting that $\dfrac{Kv_s}{v_m} = k'$,

$$\frac{y - y_o}{x} = \frac{\sqrt{n}\, k'}{2(1+k')}$$

Consequently,
$$4\left(\frac{(y-y_o)}{x}\right)^2 = n\left(\frac{k'}{(1+k')}\right)^2 = N_E \qquad (18)$$

Equation (18) displays the relationship between the column efficiency defined in theoretical plates and the column efficiency given in 'effective plates'. It is clear that the number of 'effective plates' in a column is not an arbitrary measure of the column performance, but is directly related to the column efficiency as derived from the plate theory. Equation (18) clearly demonstrates that, as the capacity ratio (k') becomes large, (n) and (N_E) will converge to the same value.

In 1965, Giddings [7] proposed the function $\left(\dfrac{k'}{\Delta k'}\right)$ as a means of defining the resolving power (R) of a column. This function is analogous to a very similar expression used in spectroscopy to define spectroscopic resolution, i.e., $\left(\dfrac{\lambda}{\Delta \lambda}\right)$, where ($\Delta\lambda$) is the minimum wavelength increment that can be resolved from an adjacent wavelength, at a mean wavelength (λ). The value that Giddings used that would be analogous to ($\Delta\lambda$) in spectroscopy was ($\Delta k'$), the band width at the base of the eluted peak. Consequently, ($\Delta k'$) will be equivalent to twice the peak width at the points of inflexion or four times the standard deviation (4σ). The function $\left(\dfrac{k'}{\Delta k'}\right)$ has an interesting relationship to the number of effective plates in the column.

Thus, from the plate theory, (R_r) the concept of resolution as introduced by Giddings, will be given by

$$R_r = \frac{k'}{\Delta k'} = \frac{n\, k v_s}{4\sqrt{n}\,(v_m + K v_s)}$$

Dividing through by (v_m), and noting that $\dfrac{Kv_s}{v_m} = k'$,

$$R_r = \frac{\sqrt{n}\, k'}{4(1+k')} = \frac{\sqrt{N_E}}{4} \qquad (19)$$

Equation (19) shows that the resolving power of the column (using the definition proposed by Giddings) is directly proportional to the square root of (N_E), the number of effective plates. It follows that (R_r) can be used to directly compare the resolving power of different columns of any size or type. Unfortunately, the magnitude of (R_r) will also depend on the value of (k') for the respective solute used for measurement and, consequently, any comparison between columns must be made using data from solutes that have the same, or closely similar, (k') values.

As a secondary consideration, the chromatographer may also need to know the minimum value of the separation ratio (α) for a solute pair that can be resolved by a particular column. The minimum value of (α) has also been suggested [8] as an alternative parameter that can be used to compare the performance of different columns. There is, however, a disadvantage to this type of criteria, due to the fact that the value of (α) becomes *less* as the resolving power of the column becomes *greater*. Nevertheless, a knowledge of the minimum value of ($\alpha_{A/B}$) can be important in practice, and it is of interest to determine how the minimum value of ($\alpha_{A/B}$) is related to the effective plate number.

The minimum value of ($\alpha_{A/B}$) of a pair of solutes that are to be separated on a specific column can be assumed to be equal to the ratio of the retention distance of the first peak plus its width at the base, to its normal retention distance. This assumption premises that resolution is satisfactory when the peak maxima are separated by 4σ.

Thus,
$$\alpha_{A/B} = \frac{nKv_s + 4\sqrt{n}(v_m + Kv_s)}{nKv_s} = 1 + \frac{4(1+k')}{\sqrt{n}\,k'}$$

Again, bearing in mind that $\dfrac{Kv_s}{v_m} = k'$

Therefore,
$$\alpha_{A/B} = 1 + \frac{4}{\sqrt{N_E}} = 1 + \frac{1}{R_r} \tag{20}$$

It is seen that the chromatographer can arrive at the minimum ($\alpha_{A/B}$) value for a pair of solutes that the column can resolve directly, from either the resolution, as defined by Giddings, or from a simple function of the number of effective plates. However, again it must be emphasized that this will not be a *unique* value for any column, as it will also depend on the (k') of the eluted solute.

Elution Curve of a Finite Charge

In most mathematical processes, including the derivation of the plate theory, the assumption is made that the initial charge is placed on the first plate of the column. This is difficult to achieve in practice, as the charge must occupy a finite portion of

Extension of the Plate Theory

the total column volume and, consequently, also occupy a specific number of theoretical plates. This situation will now be examined theoretically. Consider the situation shown in Figure 15, where it is assumed that the initial charge is distributed over (r) theoretical plates.

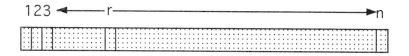

Figure 15. The Injection of a Finite Volume of Charge on to an LC Column

Let the mobile phase flow be interrupted, and a sample placed on the column such that it is dispersed over (r) theoretical plates. The mobile phase is not actually stopped in practice and thus, in reality, the injection is carried out while the flow of mobile phase continues. Now, after injection, the contents of each plate will be eluted through the column and produce its own unique elution curve as though each were the only sample injected. As a result, there will be (r) elution curves and the concentration profile of each curve will be separated from its neighbor by one plate volume of mobile phase. The first peak will result from the elution of the contents of the (r)th plate and the last peak from the elution of the contents of the first plate. The composite peak actually monitored by the detector will be a concentration profile formed by the addition of all the individual peaks. Due to the dispersion of the sample over (r) plates on injection, the last peak, which was eluted from the first plate, will pass through all the plates in the column, (*i.e.*, (n) plates). Whereas the sample originally situated in plate (r) after injection, will only pass through (n-r) plates of the column. It follows that the contents of plate (r) will start eluting as though (r) plate volumes of mobile phase had already passed through the column. From the elution curve equation, the elution curve resulting from that portion of the sample on plate one will be given by

$$X_{n(1)} = X_0 \frac{e^{-v} v^n}{n!}$$

and for that portion of the sample on plate 2, $X_{n(2)} = X_0 \dfrac{e^{-(v+1)}(v+1)^n}{n!}$

and similarly for that portion on plate 3, $X_{n(3)} = X_0 \dfrac{e^{-(v+2)}(v+2)^n}{n!}$

Consequently, the composite elution curve for the total sample originally dispersed over the first three plates will be given by,

$$X_{n(1)} + X_{n(2)} + X_{n(3)} = X_0 \left[\frac{e^{-v} v^n}{n!} + \frac{e^{-(v+1)}(v+1)^n}{n!} + \frac{e^{-(v+2)}(v+2)^n}{n!} \right]$$

It follows that the elution curve for a sample, initially dispersed over (r) plates, will be

$$X_n = \sum_{p=0}^{p=r} \frac{e^{-(v+p)}(v+p)^n}{n!} \qquad (21)$$

The sum expressed by equation (21) lends itself to a digital calculation and can be employed in an appropriate computer program to calculate actual peak profiles. In doing so, however, as (v) is measured in plate volumes and sample volumes are usually given in milliliters, they must be converted to plate volumes to be used with equation (21). To demonstrate the effect of a finite charge and the use of equation (21), the peak profiles resulting from a sample dispersed over the twenty-one consecutive plates of a column are shown in Figure 16.

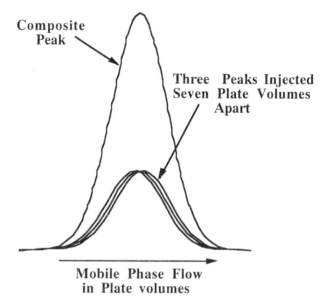

Figure 16. Elution Curves Resulting from the Injection of a Charge that Occupied the First Twenty-One Plates of a Column

In Figure 16 the elution profiles for samples from each group of seven plates are included together with the overall composite peak from the total charge. The calculations assumed a column efficiency of 5000 theoretical plates. The elution

Extension of the Plate Theory

curves from each seventh plate are seen to be identical, having the same height and width. The height of the composite curve is three times that of the individual curves from each plate. In addition, it is seen that the width of the composite curve will be twenty-one plate volumes greater than a single curve from a single plate. Figure 16 also shows that the effect of a finite volume of charge spread over a number of theoretical plates is to broaden the peak. This, in turn, reduces the apparent efficiency of the column as measured and, thus, its resolving power. It follows that for the accurate measurement of efficiency, not only must the sample *mass* be small to ensure the adsorption isotherm is linear (solute concentration must be reduced appropriately), but the sample *volume* must also be limited to minimize peak dispersion. It follows that peak dispersion by sample volume needs to be examined theoretically to identify the maximum sample volume that can be tolerated before the column efficiency is significantly impaired. However, before this can be considered the concept of the *summation of variances* must be understood.

Summation of Variances

Peak dispersion can happen in any part of the chromatographic system, from the sample valve to the sensing cell of the detector. The final band width will be the result of all the dispersion processes that take place, not merely in the column, but in all the other parts of the mobile phase conduit system. It follows that to determine the ultimate peak width, the contribution of all the extra-column dispersion processes must be identified and taken into account and then added to the dispersion that occurs in the column itself. The theory of statistics dictates that it is not possible to sum the standard deviations of a series of random processes, but it is possible to sum their variances. Nevertheless, it is important to understand that the various processes must be random and *non-interactive*; that is to say that the extent to which one random process proceeds does not affect the progress of another random process. It is difficult to prove a particular physical process is random, other than by analyzing its effect. The experimental evidence so far obtained for those dispersion processes that have been examined in detail indicates that all types of band spreading that take place in a chromatographic system are basically random in nature. Now, assuming there are (n_n) non-interacting, random dispersive processes occurring in the chromatographic system, then any process (p) acting alone will produce a Gaussian curve having a variance σ_p^2. Hence,

$$\sigma_1^2 + \sigma_2^2 + \sigma_3^2 + \ldots + \sigma_p^2 \ldots + \sigma_{n_n}^2 = \sigma^2$$

where (σ^2) is the variance of the peak as sensed by the detector.

This equation expresses the concept of the summation of variances in an algebraic form. The principle has fundamental significance in chromatography theory and is the basic assumption behind all forms of the Rate Theory. By identifying the individual dispersion processes that occur in a column and deriving an expression for the magnitude of their resulting variances, they can be summed to provide an expression for the overall column variance. However, to demonstrate the manner by which the summation of variances can be employed for practical use, an expression for the maximum volume of charge that can be tolerated by a given column before its performance becomes degenerated will be derived.

Maximum Sample Volume that Can Be Placed on a Chromatographic Column

Having established that a finite volume of sample causes peak dispersion and that it is highly desirable to limit that dispersion to a level that does not impair the performance of the column, the maximum sample volume that can be tolerated can be evaluated by employing the principle of the summation of variances. Let a volume (V_i) be injected onto a column. This sample volume (V_i) will be dispersed on the front of the column in the form of a rectangular distribution. The eluted peak will have an overall variance that consists of that produced by the column and other parts of the mobile phase conduit system plus that due to the dispersion from the finite sample volume. For convenience, the dispersion contributed by parts of the mobile phase system, other than the column (except for that from the finite sample volume), will be considered negligible. In most well-designed chromatographic systems, this will be true, particularly for well-packed GC and LC columns. However, for open tubular columns in GC, and possibly microbore columns in LC, where peak volumes can be extremely small, this may not necessarily be true, and other extra-column dispersion sources may need to be taken into account. It is now possible to apply the principle of the summation of variances to the effect of sample volume.

Thus,
$$\sigma^2 = \sigma_i^2 + \sigma_c^2$$

where (σ^2) is the overall variance of the eluted peak,
(σ_i^2) is the variance of the sample volume,
and (σ_c^2) is the variance due to column dispersion.

Now, the acceptable degree to which the performance of the column is denigrated is basically an arbitrary decision. As long ago as 1957 Klinkenberg [9] recommended

Extension of the Plate Theory

that the maximum extra-column dispersion that should be tolerated without serious loss in resolution was a 5% increase of the standard deviation (*ca.* 10% increase in peak variance). This criterion, recommended by Klinkenberg as the limit for extra-column dispersion, is now generally accepted.

It has been established that the variance of a rectangular distribution of sample volume (V_1) will be $\left(\dfrac{V_i^2}{12}\right)$. Now, if it is assumed that the peak width can be increased by 5% as a result of the dispersing effect of the sample volume, then applying the principle of the summation the variances,

$$\frac{V_i^2}{12} + \left(\sqrt{n}(v_m + Kv_s)\right)^2 = \left(1.05\sqrt{n}(v_m + Kv_s)\right)^2$$

where, from the plate theory, dispersion due to the column alone is $\left(\sqrt{n}(v_m + Kv_s)\right)^2$

Thus,
$$\frac{V_i^2}{12} = n(v_m + Kv_s)^2(1.05^2 - 1)$$

$$= n(v_m + Kv_s)^2 \, 0.102$$

Consequently,
$$V_i^2 = n(v_m + Kv_s)^2 \, 1.22$$

or
$$V_i = \sqrt{n}(v_m + Kv_s) \, 1.1$$

Recalling that $V_r = n(v_m + Kv_s)$, then

$$V_i = \frac{1.1 \, V_r}{\sqrt{n}}. \tag{22}$$

Equation (22) allows the maximum sample volume that can be used without seriously denigrating the performance of the column to be calculated from the retention volume of the solute and the column efficiency. In any separation, there will be one pair of solutes that are eluted closest together (which, as will be seen in Part 3 of this book, is defined as the *critical pair*) and it is the retention volume of the first of these that is usually employed in equation (22) to calculate the maximum acceptable sample volume.

Vacancy Chromatography

Vacancy chromatography is a special method of development that can provide both negative and positive peaks in the chromatogram. It is not commonly used, although it will be seen that it has certain characteristics that could render it useful for process monitoring and process control. If a column is supplied with a mobile phase that contains a solute at a given concentration, and the mobile phase flow is continued until equilibrium is achieved throughout the column, then the concentration of solute in the column eluent will be the same as that in the mobile phase at the inlet.

Now consider that a sample of pure mobile phase, devoid of solute, is placed on the first plate of the column. This will cause a fall in the solute concentration in the first plate which, *mathematically*, will represent the injection of a sample having a *negative concentration*. Now a negative concentration profile will pass through a column in exactly the same way as a positive concentration profile and it will be eluted from the column and be recorded as a negative peak. In addition, the concentration profile of a negative sample concentration will be described by the same elution equation and its retention volume will be the same as that for a positive concentration sample. It is the generation of negative peaks by injecting a sample of pure mobile phase into a mobile phase containing solutes at constant and known concentrations that is understood by the term *vacancy chromatography*. This term was first introduced by Zhukhovitski and Turkel'taub [10], who used the technique for process monitoring with GC (14). Later, Reilley *et al.* [11] also explored its use in process monitoring using LC.

Vacancy chromatography has some quite unique properties and a number of potentially useful applications. Vacancy chromatography can be theoretically investigated using the equations derived from the plate theory for the elution of positive concentration profiles. As already stated, if a mobile phase contains a solute at a concentration (X_o), then, after equilibrium, the eluent from the column will also contain the solute at a concentration (X_o).

Consider a sample containing the same solute at a concentration (X_i) injected onto the column under conditions where either ($X_o < X_i$) or ($X_o > X_i$), but ($X_o \neq X_i$). Such an injection will produce a transient change in the concentration (X_o) of solute in the mobile phase.

Now, from the plate theory, this transient concentration change will be eluted through the column as a *concentration difference* and will be sensed as a negative or positive peak by the detector. The equation describing the resulting concentration profile of the eluted peak, from the plate theory, will be given by

Extension of the Plate Theory

$$X_{(n)} = (X_i - X_o)\frac{e^{-v}v^n}{n!}$$

where $X_{(n)}$ is the concentration of the solute in the mobile phase leaving the (n)th plate,

(v) is the volume passed through the column in plate volumes,

and (n) is the number of theoretical plates in the column.

If the sample consisted of pure mobile phase containing no solute, then $X_i = 0$ and

$$X_{(n)} = -X_o \frac{e^{-v}v^n}{n!}$$

It follows that the concentration leaving the (n)th plate (X_E) will be

$$X_E = X_{(n)} + X_o = X_o\left(1 - \frac{e^{-v}v^n}{n!}\right)$$

Employing the alternative Gaussian form of the elution equation,

$$X_E = X_{(n)} + X_o = X_o\left(1 - \frac{e^{\frac{-w^2}{2n}}}{\sqrt{2\pi n}}\right) \quad (23)$$

Furthermore, at the peak maximum when v=n and w=0,

$$X_E = X_o\left(1 - \frac{1}{\sqrt{2\pi n}}\right) \quad (24)$$

Equation (24) shows that when the charge is placed on the first plate, (X_n) can never equal zero and pure mobile phase free of solute will never elute from the column. However, in practice, it is almost impossible to place the sample exclusively on the first plate, and there will be a finite volume of mobile phase that will occupy a finite number of theoretical plates when it is injected onto the column.

Now, if (p) plate volumes of pure mobile phase are injected onto a column that has been equilibrated with mobile phase containing a concentration (X_o) of a solute, then on the injection of a charge of pure mobile phase:

a. For a charge where p=1, the change in concentration of solute on the first plate will be

$$X_o(e^{-1}-1)$$

b. For a charge where p=2, the change in concentration of charge on the first plate will be

$$X_0(e^{-2}-1)$$

c. For a charge where p=p, the change in concentration of charge on the first plate will be,

$$X_0(e^{-p}-1)$$

After the addition of each plate volume of charge, a new concentration of solute exists in plate (1), and its contents will be eluted through the column in the normal manner.

Consider a total of (p) plate volumes of pure mobile phase are injected onto the column followed by a further (v) plate volumes of equilibrated mobile phase. After the injection of (r) plate volumes of pure mobile phase, the new concentration of solute in plate 1 will be eluted by a further (p-r) plate volumes of sample followed by (v) plate volumes of equilibrated mobile phase. Therefore, the concentration of solute leaving the (n)th plate due to the (r)th volume of pure mobile phase will be

$$X_n = X_0\left(e^{-r}-1\right) \frac{e^{\frac{-(v+p-r-n)^2}{2n}}}{\sqrt{2\pi n}}$$

Consequently, the *actual* concentration of solute in the (n)th plate (X_E) leaving the column and entering the detector that results from the injection of (p) plate volumes of *pure* mobile phase, followed by (v) plate volumes of *equilibrated* mobile phase carrying a solute concentration (X_0) of solute, will be given by

$$X_E = X_0 + \sum_{r=1}^{r=p} X_0\left(e^{-r}-1\right) \frac{e^{\frac{-(v+p-r-n)^2}{2n}}}{\sqrt{2\pi n}}. \qquad (25)$$

The sum expressed by equation (25) also lends itself to a digital solution and can be employed in an appropriate computer program to calculate actual peak profiles for different volumes of pure mobile phase that have been injected onto an equilibrated column. The values of (X_E) were calculated for a column having 500 theoretical plates and for sample volumes of 20, 50, 100 and 200 plate volumes, respectively. The curves relating solute concentration (X_E) to plate volumes of mobile phase passed through the column are shown in Figure 17.

The curves in Figure 17 show that as the injection volume is increased, so the retention volume of the peak also increases. The retention volume of the small negative peak produced by the smallest charge will be the same as that for a sample where $X_i > X_0$ and the same as that for a solute chromatographed in the normal way with the column carrying pure mobile phase only. The significant dispersion that occurs with larger charges is clearly demonstrated.

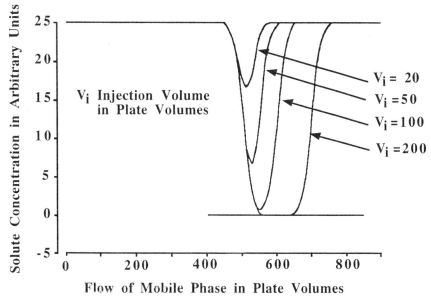

Courtesy of the Journal of Chromatography (ref. 12)

Figure 17. Vacancy Elution Curves from Different Injection Volumes on a Column of 500 Theoretical Plates

The theoretical treatment given above assumes that the presence of a relatively low concentration of solute in the mobile phase does not influence the retentive characteristics of the stationary phase. That is, the presence of a small concentration of solute does not influence either the nature or the magnitude of the solute/phase interactions that determine the extent of retention. The concentration of solute in the *eluted* peak does not fall to zero until the sample volume is in excess of 100 plate volumes and, at this sample volume, the peak width has become about five times the standard deviation of the normally loaded peak.

Equation (25) can be extended to provide a general equation for a column equilibrated with (q) solutes at concentrations $X_1, X_2, X_3, ... X_q$. For any particular solute (S), if its normal retention volume is $V_{r(S)}$ on a column containing (n) plates, then from the plate theory, the plate volume of the column for the solute (S), *i.e.*, (v_S) is given by

$$v_s = \frac{V_{r(s)}}{n}$$

Let the volume of sample injected be (V_i) ml, then the charge measured in 'plate volumes' will be ($\frac{V_i}{s}$). Similarly, if the volume of mobile phase passed through the column is (y) ml then that will be equivalent to $\frac{y}{v_s}$ plate volumes. Assuming the sample that is injected onto the column contains solute (S) at a concentration X_S^1, then from equation (25), the concentration of this solute (X_{E_S}) leaving the (n) th plate in the column will be given by:-

$$X_{E_S} = X_S + \sum_{r=1}^{r=\frac{V_i}{v_s}} \left(X_S^1 - X_S\right)\left(e^{-r} - 1\right) \frac{e^{-\left(\frac{y}{v_s} + \frac{V_i}{v_s} - r - n\right)^2 / 2n}}{\sqrt{2\pi n}}$$

Thus, for a chromatogram of (q) solutes, the elution curve equation will be given by,

$$X_{E_S} = \sum_{S=1}^{S=q} X_S + \sum_{r=1}^{r=\frac{V_i}{v_s}} \left(X_S^1 - X_S\right)\left(e^{-r} - 1\right) \frac{e^{-\left(\frac{y}{v_s} + \frac{V_i}{v_s} - r - n\right)^2 / 2n}}{\sqrt{2\pi n}} . \qquad (26)$$

If pure mobile phase is injected onto the column, then $X_S = 0$ and equation (26) becomes

$$X_E = \sum_{S=1}^{S=q} X_S + \sum_{r=1}^{r=\frac{V_i}{v_s}} X_S\left(e^{-r} - 1\right) \frac{e^{-\left(\frac{y}{v_s} + \frac{V_i}{v_s} - r - n\right)^2 / 2n}}{\sqrt{2\pi n}} \qquad (27)$$

The form of equation (27) is very similar to that obtained by Reilly et al.. [11] but the derivation is simpler, as those authors utilized the approximate binomial form of the elution curve in their procedure.

Scott et al. [12] provided some experimental evidence supporting equation (27). The mixture contained uracil, hypoxanthine, guanine and cytosine, each present in the mobile phase at a concentration of 14 mg/l. The column employed was 1m long, 1.5 mm I.D., packed with a pellicular cation exchange resin and operated at a flow rate of 0.3 ml/min.

Extension of the Plate Theory

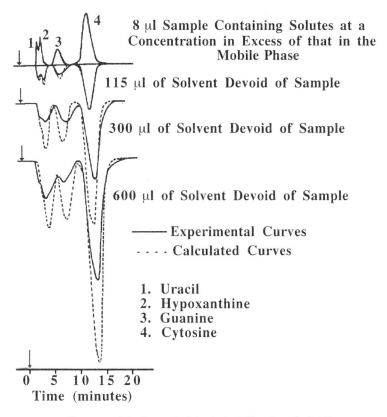

Courtesy of the Journal of Analytical Chemistry (ref. 11)

Figure 18. Vacancy Chromatograms for the Separation of a Four Component Mixture

The mobile phase was a 0.14 M potassium phosphate buffer solution adjusted to pH 4.0. The chromatograms obtained are shown in Figure 17 together with the calculated curves obtained from equation (27). Figure 18 reveals that the positions of the peaks are accurately predicted by the theory; the peak heights differ because the relative responses of the detector to the different bases were not taken into account in calculating the theoretical curves. The sample with excess concentration of solutes over that in the mobile phase, shown as a chromatogram with positive peaks, is almost exactly the mirror image of the negative chromatogram produced from the injection of 115 ml of pure mobile phase.

Although, even today, vacancy chromatography is rarely used in analytical laboratories generally, there are a number of applications where it appears it might be very useful. The technique, that was suggested by Zhukhovitski for quality control is a particularly interesting application. Consider a pharmaceutical product that contains a number of ingredients that must be kept in specified proportions. To monitor the drug for quality control purposes, a mobile phase is prepared that contains the components of the product, in the proportions specified, wa dissolved in the mobile phase, but at a low concentration suitable for LC analysis. A sample of the drug is dissolved in some pure mobile phase so that the total mass concentration is the same as that of the standards in the mobile phase. The column is brought into equilibrium with the mobile phase containing the standards and the sample injected onto the column. If the product contains the components in the specified proportion, no peaks will appear on the chromatogram, as the sample and mobile phase will have the identical composition. Any component that is in excess in the sample will give a positive peak. Conversely, any component that is present in the sample that is below specifications will give a negative peak. The peak area or peak height, for both positive and negative peaks, will give a quantitative estimation of the amount the component deviates from that specified.

The Peak Capacity of a Chromatographic Column

The peak capacity of a column has been defined as the number of peaks that can be fitted into a chromatogram between the dead point and the 'last peak', with each peak being separated from its neighbor by 4σ. The 'last peak' of chromatogram is rather a diffuse term as it depends somewhat on a number of factors such as the sensitivity of the detector and the column efficiency. Consequently, the 'last peak' can be either arbitrarily defined or defined by the properties of the column and/or the chromatograph with which it is used. Limited peak capacity can become a serious problem when multi-component mixtures are separated and the capacity of the chromatogram is insufficient to contain all the peaks discretely. When isocratic development is employed, if the early peaks of the mixture are adequately separated, then the late peaks are often broad and, consequently, are eluted at concentrations so low that they are hardly detectable from the baseline. Conversely, if the conditions are changed so that the late peaks are eluted at sufficiently low (k') values to improve detection limits, the early peaks then merge together and are not resolved.

Peak capacity can be very effectively improved by using temperature programming in GC or gradient elution in LC. However, if the mixture is very complex with a large number of individual solutes, then the same problem will often arise even under programming conditions. These difficulties arise as a direct result of the limited *peak capacity* of the column. It follows that it would be useful to derive an equation that

Extension of the Plate Theory

describes the peak capacity in terms of column and solute properties and which would provide information on how the peak capacity can be controlled.

From the Plate Theory, the peak width at the base is given by

$$4\sqrt{n}(v_m + Kv_s)$$

and the retention volume (V_r) given by $V_r = n(v_m + Kv_s)$.

Thus, a simple form of the peak capacity (S) could be given by

$$S = \frac{n}{4\sqrt{n}} \frac{(v_m + Kv_s)}{(v_m + Kv_s)} = \frac{\sqrt{n}}{4}$$

Consequently, (S) is the number of peaks, assumed to have the same width as the last peak, that can be fitted into the chromatogram from the dead volume up to, and including, the last peak. This expression does not give a valid number for the peak capacity, as all those peaks eluted before the last will have significantly smaller widths. It follows that a considerably greater number of peaks can be fitted into the chromatogram than the value of (S) calculated in this way might suggest. To obtain a more accurate number for the peak capacity of a column, a completely different approach must be employed.

The chromatogram in Figure 19 diagrammatically represents (r) resolved peaks, each peak being separated from its neighbor by 4σ.

Thus, the base width of the (r)th peak will be, $4\sqrt{n}(v_m + K_r v_s)$.

Consider the point where the last two peaks merge. At this point the retention volume of the last peak, minus half the peak width at the base, will equal the retention volume of the penultimate peak, plus half its peak width at the base.

Consequently,

$$n(v_m + K_{(r-1)}v_s) + 2\sqrt{n}(v_m + K_{(r-1)}v_s) = n(v_m + K_r v_s) - 2\sqrt{n}(v_m + K_r v_s)$$

and, $$(n + 2\sqrt{n})(v_m + K_{(r-1)}v_s) = (n - 2\sqrt{n})(v_m + K_r v_s)$$

Thus, $$(v_m + K_{(r-1)}v_s) = \frac{(n - 2\sqrt{n})}{(n + 2\sqrt{n})}(v_m + K_r v_s) \qquad (28)$$

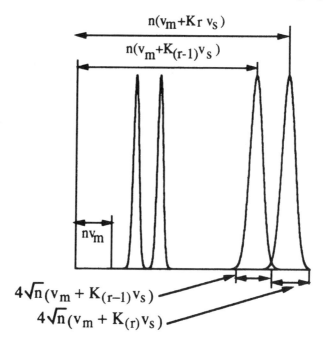

Figure 19. Diagram of the Separation of (r) Resolved Peaks

Therefore, the peak with a *base* width ($w_{(r-1)}$) of peak (r -1),

$$w_{(r-1)} = 4\sqrt{n}\left(v_m + K_{(r-1)}v_s\right) = 4\sqrt{n}\,\frac{\left(n - 2\sqrt{n}\right)}{\left(n + 2\sqrt{n}\right)}\left(v_m + K_r v_s\right)$$

In a similar way it can be shown that the base width ($w_{(r-2)}$) of peak (r -2) will be

$$w_{(r-2)} = 4\sqrt{n}\left(v_m + K_{(r-2)}v_s\right) = 4\sqrt{n}\,\frac{\left(n - 2\sqrt{n}\right)}{\left(n + 2\sqrt{n}\right)}\left(v_m + K_{(r-1)}v_s\right)$$

Substituting for $\left(v_m + K_{(r-1)}v_s\right)$ from (x),

$$w_{(r-2)} = 4\sqrt{n}\left(v_m + K_{(r-2)}v_s\right) = 4\sqrt{n}\left(\frac{n - 2\sqrt{n}}{n + 2\sqrt{n}}\right)^2 \left(v_m + K_r v_s\right)$$

and

$$w_{(r-3)} = 4\sqrt{n}\left(v_m + K_{(r-3)}v_s\right) = 4\sqrt{n}\left(\frac{n - 2\sqrt{n}}{n + 2\sqrt{n}}\right)^3 \left(v_m + K_r v_s\right)$$

Thus, assuming that the number of peaks that can be fitted into the chromatogram between the 'dead time' and the time for the complete elution of the last peak is (r), then

Extension of the Plate Theory

$$n(v_m + K_r v_s) + 2\sqrt{n}(v_m + K_r v_s) =$$

$$4\sqrt{n}\left[\left(\frac{n-2\sqrt{n}}{n+2\sqrt{n}}\right)^{(r-1)} + \left(\frac{n-2\sqrt{n}}{n+2\sqrt{n}}\right)^{(r-2)} + \ldots + \left(\frac{n-2\sqrt{n}}{n+2\sqrt{n}}\right)^2 + \left(\frac{n-2\sqrt{n}}{n+2\sqrt{n}}\right)\right](v_m + K_r v_s) + nv_m$$

and

$$n(v_m + K_r v_s) =$$

$$4\sqrt{n}\left[\left(\frac{n-2\sqrt{n}}{n+2\sqrt{n}}\right)^{(r-1)} + \left(\frac{n-2\sqrt{n}}{n+2\sqrt{n}}\right)^{(r-2)} + \ldots + \left(\frac{n-2\sqrt{n}}{n+2\sqrt{n}}\right)^2 + \left(\frac{n-2\sqrt{n}}{n+2\sqrt{n}}\right) - 0.5\right](v_m + K_r v_s) + nv_m$$

Dividing through by (n), noting that $\dfrac{Kv_s}{v_m} = k'$, and rearranging,

$$\frac{k'}{1+k'} = \frac{4}{\sqrt{n}}\left[\left(\frac{n-2\sqrt{n}}{n+2\sqrt{n}}\right)^{(r-1)} + \left(\frac{n-2\sqrt{n}}{n+2\sqrt{n}}\right)^{(r-2)} + \ldots + \left(\frac{n-2\sqrt{n}}{n+2\sqrt{n}}\right)^2 + \left(\frac{n-2\sqrt{n}}{n+2\sqrt{n}}\right) + 0.5\right]$$

Replacing the geometric series by the expression for its sum,

$$\frac{k'}{1+k'} = \frac{4}{\sqrt{n}}\left[\frac{1-\left(\frac{n-2\sqrt{n}}{n+2\sqrt{n}}\right)^r}{1-\left(\frac{n-2\sqrt{n}}{n+2\sqrt{n}}\right)} - 0.5\right] \tag{29}$$

Rearranging to provide an expression for (r)

$$r = \frac{\log\left(1-\left(\frac{\sqrt{n}\,k'}{4(1+k')}+0.5\right)\left(1-\left(\frac{n-2\sqrt{n}}{n+2\sqrt{n}}\right)\right)\right)}{\log\left(\frac{n-2\sqrt{n}}{n+2\sqrt{n}}\right)} \qquad (30)$$

Giddings [14] produced a very similar equation for the peak capacity of a column in 1967, but in Giddings's derivation, certain assumptions were made that rendered the peak capacities calculated from his expression somewhat less than those given by equation (30). The difference, however, has little practical significance. It is clear from equation (30) that the peak capacity is controlled by the column efficiency and the capacity ratio of the last eluted peak. The peak capacities of a series of columns having different efficiencies were calculated for a range of peak capacity ratios employing equation (30). The results are shown as curves relating peak capacity to capacity ratio in Figure 20.

The curves show that the peak capacity increases with the column efficiency, which is much as one would expect, however the major factor that influences peak capacity is clearly the capacity ratio of the last eluted peak. It follows that any aspect of the chromatographic system that might limit the value of (k') for the last peak will also limit the peak capacity. Davis and Giddings [15] have pointed out that the theoretical peak capacity is an exaggerated value of the true peak capacity. They claim that the individual (k') values for each solute in a realistic multi-component mixture will have a statistically irregular distribution. As they very adroitly point out, the solutes in a real sample do not array themselves conveniently along the chromatogram four standard deviations apart to provide the maximum peak capacity.

However, with practical samples the way the (k') values of the individual components for any given complex solute mixture are distributed is not predictable, and will vary very significantly from mixture to mixture, depending on the nature of the sample. Nevertheless, although the values for the theoretical peak capacity of a column given by equation (26) can be used as a reasonable practical guide for comparing *different* columns, the theoretical values that are obtained will always be in excess of the peak capacities that are actually realized in practice.

It is also apparent from Figure 20 that any property of the chromatographic system that places a limit on the maximum value of (k') must also limit the maximum peak capacity that is attainable. One property of the system that limits the maximum value

Extension of the Plate Theory

of (k') will be the sensitivity of the detector. The longer a solute is retained in the column (the higher the (k') value), the more the peak is dispersed and, thus, the smaller will be the peak height. This process continues until, at some limiting (k') value, it will be indiscernible from the noise.

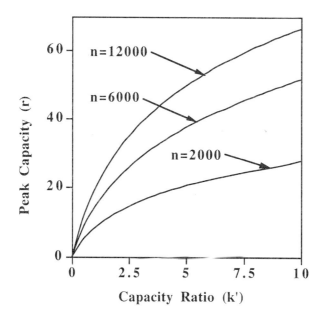

Courtesy of the Journal of Chromatogr.aphic Science (ref. 13)

Figure 20. Graph of Peak Capacity against Capacity Ratio

To unambiguously identify the presence of a peak and, in addition, be able to give some proximate estimation of its size for quantitative purposes, the peak height needs to be at least 5 times the noise level. The detector sensitivity, or the minimum detectable concentration, (X_D), is defined as that concentration of solute that will give a signal equivalent to twice the noise level and, consequently, the concentration of solute at the limiting (k') value must be $2.5 X_D$.

The concentration of solute at the peak maximum is approximately twice the average concentration. Thus, if a mass (m) is placed on the column and the peak width is $4\sqrt{n}(v_m + Kv_s)$, then,

$$\frac{2m}{4\sqrt{n}(v_m + Kv_s)} = 2.5 X_D$$

or

$$m = 5 X_D \sqrt{n} (v_m + Kv_s) \qquad (31)$$

Now, the maximum sample volume (V_i) that can be placed on the column that would restrict the increase to less than 5% has been shown to be,

$$V_i = 1.1\sqrt{n}(v_m + Kv_s)$$

It has also been shown that the effect of sample volume on peak width will be most significant for the early peaks (the most narrow peaks). In addition, the degrading effect of sample volume will progressively decrease as the capacity ratio of the peak becomes larger. However, the resolution of both late and early peaks are equally important and, consequently, the limiting sample volume will be that which restrains the dispersion of the *first* peak to 5% or less.

Thus, for an unretained peak, $\quad V_i = 1.1\sqrt{n}(v_m)$

Consequently, if the solute concentration in the sample is (X_i),

$$m = 1.1\sqrt{n}(v_m)X_i \tag{32}$$

Equating equations (31) and (32),

$$1.1\sqrt{n}(v_m) = 5X_D\sqrt{n}(v_m + Kv_s)$$

Dividing both sides by (v_m), noting that ($\frac{Kv_s}{v_m} = k'$), and simplifying,

$$\frac{1.1 X_i}{5 X_D} = 1 + k'$$

and thus

$$k' = \frac{0.22 X_i}{X_D} - 1. \tag{33}$$

Equation (33) shows that the maximum capacity ratio of the last eluted solute is inversely proportional to the detector sensitivity or minimum detectable concentration. Consequently, it is the detector sensitivity that determines the maximum peak capacity attainable from the column. Using equation (33), the peak capacity was calculated for three different detector sensitivities for a column having an efficiency of 10,000 theoretical plates, a dead volume of 6.7 ml and a sample concentration of 1%v/v. The results are shown in Table 1, and it is seen that the limiting peak capacity is fairly large.

Table 1. Capacity Ratios and Peak Capacities for Detectors of Different Sensitivities

Detector Sensitivity	Maximum (k')	Retention Time (min.)	Peak Capacity
10^{-6} (g/ml)	220	73.6	134
10^{-7} (g/ml)	2,200	736.0	177
10^{-8} (g/ml)	22,000	7360.0	194

Table 1 shows that by increasing the minimum detectable concentration by an order of magnitude, from 10^{-7} g/ml to 10^{-8} g/ml the maximum capacity ratio is also increased by an order of magnitude, but this results in a peak capacity increase of only 10%. In addition, the small increase in peak capacity is realized at the expense of a retention time that is increased from twelve hours to about five days. It is clear that attempting to increase the peak capacity of a chromatographic system by employing a detector of higher sensitivity would be very costly in time. Increased peak capacities are best achieved by employing columns of much higher intrinsic efficiency.

So far the plate theory has been used to examine first-order effects in chromatography. However, it can also be used in a number of other interesting ways to investigate second-order effects in both the chromatographic system itself and in ancillary apparatus such as the detector. The plate theory will now be used to examine the temperature effects that result from solute distribution between two phases. This theoretical treatment not only provides information on the thermal effects that occur in a column *per se,* but also gives further examples of the use of the plate theory to examine dynamic distribution systems and the different ways that it can be employed.

Temperature Changes During the Passage of a Solute Through a Theoretical Plate in Gas Chromatography

Thermal changes resulting from solute interactions with the two phases are definitely second-order effects and, consequently, their theoretical treatment is more complex in nature. Thermal effects need to be considered, however, because heat changes can influence the peak shape, particularly in preparative chromatography, and the consequent temperature changes can also be explored for detection purposes.

The theoretical treatment of temperature perturbations that result from solute phase interactions also affords an excellent example of the use of the plate concept in a wider sense of chromatography theory. When a solute passes through a theoretical plate in a gas/liquid chromatographic column, the solute is first adsorbed and then desorbed from the stationary phase. This adsorption/desorption process is accompanied by the evolution and adsorption of the heat of solution of the solute. Furthermore, this heat exchange occurs as it enters and leaves the stationary phase during its progress in and out of the plate. This reversible heat exchange, between solute and solvent and the plate surroundings, causes the temperature of the contents of the plate to rise and fall. Ray [16] suggested that this temperature effect might form the basis of a GC detection system and Klinkenberg [17] thought it might contribute to peak asymmetry. The effect was used by Claxton [18] and Grosek [19] as a detecting system in liquid/solid chromatography and Smith [20] demonstrated that, under certain circumstances, the temperature changes could adversely affect a separation in preparative chromatography.

In order to examine the thermal changes that take place in a column, it is necessary to derive an equation that describes the temperature change in a theoretical plate, in terms of its physical properties of the plate and the volume flow of mobile phase that passes through it.

Consider the (n)th theoretical plate in a GC column, as depicted in Figure 21. The properties of the plate are defined as follows,

v_g is the volume of gas in the plate,
v_l is the volume of liquid (stationary phase) in the plate,
v_s is the volume of support in the plate,
S_l is the specific heat of the stationary phase,
S_s is the specific heat of the support,
ρ_l is the density of the stationary phase,
ρ_s is the density of the support,
$X_{l(n)}$ is the concentration of solute in the stationary phase in plate (n),
$X_{g(n)}$ is the concentration of solute in the mobile phase (gas) in plate (n),
θ is the excess temperature of the plate above its surroundings,
Q is the mobile phase (gas) flow rate in ml/sec.

In the derivation of the plate-temperature equation the following assumptions will be made.

1. The flow of gas through the plate is constant.

2. The temperature of the plate surroundings is constant.

Extension of the Plate Theory

3. The concentration of solute is maintained at levels where the adsorption isotherm is linear (*i.e.*, $X_l(n) = KX_g(n)$).

4. The heat capacity of the gas is small compared with those of the stationary phase and the support.

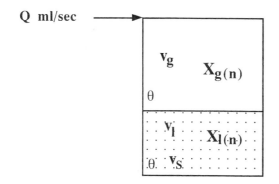

Figure 21. The Last Theoretical Plate in a GLC Column

Condition (4) assumes that the heat *convected* from the plate by the flow of gas through it is negligible compared with that *conducted* from the plate to the surroundings. This assumption is reasonable for a GC system but, as will be seen when the heat of adsorption detector is considered, this will certainly not be true when the mobile phase is a liquid.

As the plate concept is to be employed the volume flow will be measured in plate volumes, *i.e.*,

$$v = \frac{Qt}{v_g + Kv_l}$$

or

$$dt = \frac{v_g + Kv_l}{Q} dv \qquad (34)$$

Consider the heat balance of the plate,

Heat Capacity of Plate x Rise in Temperature =
    Heat Evolved in Plate (H_C) − Heat Conducted from Plate (G_C)

Consider the heat balance over a time (dt) during which there is a flow of mobile phase (dv) and a temperature change (dθ), then

$$(v_l \rho_l S_l + v_s \rho_s S_s) d\theta = \text{Heat Evolved in Plate } (H_C) \text{ in Time (dt)}.$$

− Heat Conducted Radially from Plate (G_C) in Time (dt) \qquad (35)

The solute is distributed largely over $4\sqrt{n}$ plates in the column and thus, as (n) is large, the differential temperature between plates is negligible, and so the heat conducted *axially* along the column will be very small compared with that conducted radially to the walls and from the system.

Now, the change in concentration in the mobile phase (gas) during the passage of a volume (dv) will be $\dfrac{dX_{g(n)}}{dv}dv$

Thus, the change in concentration in the liquid phase during the passage of a volume of mobile phase (dv) will be $\dfrac{K\,dX_{g(n)}}{dv}dv$.

Consequently the heat evolved in plate (n) (H$_C$) will be given by

$$H_C = \frac{hK\,dX_{g(n)}}{dv}dv \tag{36}$$

where (h) is the heat of solution of the solute in the liquid phase, given in appropriate units.

The heat conducted from the plate (G$_C$) in time (dt) will be given by

$$G_C = AZ\theta dt$$

where (A) is the surface area of the plate and (Z) is the thermal conductivity of the plate and its contents, given in appropriate units.
Substituting for (dt) from equation (34),

$$G_C = \frac{AZ}{Q}\left(v_g + Kv_l\right)\theta dv \tag{37}$$

Substituting in equation (35) for (H$_C$) from (36) and (G$_C$) from (37),

$$\left(v_l\rho_l S_l + v_s\rho_s S_s\right)d\theta = \frac{hK\,dX_{g(n)}}{dv}dv - \frac{AZ}{Q}\left(v_g + Kv_l\right)\theta dv$$

$$\frac{d\theta}{dv} = \frac{hK}{\left(v_l\rho_l S_l + v_s\rho_s S_s\right)}\frac{dX_{g(n)}}{dv} - \frac{AZ\left(v_g + Kv_l\right)}{Q\left(v_l\rho_l S_l + v_s\rho_s S_s\right)}\theta \tag{38}$$

or

$$\frac{d\theta}{dv} = \alpha\frac{dX_{g(n)}}{dv} - \beta\theta$$

Extension of the Plate Theory

where $\quad \alpha = \dfrac{hK}{(v_1\rho_1 S_1 + v_s\rho_s S_s)} \quad$ and $\quad \beta = \dfrac{AZ(v_g + Kv_1)}{Q(v_1\rho_1 S_1 + v_s\rho_s S_s)}$

or
$$\frac{d\theta}{dv} + \beta\theta = \alpha\frac{dX_{g(n)}}{dv} \tag{39}$$

Now equation (39) is a standard integral, the solution of which can be seen by differentiation to be

$$e^{\beta v}\theta = \int e^{\beta v}\alpha \frac{dX_{g(n)}dv}{dv} + E$$

Now,
$$\int u\,dv = u\int dv - \int v\,du = uv - \int v\,du$$

Integrating by parts, $\quad e^{\beta v}\theta = \left[X_{g(n)}e^{\beta v}\alpha\right] - \int\left(X_{g(n)}\alpha\beta e^{\beta v}\right)dv + E \tag{40}$

Now from the elution curve equation, $\quad X_{g(n)} = X_{g(o)}\dfrac{e^v v^n}{n!}$

Substituting for ($X_{g(n)}$) in equation (40),

$$e^{\beta v}\theta = \alpha X_{g(o)}\frac{e^{-(1-\beta)v}v^n}{n!} - \int \alpha\beta X_{g(o)}\frac{e^{-(1-\beta)v}v^n}{n!}\,dv + E \tag{41}$$

Now, from Stirling's theorem, $\quad n! = e^{-n}n^n\sqrt{2\pi n}$

Substituting for (n!), $\quad e^{\beta v}\theta = \alpha X_{g(o)}\dfrac{e^{-(1-\beta)v+n}v^n}{\sqrt{2\pi n}\,n^n} - \int \alpha\beta X_{g(o)}\dfrac{e^{-(1-\beta)v+n}v^n}{\sqrt{2\pi n}\,n^n}\,dv + E$

For algebraic convenience $\dfrac{(w+n)}{(1-\beta)}$ will be substituted for (v) in functions other than in ($e^{\beta v}$) which to simplify the algebra will be replaced later. It follows that,,

$$dv = \frac{dw}{(1-\beta)}$$

$$e^{\beta v}\theta = \alpha X_{g(o)}\frac{e^{-(1-\beta)\frac{(w+n)}{(1-\beta)}+n}\left(\dfrac{(w+n)}{(1-\beta)}\right)^n}{\sqrt{2\pi n}\,n^n} -$$

$$\int \alpha\beta X_{g(o)}\frac{e^{-(1-\beta)\frac{(w+n)}{(1-\beta)}+n}\left(\dfrac{(w+n)}{(1-\beta)}\right)^n}{\sqrt{2\pi n}\,n^n(1-\beta)}\,dw + E$$

Simplifying,

$$e^{\beta v}\theta = \alpha X_{g(o)} \frac{e^{-w}\left(\frac{w}{n}+1\right)^n}{\sqrt{2\pi n}(1-\beta)^n} - \int \alpha \beta X_{g(o)} \frac{e^{-w}\left(\frac{w}{n}+1\right)^n}{\sqrt{2\pi n}(1-\beta)^{n+1}} dw + E \quad (42)$$

Now, if x<1 (and $\frac{w}{n}$ will always be much less than unity for all significant values of $(X_{g(n)})$), then $\log_e (1+x)$ is given by the series

$$\text{Log}_e (1+x) = x - \frac{x^2}{2} + \frac{x^3}{3} - \frac{x^4}{4} + \ldots$$

Now,
$$\text{Log}\left(e^{-w}\left(\frac{w}{n}+1\right)^n\right) = -w + n\left(\frac{w}{n}\right) - n\left(\frac{w^2}{2n^2}\right) + n\left(\frac{w^3}{3n^3}\right) + \ldots$$

$$= -\frac{w^2}{2n} + \frac{w^3}{3n^2} + \frac{w^4}{4n^3} + \ldots$$

In addition, (n) is large and the whole elution curve of a given solute is practically contained between $w = -2\sqrt{n}$ and $w = 2\sqrt{n}$ (i.e., contained within four standard deviations of the Gaussian curve) thus $\frac{w^2}{2n}$ will always be very much greater than $\frac{w^3}{3n^2}$ and, thus, $\frac{w^3}{3n^2}$ and all higher terms can be ignored with respect to $\frac{w^2}{2n}$.

Thus, $\left(e^{-w}\left(\frac{w}{n}+1\right)^n\right) = e^{-\frac{w^2}{2n}}$ and substituting for $e^{-w}\left(\frac{w}{n}+1\right)^n$ in equation (42),

$$e^{\beta v}\theta = \alpha X_{g(o)} \frac{e^{-w}\left(\frac{w}{n}+1\right)^n}{\sqrt{2\pi n}(1-\beta)^n} - \int \alpha \beta X_{g(o)} \frac{e^{-w}\left(\frac{w}{n}+1\right)^n}{\sqrt{2\pi n}(1-\beta)^{n+1}} dw + E$$

Thus, integrating from $(-\infty)$ to the point of measurement (w),

$$\theta = \frac{\alpha X_{g(o)}}{e^{\beta v}(1-\beta)^n} \left[\frac{e^{-\frac{w^2}{2n}}}{\sqrt{2\pi n}} - \beta \int_{-\infty}^{w} \frac{e^{-\frac{w^2}{2n}}}{\sqrt{2\pi n}(1-\beta)} dw + E \right]$$

Now, as (n) is large when v=0, w=-n and $\theta = 0$ Thus, in addition, as

$$e^{-\frac{w^2}{2n}} = 0 = \int_{-\infty}^{w} \frac{e^{-\frac{w^2}{2n}}}{\sqrt{2pn}} dw \quad \text{then } E = 0$$

Extension of the Plate Theory

$$\theta = \frac{\alpha X_{g(o)}}{e^{\beta v}(1-\beta)^n} \left[\frac{e^{\frac{-w^2}{2n}}}{\sqrt{2\pi n}} - \frac{\beta}{(1-\beta)} \int_{-\infty}^{w} \frac{e^{\frac{-w^2}{2n}}}{\sqrt{2\pi n}} dw \right]$$

Now, again substituting (v) for $v = \frac{w+n}{(1-\beta)}$ in ($e^{\beta v}$), then

$$\beta v = \frac{\beta}{(1-\beta)}(w+n)$$

and let $\quad e^{\beta v}(1-\beta)^n = e^{\frac{\beta}{(1-\beta)}(w+n)}(1-\beta)^n = \psi$

$$\theta = \alpha X_{g(o)} e^{\frac{-\left(\beta w + \frac{n\beta^2}{2}\right)}{(1-\beta)}} \left[\frac{e^{\frac{-w^2}{2n}}}{\sqrt{2\pi n}} - \frac{\beta}{(1-\beta)} \int_{-\infty}^{w} \frac{e^{\frac{-w^2}{2n}}}{\sqrt{2\pi n}} dw \right]$$

Then

$$\log(\psi) = \frac{\beta}{(1-\beta)}(w+n) - n\mathrm{Log}(1-\beta)$$

$$\log(\psi) = \frac{\beta w}{(1-\beta)} + \frac{n\beta}{(1-\beta)} - n\left(-\beta - \frac{\beta^2}{2} - \frac{\beta^3}{3} + \ldots\right)$$

$$= \frac{\beta w}{(1-\beta)} + \frac{1}{(1-\beta)}\left(\frac{n\beta^2}{2} + \frac{n\beta^3}{6} + \frac{n\beta^4}{12} + \ldots\right)$$

Thus, if $\beta < 1$, which will be so in practice, it is seen then the expression approximates to

$$\log(\psi) = \frac{\beta w + \frac{n\beta^2}{2} + \frac{n\beta^3}{6}}{(1-\beta)} \text{ and } \psi = e^{\frac{\beta w + \frac{n\beta^2}{2} + \frac{n\beta^3}{6}}{(1-\beta)}}$$

Thus,

$$\theta = \alpha X_{g(o)} e^{\frac{-\left(\beta w + \frac{n\beta^2}{2} + \frac{n\beta^3}{6}\right)}{(1-\beta)}} \left[\frac{e^{\frac{-w^2}{2n}}}{\sqrt{2\pi n}} - \frac{\beta}{(1-\beta)} \int_{-\infty}^{w} \frac{e^{\frac{-w^2}{2n}}}{\sqrt{2\pi n}} dw \right] \quad (43)$$

Under adiabatic conditions, $\beta=0$ and

$$\theta = \alpha X_{g(o)} \frac{e^{\frac{-w^2}{2n}}}{\sqrt{2\pi n}}$$

Thus, when there is no heat lost from the plate, the temperature profile of the plate will take the same form as its concentration profile.

When $w > 3\sqrt{n}$, the solute band will have virtually passed through the plate and

$$\frac{e^{\frac{-w^2}{2n}}}{\sqrt{2\pi n}} - \frac{\beta}{(1-\beta)} \to 0 \text{ and } -\frac{\beta}{(1-\beta)} \int_{-\infty}^{w} \frac{e^{\frac{-w^2}{2n}}}{\sqrt{2\pi n}} dw \to -\beta$$

Thus, when $w > 3\sqrt{n}$ and the temperature of the plate is recovering, equation (39) becomes

$$\theta = -\alpha\beta X_{g(o)} e^{\frac{-\left(\beta w + \frac{n\beta^2}{2} + \frac{n\beta^3}{6}\right)}{(1-\beta)}}$$

or

$$\text{Log}(\theta) = \text{Log}(\xi) - \frac{\beta}{(1-\beta)} w$$

where

$$\xi = -\alpha\beta X_{g(o)} e^{\frac{-\left(\frac{n\beta^2}{2} + \frac{n\beta^3}{6}\right)}{(1-\beta)}}$$

Consequently, by plotting the logarithm of the plate temperature against (w), a straight line will be produced, the slope of which will be $\frac{\beta}{(1-\beta)}$ and, thus, will allow the heat loss factor to be evaluated. In practice, (β) ranges from 0 to about 0.2 and, thus, using equation (43) and values for (β) of 0, 0.5, 0.1, 0.15, 0.17, and 0.2, the temperature curves for the (n) th plate of a column having 2500 plates can be calculated. The results obtained are shown in Figure 22. It is seen that the expected S-shaped curve is produced. As the solute dissolves in the stationary phase, and the heat of solution is evolved, the temperature rises. After the peak maximum is reached, the solute desorbs from the plate, the heat of solution is absorbed, and the temperature falls below that of its surroundings. This effect can be simply demonstrated by inserting a thermocouple into a column and monitoring the temperature as the solute band passes. An example of a set of such curves [21] is given in Figure 23. It is seen that the expected temperature profiles are realized. It is

Extension of the Plate Theory

also interesting to note that the front of the peak is eluted at a higher temperature than the back of the peak throughout the whole length of the column.

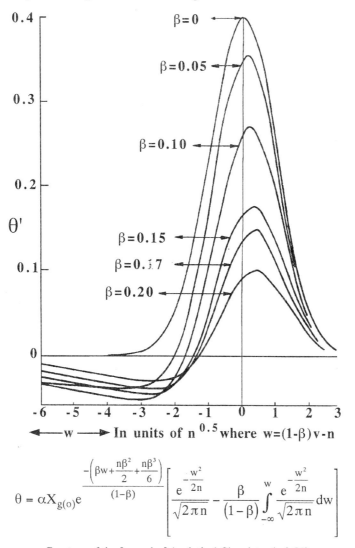

$$\theta = \alpha X_{g(o)} e^{-\left(\beta w + \frac{n\beta^2}{2} + \frac{n\beta^3}{6}\right)/(1-\beta)} \left[\frac{e^{-\frac{w^2}{2n}}}{\sqrt{2\pi n}} - \frac{\beta}{(1-\beta)} \int_{-\infty}^{w} \frac{e^{-\frac{w^2}{2n}}}{\sqrt{2\pi n}} dw \right]$$

Courtesy of the Journal of Analytical Chemistry (ref. 21)

Figure 22. Relative Changes in Plate Temperature with Column Flow, Measured in Units of $\sqrt{n}$ for Different Values of Heat Loss Factor (β)

This, as is shown by the theory, is due to the *evolution* of the heat of *absorption*, during solute adsorption at the front part of the peak. Conversely, the back of the peak is eluted at a lower temperature than the surroundings throughout the length of the column due to the *absorption* of the heat of solute *desorption*. As a result, the distribution coefficient of the solute at the front of the peak, and at a higher temperature, will be less than the distribution coefficient at the back of the peak, at the lower temperature. Consequently, the front of the peak will travel more rapidly through the column than the back of the peak, which will result in an asymmetric peak. Such asymmetry is shown by the concentration profiles given in Figure 23. It must be said, however, that although the temperature changes in the plate as the solute passes through it will contribute to peak asymmetry, it may not be the only source of asymmetry, as other factors may also be present.

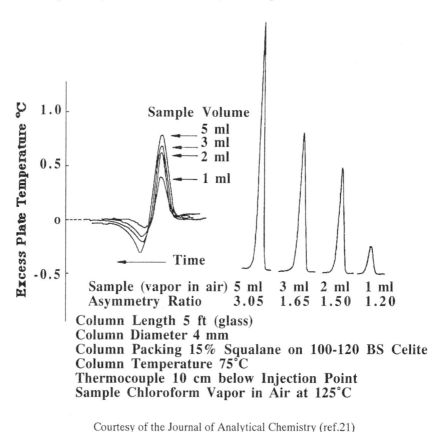

Courtesy of the Journal of Analytical Chemistry (ref.21)

Figure 23. Temperature and Concentration Curves for the Elution of Quantities of Chloroform Injected as a Saturated Vapor in Air

A Theoretical Treatment of the Heat of Absorption Detector

Thermal changes in a distribution system, although a second-order effect and, thus, more complex to deal with theoretically, can nevertheless sometimes be used to practical ends. The temperature changes that occurred in a dynamic distribution system were used, in the early days of LC, for detection purposes. Ultimately, the system proved to be ineffectual as a detector, but this could have been deduced

Extension of the Plate Theory

theoretically, if properly examined. Although, presently, it is not a practical detector, the theory behind the heat of adsorption detector will be considered for two reasons. The primary reason is to demonstrate another physical chemical process that can be examined very successfully using the plate theory, and perhaps invoke its use in the examination of other dynamic distribution phenomena. Secondly, the theory of the system will also explain how the detector functions and either deter research activities in this area based on established concepts or stimulate new thoughts on its use and, perhaps, an alternative novel system.

The heat of adsorption detector, was originally described by Claxton [18] in 1958 and was subsequently examined by a number of workers [19,20]. It was once commercially available but, as a result of its limitations which will be understood in due course, it has not been employed as an LC detector for many years. The main reason for its failure as a useful detector was the curious and apparently unpredictable shape of the temperature-time curve that resulted from sensing the usual Gaussian concentration profile of an eluted solute. These types of curves were shown in the previous theoretical treatment of temperature changes in a GC column. The detector response was further confused by the unexpected effect of changes in detector geometry, the operating conditions of the chromatograph, and the retention volume of the solute. Furthermore, for closely eluting peaks (particularly if the peaks were incompletely resolved), the sensor produced a very complex curve that was extremely difficult to interpret.

The actual sensor consists of a small plug of adsorbent, usually silica gel or alumina, through which the mobile phase passes after leaving the column. Embedded in the silica gel or alumina plug is a small volume temperature sensor (*e.g.*, a thermocouple or a thermistor) which continuously monitors the temperature of the adsorbent during the development of the chromatogram. On the elution of a solute, some is adsorbed onto the surface of the plug, causing the evolution of heat and, consequently, the temperature rises. After the peak maximum has passed through the cell, the solute is desorbed from the plug, a process that absorbs heat and consequently the temperature of the cell now falls. The sensor will, thus, first record a rise in temperature (during adsorption) and, subsequently, a fall in temperature (during desorption) relative to its surroundings. The form of the temperature profile will be determined by the heat loss during the adsorption/desorption process and the relative plate capacities of the detector cell and the associated column. In this sense, the detector differs considerably from that of the temperature change in the GC column.

Let a small cell, containing the adsorbent, be placed immediately after the column and through which the column eluent passes. Assume that the eluent is at a constant temperature due to a suitable heat exchanger being placed between the cell and the column. In addition, it is also assumed that no significant peak dispersion takes place during passage through the heat exchanger.

Let the cell have internal and external radii of (r_1) and (r_2), respectively, and length (l) and the following assumptions will be made:

  1. There is constant flow of mobile phase through the cell.

  2. The temperature of the cell surroundings is constant.

  3. The size of the cell is such that the solute is in equilibrium with the two phases.

  4. The cell does not contribute significantly to band dispersion.

Consider the heat balance of the cell. However, as opposed to the GC column, the heat capacity of the liquid mobile phase in an in LC column is relatively large. Consequently, heat *convected* from the cell by the mobile phase must also be taken into account. It follows that

(Heat Capacity of Cell) x (Change in Temperature) =

(Heat Evolved in Cell) - (Heat Convected from Cell by Mobile Phase)

- (Heat Conducted from Cell)

The same basic approach of the plate theory will be employed to develop an equation for (θ), the cell temperature and, thus, the volume flow of mobile phase will be measured in plate volumes (v) of the attached column and the column plate volume will be designated as (c_a) for solute (a).

Let a volume (dv) of mobile phase pass through the cell, carrying solute that is absorbed onto the surface of the adsorbent with the evolution of heat, and let the resulting temperature change be ($d\theta$).

Extension of the Plate Theory

Then, assuming the heat capacity of the solute is negligible,

$$H_C d\theta = \text{(Heat Evolved in Cell)} - \text{(Heat Convected from Cell)}$$
$$- \text{(Heat Conducted from Cell)} \qquad (44)$$

where, (H_C) is the heat capacity of the cell, *i.e.*,

$$i \quad H_C = V_m \rho_m S_m + V_s \rho_s S_s + V_g \rho_g S_g$$

where (V_m) is the volume of mobile phase in the cell,
(V_s) is the volume of adsorbent in the cell,
(V_g) is the volume of the wall of the cell,
(ρ_m) is the density of the mobile phase,
(ρ_s) is the density of the adsorbent,
(ρ_g) is the density of the cell wall material,
(S_m) is the specific heat of the mobile phase,
(S_s) is the specific heat of the adsorbent,
and (S_g) is the specific heat of the wall material.,

Note the italic form, (*V*), is used to distinguish the volumes involved in the cell from volume, (V), which will refer to the mobile phase, and the plate volume, (v).

Developing the arguments in the same manner as the plate theory, let a volume (dv) of mobile phase, equivalent to ($c_a$dv) ml, enter the cell, and let the concentration of solute (a) in the incremental volume be (X_n). Let an equivalent volume ($c_a$dv) of mobile phase be displaced from the cell, and let the concentration of solute in the mobile phase contained by the cell prior to the introduction of the volume (dv) be (X_m).

Now, the net change of mass of solute (dm) in the cell will be

$$dm = (X_n - X_m) c_a dv \qquad (45)$$

Equilibrium is assumed to occur in the detector cell, so the introduction of the mass of solute (dm) will result in a change in concentration of solute in the mobile phase and adsorbent of (dX_m) and (dX_s), respectively, where (X_s) is the concentration of solute in the adsorbent. Thus,

$$dm = V_s dX_s + V_m dX_m$$

and, assuming the distribution isotherm is linear over the concentrations range employed,

$$dX_s = KdX_m$$

$$dm = V_s K dX_m + V_m dX_m$$

$$= (V_s K + V_m) dX_m \tag{46}$$

Equating equations (45) and (46) and rearranging,

$$(X_n - X_m) c_a dv = (V_s K + V_m) dX_m.$$

Rearranging,

$$\left(\frac{V_s K + V_m}{c_a}\right) \frac{dX_m}{dv} + X_m = X_n$$

Now, $V_s K + V_m$ is the 'effective cell volume' of the detector in much the same way that (c_a) is the column 'plate volume'.

Let

$$\frac{(KV_s + V_m)}{c_a} = C_a = \frac{\text{"effective detector cell volume"}}{\text{"column plate volume"}} \text{ for solute (a)}$$

Thus,

$$C_a \frac{dX_m}{dv} + X_m = X_n \tag{47}$$

Multiplying throughout by ($e^{\frac{v}{C_a}}$),

$$C_a e^{\frac{v}{C_a}} \frac{dX_m}{dv} + X_m e^{\frac{v}{C_a}} = X_n e^{\frac{v}{C_a}}$$

or

$$\frac{d\left(C_a e^{\frac{v}{C_a}} X_m\right)}{dv} = X_n e^{\frac{v}{C_a}}$$

Integrating,

$$C_a e^{\frac{v}{C_a}} X_m = \int X_n e^{\frac{v}{C_a}} dv + R$$

Now, when $v = 0$, the solute has not moved from the point of injection on the column and $X_m = X_n = 0$ and consequently, $R = 0$ and

$$X_m = \left(\frac{e^{-\frac{v}{C_a}}}{C_a}\right) \int_0^V X_n e^{\frac{v}{C_a}} dv \tag{48}$$

Extension of the Plate Theory

At this stage, equation (48) provides an expression for (X_m). Continuing, if the change in mass of solute on the absorbent due to a volume flow of mobile phase $c_a dv$ is (dm_s), then the consequent heat evolved (dG) in the cell will be given by:

$$dG = g\left(\frac{dm_s}{dv}\right) dv$$

where (g) is the heat of adsorption of the solute in cal. per gram of solute.

Hence, $dG = gV_s\left(\frac{dX_s}{dv}\right) dv$ and, as $dX_s = K dX_m$. Hence,

$$dG = KgV_s\left(\frac{dX_m}{dv}\right) dv$$

From equation (47), $\quad \dfrac{dX_m}{dv} = \left(\dfrac{X_n - X_m}{C_a}\right)$

Substituting for $\left(\dfrac{dX_m}{dv}\right)$, $\quad dG = KgV_s\left(\dfrac{X_n - X_m}{C_a}\right) dv$

Thus, substituting for (X_m) from equation (48) and rearranging,

$$dG = \left(\frac{KgV_s}{C_a}\right)\left(X_n - \left(\frac{1}{C_a}\right) e^{-\frac{v}{C_a}} \int e^{\frac{v}{C_a}} X_n\, dv\right) dv$$

Now, $\quad X_n = X_o \dfrac{e^{-v} v^n}{n!}$,

Thus, $\quad dG = \dfrac{KgV_s}{C_a}\left(X_o \dfrac{e^{-v} v^n}{n!} - \left(\dfrac{X_o}{C_a}\right) e^{\frac{-v}{C_a}} \int e^{\frac{v}{C_a}} \left(\dfrac{e^{-v} v^n}{n!}\right) dv\right) dv.$

Let $\quad X_o \dfrac{e^{-v} v^n}{n!} - \left(\dfrac{X_o}{C_a}\right) e^{\frac{-v}{C_a}} \int e^{\frac{v}{C_a}} \left(\dfrac{e^{-v} v^n}{n!}\right) dv = f(v).$ (49)

Hence, $\quad dG = \dfrac{KgV_s}{C_a} f(v)\, dv$ (50)

Now, the heat conducted from the cell will be considered to be controlled by the radial conductivity of the total cell contents and not by the cell walls alone. Furthermore, the axial conductivity of the cell will be ignored as its contribution to heat loss will be several orders of magnitude less than that lost by radial convection.

Consequently, as the cell is cylindrical, the heat conducted radially from the cell has been shown to be [22]

$$\frac{2\pi l E \theta\, dt}{\mathrm{Log}\,\frac{r_1}{r_t}}$$

where, (E) is the thermal conductivity of the cell contents,
(θ) is the excess temperature of the cell above its surroundings,
(r_t) is the radius of the sensing element,
and (dt) is the time taken for a volume ($c_a dv$) of mobile phase to pass through the cell.

Now, (dt) refers to the time interval during the introduction of the volume ($c_a dv$) of mobile phase and thus, if the flow rate is (Q),

$$\frac{dv}{dt} = \frac{Q}{c_a} \quad \text{or} \quad dt = \frac{c_a dv}{Q}$$

Thus, heat conducted from the cell is $\quad \dfrac{2\pi l E c_a \theta\, dv}{Q \, \mathrm{Log}_e \frac{r_1}{r_t}}$

And the heat convected from the cell is $\quad d_m S_m \theta\, c_a\, dv$

Thus, inserting the above expressions for the heat conducted from the cell and the heat convected from the cell, together with the heat evolved from the cell, from equation (50) in equation (44),

$$H\, d\theta = \frac{K g V_s}{C_a} f(v)\, dv - \frac{2\pi l E c_a \theta\, dv}{Q\, \mathrm{Log}_e \frac{r_1}{r_t}} - d_m S_m \theta c_a dv.$$

or $\qquad \dfrac{d\theta}{dv} = \dfrac{A}{H} f(v) - \dfrac{\beta c_a \theta}{H} \qquad\qquad (51)$

where $\qquad A = \dfrac{K g V_s}{C_a} \quad \text{and} \quad \beta = \dfrac{2\pi l E c_a dv}{Q\, \mathrm{Log}_e \frac{r_1}{r_t}} - d_m S_m$

Multiplying equation (51) throughout by $e^{\beta c_a \frac{v}{H}}$ and rearranging,

$$e^{\beta c_a \frac{v}{H}} d\theta + \frac{\beta c_a e^{\beta c_a \frac{v}{H}} \theta}{H} = \frac{A}{H} f(v) e^{\beta c_a \frac{v}{H}} dv$$

Extension of the Plate Theory

or

$$\frac{d\left(e^{\beta c_a \frac{v}{H}} \theta\right)}{dv} = \frac{A}{H} f(v) e^{\beta c_a \frac{v}{H}} dv$$

Thus, integrating,

$$e^{\beta c_a \frac{v}{H}} \theta = \int e^{\beta c_a \frac{v}{H}} \frac{A}{H} f(v)\, dv + R$$

Now, on sample injection, $v=0$, $\theta = 0$ and $f(v) = 0$ and thus, $R = 0$.

Therefore,

$$\theta = e^{-\beta c_a \frac{v}{H}} \theta \int_0^v e^{\beta c_a \frac{v}{H}} \frac{A}{H} f(v)\, dv, \qquad (52)$$

Letting

$$\frac{A}{H} = \varphi \quad \text{and} \quad \frac{\beta c_a}{H} = \phi,$$

Then,

$$\theta_v = \varphi e^{-\phi v} \int_0^v e^{\phi v} f(v)\, dv. \qquad (53)$$

It is clear from equation (53) that the constant (φ) determines only the *magnitude* of (θ), but the constant (ϕ) and $f(v)$ control the *shape* of the temperature profile. (ϕ) and $f(v)$ are the source of the curiously shaped peaks produced by the detector. The constant (ϕ) is the heat loss factor of the cell. It should be noted that the magnitude of $f(v)$ will depend on the value of (C_a), the ratio of the 'effective volume of the cell' to the 'plate volume' of the column. Inserting the full expression for $f(v)$ in equation (53),

$$\theta_v = \varphi e^{-\phi v} \int_0^v e^{\phi v} \left(X_o \frac{e^{-v} v^n}{n!} - \left(\frac{X_o}{C_a}\right) e^{\frac{-v}{C_a}} \int_0^v e^{\frac{v}{C_a}} \left(\frac{e^{-v} v^n}{n!}\right) dv \right) dv \qquad (54)$$

Equation (54) is an explicit expression that defines the temperature change of the detector in terms of the initial concentration of the solute placed on the column and the volume of mobile phase that passes through it. It can be used, with the aid of a computer, to synthesize the different shaped curves that the detector can produce. Employing a computer in the manner of Smuts *et al.* [23], Scott [24] calculated the relative values of (θ) for ($v= 74$ to 160) with a column of 100 theoretical plates, and for (C_a) ranging from 0.25 to 4 and (ϕ) ranging from 0.01 to 1.25. The curves are shown in Figure 24.

The twenty curves shown in Figure 24 are graphs of (θ) versus (v) together with the integral of (θ) versus (v) for different values of (C_a) and (ϕ). They are all normalized to the same peak height. The curves include the practical range of heat loss factors that might be expected from an heat of adsorption detector cell.

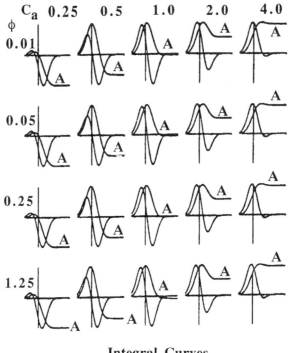

Integral Curves

Courtesy of the Journal of Chromatographic Science (ref. 24)

Figure 24. Theoretical Temperature and Integral Temperature Curves Obtained from the Heat of Absorption Detector

The curves show the effect of changes in $\frac{\text{detector cell capacity}}{\text{column plate capacity}}$ ratios that would result from different cell designs on the shape of the (θ) versus (v) curves when detecting a peak of constant width. The curves for different (C_a) values also show the effect of solutes having different capacity ratios and, thus, different peak widths, on the peak shape when passing through a cell of fixed dimensions. It is clear that the major factor affecting peak shape is the detector cell $\frac{\text{detector cell capacity}}{\text{column plate capacity}}$, ($C_a$).

Under conditions where the capacity of the detector cell is less than the plate capacity of the column ($C_a < 1$), then the negative part of the signal dominates. Conversely,

Extension of the Plate Theory

when the detector cell capacity exceeds the column plate capacity ($C_a>1$), the positive part of the signal dominates. As a result, when ($C_a>1$), the integral curve *rises to a maximum but does not return to the baseline*. In contrast, when ($C_a<1$), the integral curve first rises and then *falls below the baseline and does not return*.

Only when ($C_a=1$) does the detector signal simulate the differential form of the elution curve and, consequently, the integral curves describe a true Gaussian peak. Although it is not apparent from the curves in Figure 24, because the curves are normalized, the magnitude of the signal does vary inversely to (ϕ), albeit the overall effect of the heat loss factor (ϕ) on peak shape is small. For low values of (β), where the maximum sensitivity is realized, the peak maximum is significantly displaced from the true maximum of the elution curve. However, the maximum of the integral curves for ($C_a=1$) is almost coincident with the maximum of the elution curve for large values of (β).

It is clear that for this type of detector to be effective and produce the true Gaussian form of the eluted peak, then (C_a) must, at all times, be unity and, consequently, the detector must have the same plate capacity as that of the column. In order for the column and sensor cell to have the same capacity, the detector and column must employ the same adsorbent, have the same geometry and be packed to give the same plate height as the column. This can be achieved by making the end of the column the sensor cell and by inserting the temperature sensing element in the column packing itself. This detector arrangement will now be considered theoretically. Restating the expression for f(v), from equation (49),

$$f(v) = X_o \frac{e^{-v}v^n}{n!} - \left(\frac{X_o}{C_a}\right) e^{\frac{-v}{C_a}} \int e^{\frac{v}{C_a}} \left(\frac{e^{-v}v^n}{n!}\right) dv$$

Now, if the sensor is in the column packing $C_a=1$.

Consequently,
$$f(v) = X_o \frac{e^{-v}v^n}{n!} - X_o e^{-v} \int e^{v} \left(\frac{e^{-v}v^n}{n!}\right) dv$$

Integrating,
$$f(v) = X_o \frac{e^{-v}v^n}{n!} - X_o \frac{e^{-v}}{n!} \left[\frac{v^{n+1}}{n+1}\right]$$

$$= X_o \frac{e^{-v}v^n}{n!} - X_o \frac{e^{-v}v^{(n+1)}}{(n+1)!}$$

$$= X_n - X_{(n+1)}$$

Recalling the basic differential equation for the elution curve given in chapter 2 is,

$$f(v) = \frac{dX_{(n+1)}}{dv} \qquad (55)$$

Then, as there is no (n+1) plate that constitutes the detector, the sensing element can be considered to be placed in the (n)th plate of the column. Thus,

$$f(v) = \frac{dX_n}{dv} \qquad (56)$$

As the (n)th plate of the column acts as the detecting cell, there can be no heat exchanger between the (n-1)th plate and the (n)th plate of the column. As a consequence, there will be a further convective term in the differential equation that must account for the heat brought into the (n)th plate from the (n-1)th plate by the flow of mobile phase (dv). Thus, heat convected from the (n-1)th plate to plate (n) by mobile phase volume (dv) will be

$$d_m S_m \theta_{(n-1)} c_a dv \qquad (57)$$

Substituting for f(v) from equation (53) in equation (47) and inserting the extra convection term from (54),

$$\frac{d\theta_n}{dv} = \left(\frac{A_p}{H_p}\right)\frac{dX_n}{dv} - \left(\frac{\beta_p c_a}{H_p}\right)\theta_n + \left(\frac{d_m S_m c_a}{H_p}\right)\theta_{(n-1)}$$

where the subscript (p) accounts for the change from the already defined physical characteristics of the detecting cell to those of the last plate of the column. Thus,

$$\frac{d\theta_n}{dv} = \alpha \frac{dX_n}{dv} - B\theta_n + \gamma\, \theta_{(n-1)} \qquad (58)$$

where $\qquad \alpha = \dfrac{A_p}{H_p} \qquad B = \dfrac{\beta_p c_a}{H_p} \qquad \gamma = \dfrac{d_m S_m c_a}{H_p}$

A solution to the differential equation (58) is given by

$$\theta_n = \frac{\alpha}{(B-1)} \sum_{r=0}^{r=n} \left[\frac{(\gamma-1)}{(B-1)}\right]^r \frac{dX_{(n-r)}}{dv} \qquad (59)$$

The validity of this solution can be confirmed by differentiation as follows:

$$\frac{d\theta_n}{dv} = \frac{\alpha}{(B-1)} \sum_{r=0}^{r=n} \left[\frac{(\gamma-1)}{(B-1)}\right]^r \frac{d^2 X_{(n-r)}}{dv^2} \qquad (60)$$

Extension of the Plate Theory

Now, if
$$X_n = X_o \frac{e^{-v} v^n}{n!},$$

then
$$\frac{dX_n}{dv} = X_o \frac{e^{-v} v^{(n-1)}}{(n-1)!} - X_o \frac{e^{-v} v^n}{n!} = X_{(n-1)} - X_n$$

Thus,
$$\frac{d^2 X_n}{dv^2} = \frac{dX_{(n-1)}}{dv} - \frac{dX_n}{dv}$$

Substituting for $\frac{d^2 X_{(n-r)}}{dv^2}$ in equation (56)

$$\frac{d\theta_n}{dv} = \frac{\alpha}{(B-1)} \sum_{r=0}^{r=n} \left[\frac{(\gamma-1)}{(B-1)}\right]^r \left(\frac{dX_{(n-1-r)}}{dv} - \frac{dX_{(n-r)}}{dv}\right)$$

or,

$$\frac{d\theta_n}{dv} = \frac{\alpha}{(B-1)} \sum_{r=0}^{r=n} \left[\frac{(\gamma-1)}{(B-1)}\right]^r \frac{dX_{(n-1-r)}}{dv} - \frac{\alpha}{(B-1)} \sum_{r=0}^{r=n} \left[\frac{(\gamma-1)}{(B-1)}\right]^r \frac{dX_{(n-r)}}{dv}$$

Now, for the series, $\frac{dX_{(-1)}}{dv} = 0$. Thus, $\frac{d\theta_n}{dv} = \theta_{(n-1)} - \theta_n$.

Consequently,

$$\frac{d\theta_n}{dv} + B\theta_n - \gamma \theta_{(n-1)} = \theta_{(n-1)} - \lambda \theta_{(n-1)} - \theta_n + \beta \theta_n$$
$$= (1-\gamma)\theta_{(n-1)} + (B-1)\theta_n \quad (61)$$

From equation (56) and using its expanded form,

$$(B-1)\theta_n = \alpha \frac{dX_n}{dv} + \frac{\alpha(\gamma-1)}{(B-1)} \frac{dX_{(n-1)}}{dv} + \alpha\left[\frac{(\gamma-1)}{(B-1)}\right]^2 \frac{dX_{(n-2)}}{dv} + \ldots$$

$$(1-\gamma)\theta_{(n-1)} = -\frac{\alpha(\gamma-1)}{(B-1)} \frac{dX_{(n-1)}}{dv} - \alpha\left[\frac{(\gamma-1)}{(B-1)}\right]^2 \frac{dX_{(n-2)}}{dv} - \ldots$$

Thus,
$$(1-\gamma)\theta_{(n-1)} + (B-1)\theta_n = \alpha \frac{dX_n}{dv} \quad (62)$$

Substituting $\alpha \frac{dX_n}{dv}$ from equation (62) for $(1-\gamma)\theta_{(n-1)}+(B-1)\theta_n$ in equation (61)

and rearranging,
$$\frac{d\theta_n}{dv} = \alpha \frac{dX_n}{dv} - B\theta_n + \gamma \theta_{(n-1)} \quad (63)$$

Equations (63) and (58) are seen to be identical, which substantiates the validity of equation (60) for (θ_n). It is clear from equation (59) that (α) only affects the magnitude of the curve, while (γ) and (B) affect both its shape as well as its magnitude.

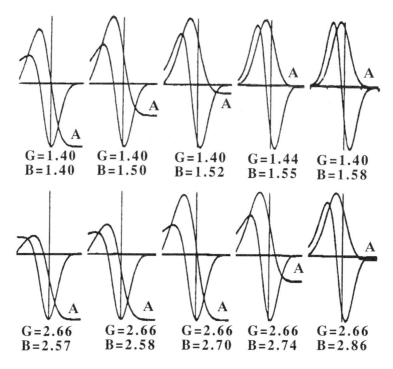

Courtesy of the Journal of Chromatographic Science (ref. 24)

Figure 25. Theoretical Temperature and Integral Temperature Curves Generated by a Temperature Sensor Situated in the Column Packing

Scott [24], assumed practical values for the various physical parameters of the system and calculated the temperature and integral temperature curves for a series of different practical values of (B) and (γ). The results are shown in Figure 25. In the figure, the values of (γ) are represented as (G). The curves in Figure 25 show that the heat convected into the detector cell or plate also distorts the curves. It would seem that, unless the heat lost radially is extremely high so that little heat is *convected to* the sensor, it would not be possible to obtain symmetrical integral peaks. It would also appear that this level of heat transfer would be impossible to achieve in practice. It must be concluded that the heat of absorption detector designed in the forms that have been theoretically examined does not seem viable for LC. The examples given demonstrate how the plate theory can be used in a number of different ways to theoretically examine dynamic distribution systems. The plate theory has also been

Extension of the Plate Theory

used to investigate pressure changes that take place in a GC column, [23] the influence of solute decomposition on band profiles and other similar effects that can take place in a chromatographic system.

Synopsis

By changing the origin from the injection point, which is appropriate for the Poisson form of the elution equation, to the position of the peak maximum, the Gaussian or Error function form of the elution equation can be derived. The Error function form of the elution equation can be useful for certain theoretical treatments and, particularly, for the computer simulation of elution curves. Due to interaction between the concentration profiles of closely eluting peaks, the apparent retention times, or retention volumes, as measured from the position of the peak maxima, may differ significantly from the true retention times or volumes. This error will become greater as the resolution is less and the column efficiency reduced. If the peaks are asymmetrical, the effect is exacerbated and will even give a false indication of the relative peak heights. The interaction between the concentration profiles of closely eluting peaks can, however, be employed usefully for the analysis of completely unresolved peaks. The positions of the peak maxima can be related to the relative proportions of the two unresolved components. Thus, the composition of the mixed peak can be determined from the retention time of the composite peak. Peak asymmetry normally results from the separation being carried out under conditions where the adsorption isotherm is nonlinear. There can be a number of causes for this to happen, such as stationary phase saturation, or solute/solute interaction. If the isotherm is Langmuir in type, large samples will produce peaks with a sharp front and a sloping tail. If the isotherm is Freundlich in type, then the peak will have a sloping front and a sharp back. By equating the second differential of the elution equation to zero and solving, the peak width at the points of inflexion can be determined. From the peak width, by simple proportion, the efficiency of the column can be calculated from measurements made on the chromatogram. In order to do this, the position of the points of inflection, relative to the peak height, must be identified. This can be achieved, again, from the elution equation, and the height of the points of inflection are found to be 0.6065 of the peak height. Once more employing the elution curve equation, an expression for the resolution of a column can be derived, and the efficiency required to separate a pair of solutes can be calculated from the capacity factor of the first peak and their separation ratio. The efficiency of a column does not reflect its resolution for solutes eluted at low capacity ratios, the 'effective plate number' was introduced. The effective plate number and the theoretical plate number converge at high capacity ratios but, at low capacity ratios, the effective plate number is much less and more nearly relates to the resolution of the column. The plate theory assumes that the sample is placed on the first plate but, in practice, this does not occur and, thus, the ultimate peak consists of a composite of all the peaks eluted from each plate that contained sample. Samples of finite volume will always cause the peak to be broader than the theoretical width that would result from the sample being placed only on the first plate. Peak dispersion can occur in many parts of the chromatograph other than the column, and the total dispersion is a result of all the dispersion processes that occur. The net effect can be obtained by adding the individual variances from each dispersion process and the total will be the variance of the final peak. This procedure is called the *summation of variances*. By applying this principle to the sample volume, it can be shown that the maximum sample that can be placed on a column, without increasing the peak width by more than 5%, is a simple function of the retention volume of the solute and the column efficiency. If a column is supplied with a mobile phase containing a solute at a given concentration, and equilibrium is allowed to take place, the concentration of solute in the eluent will be the same as that at the inlet. A sample of pure mobile phase placed on the equilibrated column will cause a fall in the solute concentration which, *mathematically*, will represent the injection of a sample having a *negative concentration*. A negative concentration profile passes through a column in exactly the same way as a positive concentration profile and, on elution, will be recorded as a negative peak. This type of chromatography is called *vacancy* chromatography. Vacancy chromatography can be described by the same elution equation as that for a positive concentration of sample, and vacancy chromatograms can be synthesized by means of a computer to demonstrate the different properties of the technique, again employing the same elution equation. One limitation of a chromatographic system, and this is particularly important in the separation of multi-component mixtures, is the peak capacity of the column. Peak capacity is defined as the number of peaks that can be fitted into a chromatogram between the dead point and the last eluted peak. An explicit equation for the peak capacity can be derived from the plate theory and is shown to be a function of the capacity ratio of the last eluted solute and the column efficiency. Any property of the chromatographic system that places a limit on the maximum capacity of the solute, such as detector sensitivity, also places a limit on the peak capacity. During the passage of a solute through a column, it is continuously adsorbed into the stationary phase and then desorbed from the stationary phase. In fact, as already discussed in an earlier chapter, at the front of the peak there is a net transfer of solute from the mobile phase to the stationary phase, and at the rear of the peak there is net transfer of solute from the stationary phase to the mobile phase. During transfer to the stationary phase, the heat of adsorption or solution is *evolved*, and at the rear of the peak during transfer to the mobile phase, the heat of desorption or vaporization is *absorbed*. This results in the stationary phase at the front of the peak being at a temperature above its surroundings and at the back of the peak, at a temperature below that of its surroundings. This temperature change can be examined by use of the plate theory, and an equation that describes the temperature of a theoretical plate in terms of the flow of mobile phase though it can be derived. The same technique can be used to investigate the characteristics of the heat of adsorption detector. It can be shown that, even when the column itself is used as the detector cell, the difficulties in achieving the necessary thermal conditions renders the heat of adsorption detector, in its present accepted form, an impractical detecting device for liquid chromatography.

References

1. R. P. W. Scott and C. E. Reese, *J. Chromatogr.*, **138**(1977)283.
2. R. P. W. Scott and P. Kucera, *J. Chromatogr.*, **149**(1978)93.
3. J. H. Purnell, *Nature* (London),**184, Suppl. 26**(1959)2009.
4. J. H. Purnell and J. Bohemen, *J. Chem. Soc.* (1961)2030.
5. D. H. Desty and A. Goldup, "Gas *Chromatography 1960*" (Ed. R. P. W. Scott), Butterworths Scientific Publications, London (1960)162.
6. R. P. W. Scott, *Nature (London)*, **183**(1959)1753.
7. J. C. Giddings, "The *Dynamics of Chromatography* ", Marcel Dekker, New York (1965)265.
8. J. C. Giddings, *J. Chromatogr. Sci* , **12**(1974)1753.
9. A. Klinkenberg, in "*Gas Chromatography 1960* (Ed. R. P. W. Scott), Butterworths Scientific Publications, London (1960)194.
10. A. A. Zhukhovitski and Turkel'taub, *Dokl. Acad. Nauk. USSR.*, **143**(1961)646.
11. C. N. Reilley, C. P. Hildebrand and J.W.Ashley, *Anal. Chem.*, **34**(1962)1198.
12. R. P. W. Scott, C. G. Scott and P. Kucera, *Anal. Chem.*, **44 No.1**(1972)100.
13. R. P. W. Scott, *J. Chromatogr. Sci.*, **9**(1971)449.
14. J. C. Giddings, *Anal. Chem.*, **39 No.8**(1967)1027.
15 J. M. Davis and J. C. Giddings, *Anal Chem.*, **55**(1983)418.
16. N. H. Ray, private communication, *Symposium on Vapor Phase Chromatography*, London (1956).
17. A. Klinkenberg, *"Vapour Phase Chromatography"* (Ed. D. H. Desty),

Butterworths Scientific Publications, London (1957).

18. G. C. Claxton, *J. Chromatogr.*, **2**(1959)136.
19. A. J. Groszek, *Nature (London)*, **182**(1958)1152.
20. J. Smith, Rubber Research Association, Welyn, England, private communication (1958).
21. R. P. W. Scott, *Anal. Chem.*, **35 No. 4**(1963)481.
22 R. H. Perry, C. H. Chilton and S. D. Kirkpatrick, Chemical Enginerring Handbook, McGraw Hill, NewYork (1970)10.
23. T. W. Smuts, P. W. Richter and V. J. Pretorious, *J. Chromatogr. Sci.*, **9**(1971)457.
24. R. P. W. Scott, *J. Chromatogr. Sci.*, **July**(1973)349.

Part 2

The Mechanism of Dispersion

Dispersion in Columns and Mobile Phase Conduits, the Dynamics of Chromatography, the Rate Theory and Experimental Support of the Rate Theory

Chapter 7

The Dynamics of Peak Dispersion

In Part 1 of this book, the mechanism of retention was explained and it was established that retention is one essential requirement for chromatographic resolution. The individual components of a mixture must be moved apart in the chromatographic system in order that they may be eluted discretely. As already discussed, however, the discrete elution of each solute stipulates that the concentration profile of each substance (the respective peak) must be sufficiently narrow to ensure that it does not merge with its neighbors. Thus, the constraint of peak dispersion is a second essential requirement for chromatographic resolution. In Part 2 of this book, the dynamics of the chromatographic distribution system will be examined, qualitatively and quantitatively, to determine those factors that control peak dispersion both within and external to the column. In addition, experimental support for the equations that describe peak dispersion will be given and the impact of dispersion control on instrument design will also be discussed.

The dynamics of chromatography are not easily discussed using the dependence of solute concentration on volume flow of mobile phase, which are the variables used in the plate theory. The variables used in dynamic studies usually involve the dependence of solute concentration on the distance traveled by the solute along the column. Now, the elution curve of a chromatogram can be readily expressed using parameters other than the *volume flow* of mobile phase as the independent variable. Instead of using milliliters of mobile phase, solute concentration in the mobile phase leaving the column can also be related to *elapsed time* or *distance* traveled by the solute band along the column and, proportionally, the same chromatogram will be obtained. The different chromatograms obtained using the variables volume flow of mobile phase (v), elution time (t) and distance traveled along the column (x) are illustrated in Figure 1.

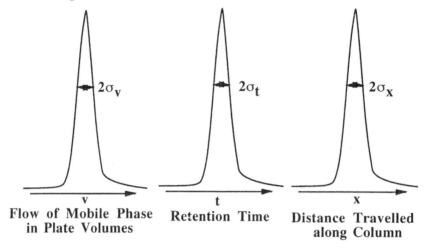

Figure 1.   The Alternative Axes of a Chromatogram

Now, all the curves are describing the same chromatogram; thus, by simple proportion, the ratios of the variance of each elution curve to the square of the retention (in the respective units in which the variables are defined) will all be equal.

Consequently,
$$\frac{\sigma_v^2}{v_r^2} = \frac{\sigma_x^2}{l^2} = \frac{\sigma_t^2}{t_r^2}$$

where (σ_v), (σ_x) and (σ_t) are the standard deviations of the elution curves when related to the volume flow of mobile phase, the distance traveled by the solute along the column and time, respectively.

and (V_r), (l) and (t_r) refer to the retention volume, column length and retention time, respectively.

From the Plate Theory it has been shown that

$$\sigma_v = \sqrt{n}(v_m + Kv_s) \quad \text{and} \quad V_r = n(v_m + Kv_s)$$

Thus,
$$\frac{\sigma_v^2}{V_r^2} = \frac{n(v_m + Kv_s)^2}{n^2(v_m + Kv_s)^2} = \frac{1}{n} = \frac{\sigma_x^2}{l^2}$$

Therefore,
$$\frac{1}{n} = \frac{\sigma_x^2}{l}$$

Mechanism of Dispersion

The ratio $\left(\dfrac{\sigma_x^2}{l}\right)$ is the variance per unit length of the column and will include all the variance contributions of the different dispersion processes that can be summed to give the final variance. Now, the ratio $\left(\dfrac{l}{n}\right)$ is the length of the column divided by the number of theoretical plates and, thus, has been given the obvious term *Height of the Theoretical Plate* (H). Furthermore, it is clear that the value of (H) can be easily calculated from the column length and the column efficiency and, thus, due to the equivalence, the variance per unit length of any column can also be calculated. It follows that, if the dynamics of the distribution system are examined, and the different dispersion processes identified and their variances summed then the theoretically predicted relationships can be compared with those obtained experimentally and the validity of the theory confirmed.

The different dispersion processes (1, 2, 3,...) that occur in a column will now be considered theoretically, their individual contributions to the variance per unit length of the column (H_1, H_2, H_3...) evaluated and then summed to provide an expression for the total variance per unit length of the column (H), *i.e.*,

$$H = H_1 + H_2 + H_3 + ...$$

The theory that results from the investigation of the dynamics of solute distribution between the two phases of a chromatographic system and which allows the different dispersion processes to be qualitatively and quantitatively specified has been designated the Rate Theory. However, historically, the Rate Theory was never developed as such, but evolved over more than a decade from the work of a number of physical chemists and chemical engineers, such as those mentioned in chapter 1.

Various mathematical concepts and techniques have been used to derive the functions that describe the different types of dispersion and to simplify further development of the rate theory; two of these procedures will be discussed in some detail. The two processes are, firstly, the Random Walk Concept [1] which was introduced to the rate theory by Giddings [2] and, secondly, the mathematics of diffusion which is both critical in the study of dispersion due to longitudinal diffusion and that due to solute mass transfer between the two phases. The random walk model allows the relatively simple derivation of the variance contributions from two of the dispersion processes that occur in the column and, so, this model will be the first to be discussed.

The Random Walk Model

The random walk model has been described in detail elsewhere [1], but in simple form it postulates that if each molecule of a group takes a series of steplike movements, which may be positive or negative, the direction being completely random, then after (p) steps have been taken, each step having a *mean* length (s), the average of the molecules will have moved some distance from the starting position and will form a Gaussian type distribution curve with a variance of σ^2. Now, according to the random walk model, the following simple relationship holds:

$$\sigma = s\sqrt{p} \tag{1}$$

Equation (1) can be used in a general way to determine the variance resulting from the different dispersion processes that occur in an LC column. However, although the application of equation (1) to physical chemical processes may be simple, there is often a problem in identifying the average step and, sometimes, the total number of steps associated with the particular process being considered. To illustrate the use of the Random Walk model, equation (1) will be first applied to the problem of radial dispersion that occurs when a sample is placed on a packed LC column in the manner of Horne *et al.* [3].

When a stream of mobile phase carrying a solute impinges against a particle, the stream divides and flows around the particle. Part of the divided stream then joins other split streams from neighboring particles, impinges on another particle and divides again. If a sample is placed on the column at the center of the packing, it is not in radial equilibrium. However, as a result of this stream splitting process, during passage through the column, the sample is dispersed and may eventually achieve radial equilibrium some distance down the column. In the early days of liquid chromatography, relatively low inlet pressures were employed and samples were injected on the column by turning off the pump and injecting the sample with a syringe, through an appropriate septum device, into the center of the packing. This procedure often resulted in radial equilibrium *never* being achieved by the solutes before they were eluted.

The introduction of the sample valve, however, helped establish radial equilibrium early in the separation but, unless some special sample spreading device is employed at the front of the column, equilibrium will not necessarily occur at the point of injection. The stream splitting process is depicted in Figure 2.

Mechanism of Dispersion

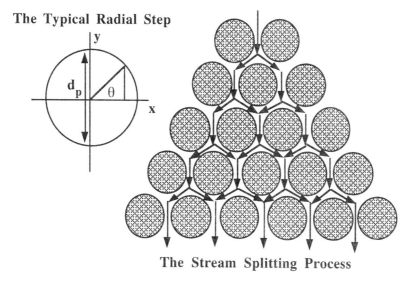

Figure 2. The Mechanism of Radial Dispersion

Consider a molecule passing round a particle; it will suffer a lateral movement which is seen, from Figure 2, to be,

$$\text{Lateral Movement/Particle} = \frac{d_p}{2}\cos\theta$$

where (d_p) is the particle diameter.

It follows that the *average* lateral step (s) will be

$$s = \frac{d_p}{2}\int_{-\frac{\pi}{2}}^{+\frac{\pi}{2}}\frac{\cos\theta}{\pi}d\theta = \frac{d_p}{2}\left[\frac{\sin\theta}{\pi}\right]_{-\frac{\pi}{2}}^{\frac{\pi}{2}} = \frac{d_p}{2}\left[\frac{\sin\left(\frac{\pi}{2}\right)-\sin\left(-\frac{\pi}{2}\right)}{\pi}\right] = \frac{d_p}{2}\left[\frac{1+1}{\pi}\right] = \frac{d_p}{\pi}$$

Now, employing the random walk relationship, the radial variance is given by

$$\sigma^2 = (\text{Number of Steps})\times(\text{Step Length})^2 = s^2 p \qquad (2)$$

Assuming one lateral step is taken by a molecule for every distance (jd_p) that it moves axially, then (p), the number of steps, is given by

$$p = \frac{1}{jd_p}$$

where (1) is the distance traveled axially by the solute band.

Now, substituting for (p) and (s) in equation (2),

$$\sigma^2 = \frac{1}{jd_p}\left(\frac{d_p}{\pi}\right)^2 = \frac{ld_p}{j\pi^2}$$

It is probable that, in practice, (j) will lie between 0.5 and 1.0 but, to simplify the argument, (j) will be taken as unity. Thus, it will be assumed that one lateral step will be taken by a given molecule for every step traveled axially that is equivalent to one particle diameter.

Consequently,
$$\sigma^2 = \frac{ld_p}{\pi^2} \tag{2}$$

or
$$\sigma = \frac{\sqrt{(ld_p)}}{\pi} \tag{3}$$

Equation (3) allows the calculation of the distance traveled axially by a solute band before the radial standard deviation of the sample is numerically equal to the column radius. Consider a sample injected precisely at the center of a 4 mm diameter LC column. Now, radial equilibrium will be achieved when (σ), the radial standard deviation of the band, is numerically equal to the radius, *i.e.*, $\sigma = 0.2$ cm.

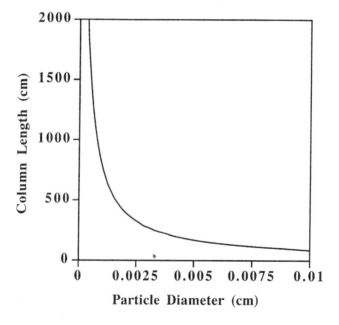

Figure 3. Graph of Length of Column Traversed by the Solute Before Radial Equilibrium Is Achieved against Particle Diameter

Mechanism of Dispersion

Thus, from equation (3)
$$0.2 = \frac{\sqrt{ld_p}}{n}$$

or,
$$l = \frac{(0.2n)^2}{d_p} = \frac{0.793}{d_p} \tag{4}$$

Equation (4) allows the distance that a solute band must travel along a column before a sample, injected at the center of the packing, is evenly spread across the column diameter to be calculated. A curve relating column length (l) in equation (4) to the particle diameter is shown in Figure 3. It is seen that radial transfer of solute can be very slow if the particle diameter of the packing is less than 20 μm, and, at the extreme, if the particle diameter is 5 μm or less, radial equilibrium is highly unlikely to occur at all as the column will not be long enough. It is, therefore, important to employ an injection technique that ensures radial equilibrium occurs rapidly after injection.

This example of the use of the Random Walk model illustrates the procedure that must be followed to relate the variance of a random process to the step width and step frequency. The model will also be used to derive an expression for other dispersion processes that take place in a column.

The Diffusion Process

Diffusion plays an important part in peak dispersion. It not only contributes to dispersion directly (*i.e.*, longitudinal diffusion), but also plays a part in the dispersion that results from solute transfer between the two phases. Consider the situation depicted in Figure 4, where a sample of solute is introduced in plane (A), plane (A) having unit cros-sectional area. Solute will diffuse according to Fick's law in both directions ($\pm$ x) and, at a point (x) from the sample point, according to Ficks law, the mass of solute transported across unit area in unit time (m_x) will be given by

$$m_x = D_m \frac{dc}{dx} \tag{5}$$

where (D_m) is the Diffusion Coefficient or the Diffusivity of the solute in the fluid and $\left(\frac{dc}{dx}\right)$ is the concentration gradient at (x).

Now, the mass leaving the slice (dx) thick at (x+dx), *i. e.*, (m_{x+dx}), will be

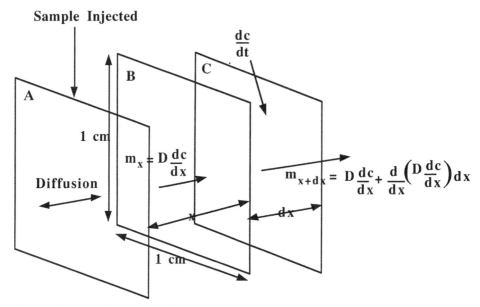

Figure 4. The Diffusion Process

$$m_{x+dx} = D_m \frac{dc}{dx} + \frac{d\left(D_m \frac{dc}{dx}\right)dx}{dx}$$

Thus, the net change in mass per unit time in the slice (dx) thick will be

$$dm = m_{x+dx} - m_x = D_m \frac{dc}{dx} + \frac{d\left(D_m \frac{dc}{dx}\right)dx}{dx} - D_m \frac{dc}{dx}$$

or

$$\frac{dm}{dt} = D_m \frac{d^2c}{dx^2} dx$$

Now, as $dc = \frac{dm}{1 \times 1 \times dx}$, then $\frac{dm}{(1 \times 1 \times dx)dt} = \frac{dc}{dt} = \frac{D_m}{(1 \times 1 \times dx)} \frac{d^2c}{dx^2} dx = D_m \frac{d^2c}{dx^2}$

or

$$\frac{dc}{dt} = D_m \frac{d^2c}{dx^2} \quad (6)$$

Now, this is a standard differential equation and one solution to this equation, which can be proved by appropriate differentiation, takes the Gaussian form as follows:

$$c = \frac{M}{\sqrt{t}} e^{\left(\frac{-x^2}{4D_m t}\right)}$$

Now, we know from the Plate Theory that $X_{m(n)} = \dfrac{X_o e^{\frac{w^2}{2n}}}{\sqrt{2\pi n}}$, where (n) is the variance of the Gaussian curve. As (n) is the volume variance of the Gaussian curve (i.e., σ_v^2), then, by comparison, $(2D_m t)$ will be the length variance $\left(\sigma_x^2\right)$ of the concentration curve where (t) is the elapsed time. Consequently, if a differential equation of the form $\dfrac{dc}{dt} = D_m \dfrac{d^2 c}{dx^2}$ is derived that describes some form of dispersion that arises from a random diffusion process, then the solution will be a Gaussian function and, more important from the point of view of the Rate Theory, the Gaussian curve will have a variance given by $(2D_m t)$.

Thus, if $\dfrac{dc}{dt} = D_m \dfrac{d^2 c}{dx^2}$, the solution of the equation is a Gaussian function, and, for that equation,
$$\sigma^2 = 2D_m t \tag{7}$$

Sources of Dispersion in a Packed Column

The dispersion of a solute band in a packed column was originally treated comprehensively by Van Deemter *et al.* [4] who postulated that there were four first-order effect, spreading processes that were responsible for peak dispersion. These the authors designated as *multi-path dispersion, longitudinal diffusion, resistance to mass transfer in the mobile phase* and *resistance to mass transfer in the stationary phase*. Van Deemter derived an expression for the variance contribution of each dispersion process to the overall variance per unit length of the column. Consequently, as the individual dispersion processes can be assumed to be random and *non-interacting,* the total variance per unit length of the column was obtained from a sum of the individual variance contributions.

Dispersion from the Multi-path Effect

The multipath dispersion has been treated in two ways. The first was by Van Deemter *et al.* who assumed it to be a simple direct dispersion contribution to the overall variance per unit length of the column. The second was by Giddings, who considered multi-path dispersion to be the limiting effect of what Giddings termed *eddy diffusion* that occurs when the mobile phase velocity became significantly greater than the effective diffusion velocity. Eddy diffusion is an important effect in liquid chromatography as it reduces the resistance to mass transfer in the mobile phase between the particles and is one factor that makes the variance per unit length of the capillary and packed columns to be very similar in magnitude. Both types of dispersion will be considered but the simple effect described by Van Deemter will be considered first.

In a packed column the individual solute molecules will describe a tortuous path through the interstices between the particles and some will randomly travel shorter routes than the average and some longer. The multi-path effect is diagramatically depicted in Figure 5.

Figure 5. Multi-path Dispersion

Those molecules taking the shorter paths will move ahead of the mean and those that take the longer paths will lag behind by the distance (dl), as shown in Figure 5, which results in band dispersion. An equation that describes the dispersion resulting from the multi-path effect can be derived using the Random Walk Model. Now, in this case, the average path length (s) will be equivalent to the diameter of the particle (d_p). It follows that the number of steps will be equivalent to the column length (l) divided by the average step, *i.e.*, $p = \dfrac{l}{d_p}$.

Thus, applying the Random Walk relationship, *i.e.*, $\sigma = s\sqrt{p}$, then

$$\text{Then,} \quad \sigma = d_p \left(\dfrac{l}{d_p}\right)^{0.5} = \sqrt{(ld_p)} \quad \text{or} \quad \sigma^2 = ld_p$$

Mechanism of Dispersion

Dividing the total variance by the column length (1), the multi-path contribution (H_M) to the overall height of the theoretical plate (H) is obtained.

$$\frac{\sigma^2}{1} = H_M = d_p \qquad (8)$$

Equation (5), however, would apply only to a perfectly packed column so Van Deemter introduced a constant (2λ) to account for the inhomogeneity of real packing (for ideal packing (λ) would take the value of 0.5). Consequently, his expression for the multi-path contribution to the total variance per unit length for the column (H_M) is

$$H_M = 2\lambda d_p \qquad (9)$$

Dispersion from Longitudinal Diffusion

Driven by the concentration gradient, solutes naturally diffuse when contained in a fluid. Thus, a discrete solute band will diffuse in a gas or liquid and, because the diffusion process is random in nature, will produce a concentration curve that is Gaussian in form. This diffusion effect occurs in the mobile phase of both packed GC and LC columns. The diffusion process is depicted in Figure 6.

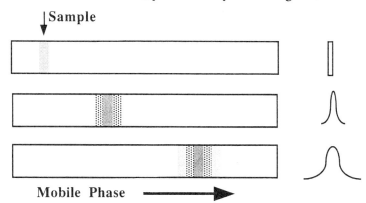

Figure 6. Peak Dispersion by Longitudinal Diffusion

Longitudinal Diffusion in the Mobile Phase

Thus, treating the diffusion process in a similar way to that shown in Figure 4 the total variance due to longitudinal diffusion in a column of length (1) is given by equation (7), *viz.*,

$$\sigma_1^2 = 2D_m t$$

From the plate theory, the retention time of the solute (t_r) is given by $t_o''(1+k'')$

where (t_o'') is the retention time of a *non-permeating, non-absorbed* solute,

and (k'') is the *dynamic* capacity ratio of the solute.

Thus,
$$\sigma_1^2 = 2D_m t_r = 2D_m t_o''(1+k'') \tag{10}$$

In addition, over the time period (t_r), only a fraction $\left(\dfrac{1}{(1+k'')}t_r\right)$ is spent by the solute in the mobile phase, due to distribution between the two phases; thus,

$$\sigma_1^2 = 2D_m t_o''(1+k'')\dfrac{1}{(1+k'')} = 2D_m t_o''$$

Now, bearing in mind that $\left(t_o'' = \dfrac{1}{u}\right)$, $\sigma_1^2 = 2D_m \dfrac{1}{u}$,

($H_{D(m)}$), the variance per unit length of the column contributed by longitudinal diffusion in the mobile phase, will be $\left(\dfrac{\sigma_1^2}{1}\right)$.

Thus,
$$H_{D(m)} = \dfrac{\sigma_1^2}{1} = 2D_m \dfrac{1}{u}\dfrac{1}{1} = \dfrac{2D_m}{u}$$

i.e.,
$$H_{D(m)} = \dfrac{2D_m}{u}. \tag{11}$$

Equation (11) accurately describes longitudinal diffusion in a capillary column where there is no impediment to the flow from particles of packing. In a packed column, however, the mobile phase swirls around the particles. This tends to increase the effective diffusivity of the solute. Van Deemter introduced a constant (γ) to account for this effect. Thus, equation (11) must be modified, for a packed column, to the following form:

$$H_{D(m)} = \dfrac{2\gamma_1 D_m}{u} \tag{12}$$

Giddings [2] estimated that, for a well-packed column, (γ) takes a value of about 0.6. Equation (11) accurately describes longitudinal dispersion in GC capillary columns and equation (12) accurately describes longitudinal dispersion in GC and LC packed columns. Experimental support for these equations will be given in a later chapter.

Longitudinal Diffusion in the Stationary Phase

Theoretically, dispersion can take place by diffusion in the stationary phase but, as will be seen, in practice, is much less in magnitude than that in the mobile phase. The theoretical treatment is similar to that for dispersion in the mobile phase using equation (10).

Mechanism of Dispersion

Reiterating equation (10),

$$\sigma_1^2 = 2D_s t_r = 2D_s t_o''(1+k'')$$

where (D_s) is now the Diffusivity of the solute in the *stationary phase*.

However, in this case, over the time period (t_r), a fraction of time $\left(\dfrac{k''}{(1+k'')} t_r\right)$ is spent by the solute in the *stationary* phase and, thus,

$$\sigma_1^2 = 2D_s t_o''(1+k'') \dfrac{k''}{(1+k'')} = 2D_s k'' t_o$$

Now, again noting that $\left(t_o'' = \dfrac{1}{u}\right)$, $\quad \sigma_1^2 = 2D_s k'' \dfrac{1}{u}$

Noting, ($H_{D(S)}$), the contribution to the variance per unit length, will be $\left(\dfrac{\sigma_1^2}{1}\right)$.

Then, the contribution to the total variance per unit length for the column from longitudinal diffusion in the *stationary* phase will be

$$H_{D(S)} = \dfrac{\sigma_1^2}{1} = 2D_s k'' \dfrac{1}{u} \dfrac{1}{1} = \dfrac{2k'' D_s}{u}$$

i.e.,
$$H_{D(S)} = \dfrac{2k'' D_s}{u}. \tag{13}$$

Equation (13) accurately describes the dispersion that results from longitudinal diffusion in the stationary phase of a capillary column, where the film of stationary phase is continuous along the length of the column. In addition, if, or when, capillary columns become a practical reality for use in LC, dispersion due to longitudinal diffusion in the stationary phase will make a significant contribution to the overall dispersion of the column. This is because the magnitude of the diffusivity of both phases will be similar in magnitude. In GC, however, the diffusivity of the solute in the mobile phase is over four orders of magnitude *less* than that in the stationary phase. As (k'') is unlikely to exceed two orders of magnitude in the practical use of capillary columns, the contribution from diffusion in the stationary phase will always be less than one percent of that in the mobile phase and, thus, from a practical point of view, can be ignored.

In a packed column, however, the situation is quite different and more complicated. Only *point contact* is made between particles and, consequently, the film of stationary phase is largely *discontinuous*. It follows that, as solute transfer between particles can only take place at the points of contact, diffusion will be severely impeded. In practice the throttling effect of the limited contact area between particles renders the dispersion due to diffusion in the stationary phase insignificant. This is true even in packed LC columns where the solute diffusivity in both phases are of the same order of magnitude. The negligible effect of dispersion due to diffusion in the stationary phase is also supported by experimental evidence which will be included later in the chapter.

In summary, equation (13) accurately describes longitudinal dispersion in the stationary phase of capillary columns, but it will only be significant compared with other dispersion mechanisms in LC capillary columns, should they ever become generally practical and available. Dispersion due to longitudinal diffusion in the stationary phase in packed columns is not significant due to the discontinuous nature of the stationary phase and, compared to other dispersion processes, can be ignored in practice.

Dispersion Due to Resistance to Mass Transfer

Resistance to Mass Transfer in the Mobile Phase

During migration through the column, the solute molecules are continually and reversibly transferring from the mobile phase to the stationary phase. This transfer process is not instantaneous; a finite time is required for the molecules to traverse (by diffusion) through the mobile phase in order to reach the interface and enter the stationary phase. Thus, those molecules close to the stationary phase enter it immediately, whereas those molecules some distance away will find their way to it some time later. However, since the mobile phase is moving during this time interval, those molecules that remain in the mobile phase will be swept along the column and dispersed away from those molecules that were close to and entered the stationary phase immediately. The dispersion resulting from the resistance to mass transfer in the mobile phase is depicted in Figure 7. The diagram shows 6 solute molecules in the mobile phase and those closest to the surface (1 and 2) enter the stationary phase immediately. During the period, while molecules 3 and 4 diffuse through the mobile phase to the interface, the mobile phase moves on. Thus, when molecules 3 and 4 reach the interface they will enter the stationary phase some distance ahead of the first

Mechanism of Dispersion

two. Finally, while molecules 5 and 6 diffuse to the interface the mobile phase moves even further down the column until molecules 5 and 6 enter the stationary phase further ahead of molecules 3 and 4. Thus, the 6 molecules, originally relatively close together, are now spread out in the stationary phase. This explanation is a little over-simplified, but gives a correct description of the mechanism of mass transfer dispersion.

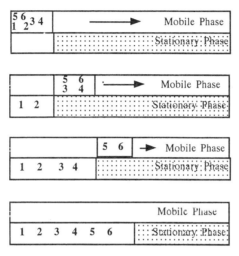

Figure 7. Resistance to Mass Transfer in the Mobile Phase

Resistance to Mass Transfer in the Stationary Phase

Dispersion caused by the resistance to mass transfer in the stationary phase is exactly analogous to that in the mobile phase. Solute molecules close to the surface will leave the stationary phase and enter the mobile phase before those that have diffused further into the stationary phase and have a longer distance to diffuse back to the surface. Thus, as those molecules that were close to the surface will be swept along in the moving phase, they will be dispersed from those molecules still diffusing to the surface. The dispersion resulting from the resistance to mass transfer in the stationary phase is depicted in Figure 8.

At the start, molecules 1 and 2, the two closest to the surface, will enter the mobile phase and begin moving along the column. This will continue while molecules 3 and 4 diffuse to the interface at which time they will enter the mobile phase and start following molecules 1 and 2. All four molecules will continue their journey while molecules 5 and 6 diffuse to the mobile phase/stationary phase interface. By the time molecules 5 and 6 enter the mobile phase, the other four molecules will have been smeared along the column and the original 6 molecules will have suffered dispersion.

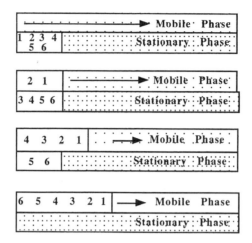

Figure 6. Resistance to Mass Transfer in the Stationary Phase

Quantitative Treatment of Resistance to Mass Transfer Dispersion

The two forms of resistance to mass transfer dispersion, *i.e.*, that in the mobile phase and that in the stationary phase, can be treated quantitatively using the random walk model. Recalling the basic principles of the Random Walk procedure, if each molecule of a group takes a series of steplike movements, which may be positive or negative, the direction being completely random, then after (p) steps have been taken, each step having an average length (s), the average of the molecules will have moved some distance from the starting position and will form a Gaussian type distribution curve with a variance of σ^2, where

$$\sigma = s\sqrt{p} \quad \text{or} \quad \sigma^2 = s^2 p$$

If it is considered, in the first instance, that the distribution is energy controlled and not diffusion controlled, a solute molecule will desorb from the stationary phase when it randomly has sufficient kinetic energy to break its association with a molecule of stationary phase, as discussed in chapter 1. Similarly, a molecule will be absorbed under the same conditions.

Now, if (k_d) is the desorption rate constant, then the mean desorption time (t_d) for the adsorbed molecule will be $\left(t_d = \dfrac{1}{k_d}\right)$. Similarly, if the adsorption rate constant is (k_a), then the mean adsorption time for a free molecule in the mobile phase will be $\left(t_a = \dfrac{1}{k_a}\right)$.

Mechanism of Dispersion

Consider a local concentration of solute migrating down a column. During this migration, adsorption and desorption steps will continuously and frequently occur. In addition, each occurrence will be a random event. Now a desorption step will be a random movement forward as it releases a molecule into the mobile phase, where it can move forward. Conversely, an adsorption step is a step backward, as it results in a period of immobility for the molecule while the rest of the zone moves forward. The total number of random steps taken as the solute mean position moves a distance (l) along the column is the number of forward steps plus the number of backward steps. Since the distribution is dynamic and is an equilibrium system, each desorption step must be followed by an adsorption step and, so, the total number of steps will be twice the number of adsorption steps that take place in the migration period.

On average, a molecule will remain in the mobile phase a time (t_a) before it is adsorbed. During this time, it will be moving at the mean velocity of the mobile phase (u) and will, thus, move a distance (ut_a). Thus, in moving a distance (l), the total number of adsorptions will be $\left(\dfrac{l}{ut_a}\right)$ and the total number of steps including the adsorption and desorption steps will be $\left(\dfrac{2l}{ut_a}\right)$.

Thus, in the Random Walk Model $p = \left(\dfrac{2l}{ut_a}\right)$, where $\sigma^2 = s^2 p$.

It is now necessary to determine the average step length (s) to obtain an expression for (H). The step length is that length moved by the molecule relative to that of the zone center and, while the molecule has move (vt_a) during time (t_a), the zone center has also moved. Now, it was shown in the Plate Theory that the zone velocity is $\left(\dfrac{u}{1+k''}\right)$ where (k'') is the dynamic capacity ratio of the solute. Thus,

$$s = ut_a - \left(\dfrac{ut_a}{1+k''}\right) = ut_a\left(1 - \dfrac{1}{1+k''}\right) = ut_a\left(\dfrac{k''}{1+k''}\right).$$

Consequently, $\sigma^2 = s^2 p = \left(ut_a\left(\dfrac{k''}{1+k''}\right)\right)^2 \left(\dfrac{2l}{ut_a}\right) = \left(\dfrac{2(k'')^2}{(1+k'')^2}\right)lut_a,$

i.e., $H_{RMT} = \dfrac{\sigma^2}{l} = \left(\dfrac{2(k'')^2}{(1+k'')^2}\right)ut_a$

In practice, it is more convenient to express (H) in terms of (t_d) as opposed to (t_a).

Now, the ratio of the mean phase residence times is the time the solute spends in the mobile phase divided by the time spent in the stationary phase. Thus,

$$\frac{t_a}{t_d} = \frac{\frac{1}{1+k''}}{1 - \frac{1}{1+k''}} = \frac{1}{k''}$$

and

$$H_{RMT} = \left(\frac{2(k'')^2}{(1+k'')^2}\right) u \frac{t_d}{k''} = \frac{2k''}{(1+k'')^2} u t_d. \quad (14)$$

Equation (14), although derived from the approximate random walk theory, is rigorously correct and applies to heterogeneous surfaces containing wide variations in properties and to perfectly uniform surfaces. It can also be used as the starting point for the random walk treatment of diffusion controlled mass transfer similar to that which takes place in the stationary phase in GC and LC columns.

Diffusion Controlled Dispersion in the Stationary Phase

The major difference between diffusion controlled dispersion and that resulting from adsorption and desorption is that the transfer process is concentration controlled. Reiterating equation (7),

$$\sigma^2 = 2D_m t$$

Thus, during solute transfer between the phases, (t) is now the average diffusion time (t_D) and (σ) is the mean distance through which the solute diffuses, *i.e.*, the depth or thickness of the film of stationary phase (d_f). Thus,

$$d_f^2 = 2D_S t_D \quad \text{or,} \quad t_D = \frac{d_f^2}{2D_S} \quad (15)$$

where (D_S) is the Diffusivity of the solute in the stationary phase.

Substituting (t_D) for (t_d) from (15) in (14), $H_{MTS} = \frac{q k''}{(1+k'')^2} \frac{d_f^2}{D_S} u \quad (16)$

where (q) is a configuration factor.

Mechanism of Dispersion

The value of (q) takes into account the precise shape of the pool of stationary phase; for a uniform liquid film as in a GC capillary column, q = 2/3. Diffusion in rod shaped and sphere shaped bodies (*e.g.*, *paper* chromatography and LC) gives q=1/2 and 2/15, respectively [2].

Thus, for a GC capillary column,

$$H_{MTS} = \frac{2k''}{3(1+k'')^2} \frac{d_f^2}{D_S} u \qquad (17)$$

and a close approximation for an LC or GC packed column (H_{MTS}) would be given by

$$H_{MTS} = \frac{2k''}{15(1+k'')^2} \frac{d_f^2}{D_S} u \qquad (18)$$

The validity of equation (18) for LC packed columns has also been experimentally demonstrated and will be discussed in a later chapter.

Diffusion Controlled Dispersion in the Mobile Phase

Dispersion in the mobile phase is again diffusion controlled and, so, again reiterating equation (7),

$$\sigma^2 = 2D_m t$$

During the movement of a solute molecule in the mobile phase of a chromatographic column, it must traverse from localities of high velocity (the center of a capillary column or the center of an inter-particle channel) to that of low velocity (the interface between the two phases at the capillary column walls or the surface of the particles in the packed column). The 'exchange time' between the two extreme velocities is designated as (t_m). Now, the distance between the extremes of velocity will depend on the geometry of the column system, but a general case will be assumed where a molecule must diffuse a distance ($\omega_m d_m$) to move from one velocity extreme to the other. Depending on whether the column is open tubular or packed, and depending on the homogeneity of the packing, the particle shape and size, the value of (ω_m) may range from considerably less than unity to significantly greater than unity. Now, from equation (7), ($\sigma = \omega_m d_m$) and ($t = t_m$). Thus,

$$\omega_m^2 d_m^2 = 2D_m t_m$$

or
$$t_m = \frac{\omega_m}{2D_m} d_p^2 \qquad (19)$$

Now, in the equation from the Random walk concept, $\sigma = s\sqrt{p}$ or $\sigma^2 = s^2 p$

In addition, $s = t_m u$ and $p = \left(\frac{1}{s}\right) = \left(\frac{1}{t_m u}\right)$, or

$$\sigma^2 = s^2 p = t_m^2 u^2 \frac{1}{t_m u} = t_m u l. \qquad (20)$$

Substituting for (t_m) from (19) in (20),

$$\sigma^2 = \frac{\omega_m}{2D_m} d_p^2 u l$$

Now,
$$\frac{\sigma^2}{l} = H_{MTM}$$

Thus,
$$H_{MTM} = \frac{\sigma^2}{l} = \frac{\omega_m}{2D_m} d_p^2 u \qquad (21)$$

Equation (21) applies to all types of columns each requiring a different constant (ω_m), This constant is partly determined by the geometry of the distribution system and partly by the capacity ratio of the specific solute being eluted. There has not been a general value for (ω_m) developed for packed LC or CC columns but due to the geometric simplicity of the capillary column Golay [4] developed the function

$$\omega_m = \frac{1 + 6k'' + 11k''^2}{24(1 + k''^2)^2}$$

and thus for a capillary column,

$$H_{MTM} = \frac{1 + 6k'' + 11k''^2}{24(1 + k''^2)^2} \frac{r^2}{2D_m} u \qquad (22)$$

The development of the function describing (t_m) for a capillary column is similar to that for the packed column but (r), the column radius, replaces (d_p), the particle diameter.

Due to the varying physical nature of the different packings, it appears that no one has developed a specific function for (ω_m) for a packed column, but it was suggested by Klinkenberg and Purnell that the function $\omega_m = \dfrac{1+6k''+11k''^2}{24(1+k'')^2}$ could also be used for packed GC and LC columns in a general expression for the resistance to mass transfer in the mobile phase contribution to the overall variance per unit length. In fact, this function has often been used in the literature. However, the flow patterns in the packed column can be very complex and the eddies and pseudo-turbulence that appears to take place between the particles greatly increases the effective diffusivity of the solute in the mobile phase. In fact, the magnitude of equation (21) can be so reduced by the resulting large increase in (D_m) that the overall contribution to the peak dispersion by resistance to mass transfer in the mobile phase can become small enough to be ignored.

Fudge Factors

The concept of the "fudge factor" was introduced by Golay to describe such constants as (λ), (γ), (ω) and (q) used by Van Deemter, Giddings and others in the derivation of the rate equations for GC and LC packed columns. Although in some ways unkind to those who evoked such factors, the inference associated with the term fudge factor has some validity. Nevertheless, Golay had the great advantage of working with the geometrically simple open tube, coated with a cylindrical film of stationary phase, which allowed him to derive, in a relatively simple manner, explicit equations that avoided the need for "fudge factors". Van Deemter and his colleagues, however, had to take into account the widely variable characteristics of the column packing. Not only did the physical form of the packing vary greatly from column to column, but the shapes of the particles could range from completely irregular to almost perfectly spherical. In addition, the diameter of the particles was never precise and the distribution of diameters could have a standard deviation that was as great as 50% of the mean. Furthermore, the pore volume, pore size and surface area of the adsorbents and bonded phases also vary widely which, together, can give the fudge factors a wide range of plausible values.

Giddings made a stalwart effort to provide values for the different constants that would apply to diverse stationary phase and support conditions [2]. However, at best, his values are the closest estimates from an assumed set of conditions that may fit, to a greater or lesser extent, the properties of the actual stationary phase or support in use. In some cases, his constants may be used in column design and to help in the choice of those operating conditions that will provide the required resolution and the minimum analysis time. However, if accuracy is essential, then the constants should be obtained *experimentally* in the manner to be described in a later chapter.

Ultimately,
the truth will only be found on the laboratory bench.

In summary, the rate theory provides the following equations for the variance per unit length (H) for four different columns.

1. The Open Tubular GC Column

$$H = \frac{2D_m}{u} + \frac{1+6k''+11k''^2}{24(1+k'')^2} \frac{r^2}{D_m} u + \frac{2k''}{3(1+k'')^2} \frac{d_f^2}{D_s} u \tag{23}$$

2. The Packed GC Column

$$H = 2\lambda d_p + \frac{2\gamma D_m}{u} + \frac{1+6k''+11k''^2}{24(1+k'')^2} \frac{d_p^2}{D_m} u + \frac{2k''}{3(1+k'')^2} \frac{d_f^2}{D_s} u \tag{24}$$

3. The Packed LC Column

$$H = 2\lambda d_p + \frac{2\gamma D_m}{u} + \omega \frac{d_p^2}{D_m} u + \frac{2k''}{3(1+k'')^2} \frac{d_f^2}{D_s} u \tag{25}$$

4. The Open Tubular LC Column

$$H = 2\lambda d_p + \frac{2\gamma_1 D_m}{u} + \frac{2\gamma_2 k'' D_s}{u} + \omega \frac{d_p^2}{D_m} u + \frac{2k''}{3(1+k'')^2} \frac{d_f^2}{D_s} u \tag{26}$$

It should be noted that all the equations assume that the mobile phase is incompressible which will not be true for equations (23) and (24). It follows that equations (23) and (24) will require modification in order to be applicable to practical situations. It will also be shown in a later chapter that, from experimental data, (ω) approaches zero in LC and the resistance to mass transfer in the *moving* phase (as opposed to the *mobile* phase) does not contribute significantly to the overall dispersion of the peak in liquid chromatography.

Synopsis

Equations that quantitatively describe peak dispersion are derived from the rate theory. The equations relate the variance per unit length of the solute concentration

Mechanism of Dispersion

profile to the physical and chromatographic properties of the solute and phase system, and the linear velocity of the mobile phase. There are two mathematical techniques used in the derivation of the pertinent equations, one based on the random walk model and the other the basic diffusion equation that evolves from Fick's law. In the random walk model, it is assumed that if each molecule of a group takes a series of steplike movements, which may be positive or negative, the direction being completely random; then, after (p) steps have been taken, each step having a mean length (s), the average of the molecules will have moved some distance from the starting position and will form a Gaussian type distribution. According to the random walk model, the following simple relationship holds, $\sigma = s\sqrt{p}$, where variance of the Gaussian curve is σ^2. Employing Fick's law, the following differential equation, $\frac{dc}{dt} = D_m \frac{d^2c}{dx^2}$, can be derived that describes the form of dispersion that arises from a random diffusion process. The solution of the differential equation will be a Gaussian function and, more important from the point of view of the Rate Theory, the Gaussian curve will have a variance given by $2D_m t$. Using the random walk model, the contribution from multi path dispersion to the overall variance per unit length of the column can be derived for a GC and an LC packed chromatographic column. Using the diffusion equation alone, equations for the contribution of longitudinal diffusion dispersion in both packed and capillary columns, and in both the mobile phase and the stationary phase, to the total variance per unit length of the column can also be derived. Finally, employing both the random walk model and the equation for diffusion, the dispersion contribution from the resistance to mass transfer in both phases and in packed and capillary columns to the overall column variance can be obtained. The simple HETP equations that are given assume that the mobile phase is incompressible and so will not be valid for those that employ a gas as the mobile phase. To render the equations applicable to compressible mobile phases they will require further development.

References

1. W. Feller, *"Probability Theory and Its Applications"*, John Wiley and Sons, New York (1961)Chapter 3.
2. J. C. Giddings, *"Dynamics of Chromatography"*, Marcel Dekker, New York (1965)Chapter 2.
3. D. S. Horne, J. H. Knox and E. McLaren, *"Separation Techniques in Chemistry and Biochemistry "*, (ed.R. A. Keller), Marcel Dekker, New York.

4. M. J. E. Golay, in "*Gas Chromatography 1958* "(ed. D. H. Desty), Butterworths, London (1958)36.

Chapter 8

The Rate Theory Equations

The Van Deemter equation was confidently used to describe the peak dispersion that took place in a packed column until about 1961 when, by the use of small particles and high pressures, very high efficiency LC columns were produced. As a result of the very narrow peaks produced by these columns, it was found that, when the Van Deemter equation was tested against experimental data obtained at high linear mobile phase velocities, very poor agreement was realized. This poor agreement between theory and experiment was eventually shown to be due to the presence of experimental artifacts arising from extra-column dispersion generated in the detector sensor, detector electronics, sample valve and connecting tubes. At the time, however, the importance of extra-column dispersion was not appreciated and certainly not fully understood. As a consequence, the apparent failure of the Van Deemter equation provoked the development of alternative HETP equations in the hope that a more exact relationship between HETP and linear mobile phase velocity could be obtained that would agree well with experimental data. As it turned out, much of this work was futile as, when appropriate precautions were taken to eliminate extra-column dispersion, it was found (and the data will be discussed in a later chapter) that the Van Deemter equation described dispersion in packed columns very accurately.

The Giddings Equation

The first alternative HETP equation to be developed was that of Giddings in 1961 [1] of which the Van Deemter equation appeared to be a special case. Giddings did not develop his equation because the Van Deemter equation did not fit experimental data, but because he was dissatisfied that the equation predicted a finite contribution to dispersion at the limit of zero mobile phase velocity. This concept, not surprisingly, appeared to him unreasonable and unacceptable. As a consequence, Giddings developed an equation of the following form to avoid this irregularity.

$$H = \frac{A}{1+\frac{E}{u}} + \frac{B}{u} + Cu \qquad (1)$$

It is seen that when $u \gg E$, equation (1) reduces to the Van Deemter equation,

$$H = A + \frac{B}{u} + Cu$$

It is also seen that, at very low velocities, where $u \ll E$, the first term tends to zero, thus meeting the logical requirement that there is no multipath dispersion at zero mobile phase velocity. Giddings also introduced a coupling term that accounted for an increase in the 'effective diffusion' of the solute between the particles. The increased 'diffusion' has already been discussed and it was suggested that a form of microscopic turbulence induced rapid solute transfer in the interparticulate spaces. When $u \gg E$, this interstitial mixing effect was considered complete, and the resistance to mass transfer in the mobile phase between the particles becomes very small and the equation again reduces to the Van Deemter equation. However, under these circumstances, the C term in the Van Deemter equation now only describes the resistance to mass transfer in the *mobile phase contained in the pores* of the particles and, thus, would constitute an additional *resistance to mass transfer in the stationary (static mobile) phase*. It will be shown later that there is experimental evidence to support this. It is possible, and likely, that this was the rationale that explains why Van Deemter *et al.* did not include a resistance to mass transfer term for the *mobile phase* in their original form of the equation.

The Huber Equation

In 1967, Huber and Hulsman [2] introduced yet another HETP equation having a very similar form to that of Giddings. Their equation included a modified multipath term somewhat similar in form to that of Giddings and a separate term describing the resistance to mass transfer in the mobile phase contained between the particles. The form of their equation was as follows:

Rate Theory Equations

$$H = \frac{A}{1+\dfrac{E}{u^{1/2}}} + \frac{B}{u} + Cu + Du^{1/2} \qquad (2)$$

It is seen that the first term differs from the Giddings equation and now contains the mobile phase velocity to the power of one-half. However, when $u^{1/2} \gg E$, the first term reduces to a constant similar to the Van Deemter equation. The additional term for the resistance to mass transfer in the mobile phase is an attempt to take into account the 'turbulent mixing' that takes place between the particles. Huber's equation, although not explicitly stated by the authors, implies that the mixing effect between the particles (that reduces the magnitude of the resistance to mass transfer in the mobile phase) only starts when the mobile phase velocity approaches the optimum velocity (as defined by the Van Deemter equation). In addition, the mixing effect is not complete until the mobile phase velocity is well above the optimum velocity. Thus, the shape of the HETP/u curve will be different from that predicted by the Van Deemter equation. The form of the HETP curve that is produced by the Huber equation is shown in Figure (1).

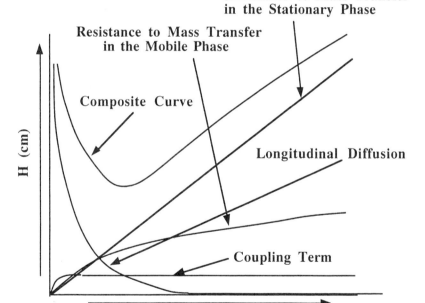

Figure 1. H Versus u Curves from the Huber Equation

The composite curve from the Huber equation is similar to that obtained from that of Van Deemter but the individual contributions to the overall variance are different. The contributions from the resistance to mass transfer in the stationary phase and longitudinal diffusion are common to both equations; the (A) term from the Huber equation increases with mobile phase flow rate and only becomes a constant value (similar to the multipath term in the Van Deemter equation) when the mobile velocity is greater than 0.05 cm/sec. The magnitude of the mobile velocity, at which the (A) term becomes a constant value is significantly below the normal range of operating velocities. That portion of the composite curve at higher velocities is not quite linear due to the nonlinear relationship for the resistance to mass transfer in the mobile phase. This becomes more obvious at the higher mobile phase velocities. At and around the optimum velocity, however, the forms of the Huber and Van Deemter curves are very similar.

The Knox Equation

In 1972-1973 Knox *et al.* [3, 4, 5] examined, in considerable detail, a number of different packing materials with particular reference to the effect of particle size on the *reduced plate height* of a column. The *reduced plate height (h)* and *reduced velocity* (v) were introduced by Giddings [6,7] in 1965 in an attempt to form a rational basis for the comparison of different columns packed with particles of different diameter. The reduced plate height is the normal plate height measured in units of particle diameter and is defined by the following expression,

$$h = \frac{H}{d_p} \qquad (3)$$

It is seen that the reduced plate height is dimensionless.

The reduced velocity is defined as

$$v = \frac{u d_p}{D_m} \qquad (4)$$

The reduced velocity compares the mobile phase velocity with the velocity of the solute diffusion through the pores of the particle. In fact, the mobile phase velocity is measured in units of the intraparticle diffusion velocity. As the reduced velocity is a ratio of velocities then, like the reduced plate height, it also is dimensionless. Employing the reduced parameters, the equation of Knox takes the following form

$$h = \frac{B}{v} + Av^{1/3} + Cv \qquad (5)$$

The equation of Knox *et al.* was not derived theoretically from a basic dispersion model, but the constants of the equation were determined by a curve fitting procedure

Rate Theory Equations

to experimental data and, as a consequence, the Knox equation has limited use in column design. It is, however, extremely valuable in assessing the quality of the packing. The curves resulting from the Knox equation are shown in Figure 2.

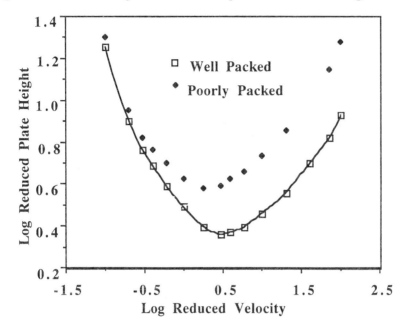

Figure 2. Graph of Log of Reduced Plate Height against Log of Reduced Velocity for Poorly and Well Packed Columns

The curves represent a plot of log (h) (reduced plate height) against log (v) (reduced velocity) for two very different columns. The lower the curve, the better the column is packed (the lower the minimum reduced plate height). At low velocities, the (B) term (longitudinal diffusion) dominates, and at high velocities the (C) term (resistance to mass transfer in the stationary phase) dominates, as in the Van Deemter equation. The best column efficiency is achieved when the minimum is about 2 particle diameters and thus, log (h) is about 0.35. The optimum reduced velocity is in the range of 3 to 5 cm/sec., that is log (v) takes values between 0.3 and 0.5. The Knox equation is a simple and effective method of examining the quality of a given column but has limited value in column design due to the empirical nature of the constants.

The Horvath and Lin Equation

In (1976) Horvath and Lin [8,9] introduced yet another equation to describe the value of (H) as a function of the linear mobile phase velocity (u). Again, it would appear largely to account for extra-column dispersion which, mistakenly, was thought to originate inside the column.

The equation they derived is as follows,

$$H = \frac{A}{1+\frac{E}{u^{1/3}}} + \frac{B}{u} + Cu + Du^{2/3} \qquad (6)$$

Horvath and Lin's equation is very similar to that of Huber and Hulsman, only differing in the magnitude of the power function of (u) in their (A) and (D) terms. These workers were also trying to address the problem of a zero (A) term at zero velocity and the fact that some form of 'turbulence' between particles aided in the solute transfer across the voids between the particles.

The Golay Equation

The Golay equation [9] for open tubular columns has been discussed in the previous chapter. It differs from the other equations by the absence of a multi-path term that can only be present in packed columns. The Golay equation can also be used to examine the dispersion that takes place in connecting tubes, detector cells and other sources of extra-column dispersion. Extra-column dispersion will be considered in another chapter but the use of the Golay equation for this purpose will be briefly considered here. Reiterating the Golay equation from the previous chapter,

$$H = \frac{2D_m}{u} + \frac{1+6k''+11k''^2}{24(1+k'')^2}\frac{r^2}{D_m}u + \frac{2k''}{3(1+k'')^2}\frac{d_f^2}{D_s}u \qquad (7)$$

where the symbols have the meanings previously defined in chapter 7.

If the solute is unretained (*i.e.*, k''=0), then the Golay equation reduces to

$$H = \frac{2D_m}{u} + \frac{1}{24}\frac{r^2}{D_m}u \qquad (8)$$

Taking a value of 2.5×10^{-5} for D_m (*e.g.*, the diffusivity of benzyl acetate in *n*-heptane), equation (8) can be employed to calculate the curve relating (H) and (u) for an uncoated capillary tube. The results are shown in Figure 3.

Rate Theory Equations

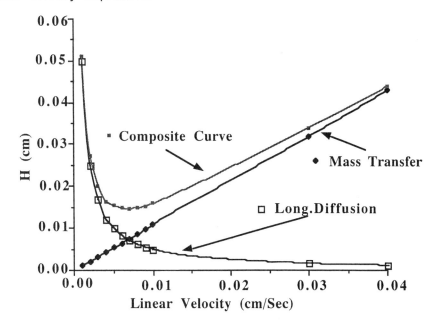

Figure 3. Capillary Column Curves Relating H against u for an Uncoated Column (Unretained Peak)

It is seen that the value of (H) is completely dependent on the diffusivity of the solute in the mobile phase, the column radius and the linear velocity of the mobile phase. The simple uncoated open tube can clearly be used to determine the diffusivity of any solute in any given solvent (the mobile phase). This technique for measuring diffusivities will be discussed in a later chapter.

Effect of Mobile Phase Compressibility On the HETP Equation for a Packed GC Column

In the treatment of peak dispersion thus far, the effect of mobile phase compressibility has been ignored. As the pressure falls along the column length, the velocity changes and, as the solute diffusivity depends on the pressure, the diffusivity of the solute will also change. Obviously, the multi-path term, which contains no parameters that depend on the velocity or gas pressure, will be unaffected and the expression that describes it will remain the same. It should be mentioned, however, that the coupling term, described by Giddings, does contain a function that includes solute diffusivity. Nevertheless, this merely changes slightly the magnitude of the mobile phase velocity at which the function becomes constant and equal to the multi-path term. In practice, separations are carried out at velocities well above that where the function becomes constant and, so, the multipath term can be considered independent of mobile phase pressure. The other terms, however, all contain parameters that are affected by gas pressure (solute diffusivity and mobile phase velocity) and, therefore, the manner in which they change with pressure needs to be examined. Reiterating the HETP equation for a packed column,

$$H = 2\lambda d_p + \frac{2\gamma D_m}{u} + \frac{f_1(k'')d_p^2}{D_m}u + \frac{f_2(k'')d_f^2}{D_S}u$$

where $f_1(k'') = \dfrac{1+6k''+11k''^2}{24(1+k'')^2}$ and $f_2(k'') = \dfrac{2k''}{3(1+k'')^2}$

It is clear that the expression must be modified to accommodate pressure and velocity changes; *i.e.*, at a point (x) along the column,

$$H_x = 2\lambda d_p + \frac{2\gamma D_{m(x)}}{u_x} + \frac{f_1(k'')d_p^2}{D_{m(x)}}u_x + \frac{f_2(k'')d_f^2}{D_S}u_x$$

It is seen that the HETP equation, as derived by Van Deemter, now applies only to a point distance (x) from the inlet of the column. Now, it has already been shown that

$$P_x u_x = P_o u_o \quad \text{or} \quad u_x = \frac{u_o P_o}{P_x}$$

where (P_x) is the pressure at point (x) along the column,
(u_x) is the linear velocity of the mobile phase at point (x),
(P_o) is the pressure at the column exit,
and (u_o) is the linear velocity of the mobile phase at the column exit.

and from the kinetic theory of gases, it is known that the diffusivity of a solute in a gas is inversely proportional to the pressure. Thus,

$$D_x P_x = D_o P_o \quad \text{or} \quad D_x = \frac{D_o P_o}{P_x}$$

where (D_o) is the solute diffusivity at the end of the column at (P_o)
and (D_x) is the solute diffusivity at point (x) and pressure (P_x)

Thus, $\quad \dfrac{u_x}{D_x} = \dfrac{u_o P_o}{P_x} \dfrac{P_x}{D_o P_o} = \dfrac{u_o}{D_o} \quad$ and $\quad \dfrac{D_x}{u_x} = \dfrac{D_o}{u_o}$

Rate Theory Equations

Substituting for $\dfrac{u_x}{D_x}$ and $\dfrac{D_x}{u_x}$ in the HETP equation,

$$H_x = 2\lambda d_p + \frac{2\gamma D_{m(o)}}{u_o} + \frac{f_1(k'')d_p^2}{D_{m(o)}}u_o + \frac{f_2(k'')d_f^2 P_0}{D_S P_x}u_o$$

It is now seen that *only* the resistance to the mass transfer term for the stationary phase is position dependent. All the other terms can be used as developed by Van Deemter, providing the diffusivities are measured at the outlet pressure (atmospheric) and the velocity is that measured at the column exit.

The resistance to the mass transfer term for the stationary phase will now be considered in isolation. The experimentally observed plate height (variance per unit length) resulting from a particular dispersion process [*e.g.*, (h_S), the resistance to mass transfer in the stationary phase] will be the sum of all the local plate height contributions (h'); *i.e.*,

$$h_s = \frac{1}{L}\int_0^L h'\, dx$$

Substituting for (h') the expression for the resistance to mass transfer in the stationary phase,

$$h_s = \frac{1}{L}\int_0^L \frac{f_2(k'')d_f^2 P_0}{D_S P_x} u_o\, dx$$

or
$$h_s = \frac{f_2(k'')d_f^2 P_0}{L D_S} u_o \int_0^L \frac{dx}{P_x} \qquad (9)$$

Now, reiterating equation (21) from Chapter 3,

$$\frac{P_x}{P_o} = \left[\gamma^2 - (\gamma^2-1)\frac{x}{L}\right]^{0.5} \quad \text{or} \quad P_x = P_o\left[\gamma^2 - (\gamma^2-1)\frac{x}{L}\right]^{0.5} \qquad (10)$$

Substituting for (P_x) from equation (10) in equation (9),

$$h_s = \frac{f_2(k'')d_f^2 P_0}{L D_S} u_o \int_0^L \frac{dx}{P_o\left(\gamma^2 - \frac{x}{L}(\gamma^2-1)\right)^{0.5}} \qquad (11)$$

Let
$$w = \gamma^2 - \frac{x}{L}(\gamma^2-1)$$

Then
$$dw = \frac{L\,dx}{-(\gamma^2-1)} \quad \text{or} \quad dx = \frac{-(\gamma^2-1)}{L}dw$$

Furthermore, when x=0, then $w=\gamma^2$ and when $w = L$, then $w=1$.

Substituting for $\left(\gamma^2 - (\gamma^2-1)\frac{x}{L}\right)$ and (dx) in equation (11) and inserting the new limits and integrating,

$$h_s = \frac{f_2(k'')\,d_f^2 P_o}{LD_S} u_o \int_{\gamma^2}^{1} \frac{L w^{-0.5}\,dx}{-P_o((\gamma^2-1))}$$

or
$$h_s = \frac{f_2(k'')\,d_f^2}{D_S} u_o \int_{\gamma^2}^{1} \frac{w^{-0.5}\,dw}{-((\gamma^2-1))} = \frac{f_2(k'')\,d_f^2}{D_S} u_o \left[\frac{2w^{0.5}}{-((\gamma^2-1))}\right]_{\gamma^2}^{1}$$

$$h_s = \frac{2f_2(k'')\,d_f^2}{D_S} u_o \left[\frac{1-\gamma}{-((\gamma^2-1))^{0.5}}\right] = \frac{2f_2(k'')\,d_f^2}{D_S} u_o \left[\frac{\gamma-1}{((\gamma^2-1))}\right]$$

Thus,
$$h_s = \frac{2f_2(k'')\,d_f^2}{D_S(\gamma+1)} u_o \tag{12}$$

Thus, the complete HETP equation for a packed GC column that takes into account the compressibility of the carrier gas will be

$$H = 2\lambda d_p + \frac{2\gamma D_{m(o)}}{u_o} + \frac{f_1(k'')d_p^2}{D_{m(o)}} u_o + 2\frac{f_2(k'')\,d_f^2}{D_S(\gamma+1)} u_o \tag{13}$$

The inlet-outlet pressure ratio (γ) can be eliminated by integrating equation (20) from chapter 2,

Reiterating equation (11)
$$u = -\frac{K}{\eta}\frac{dp}{dx}$$

or
$$u_x P_x = -\frac{K P_x}{\eta}\frac{dp}{dx} = u_0 P_o$$

Thus,
$$u_0 P_o dx = \int -\frac{K P_x}{\eta} dp$$

Rate Theory Equations

Integrating from x=0 to x = L and $P_x = P_i$ to P_o,

$$u_0 P_o L = \frac{K}{\eta}\left(P_i^2 - P_o^2\right)$$

or

$$u_0 = \frac{KP_o}{\eta L}\left(\gamma^2 - 1\right)$$

Solving for (γ),

$$\gamma = \left(\frac{u_0 \eta L}{KP_o} + 1\right)^{0.5} \qquad (14)$$

Substituting for (γ) in equation (12) from equation (14),

$$h_s = 2\frac{f_2(k'')d_f^2}{D_s\left(\left(\frac{u_0 \eta L}{KP_o}+1\right)^{0.5}+1\right)}u_o$$

Thus the HETP equation given by equation (13) becomes

$$H = 2\lambda d_p + \frac{2\gamma D_{m(o)}}{u_o} + \frac{f_1(k'')d_p^2}{D_{m(o)}}u_o + 2\frac{f_2(k'')d_f^2}{D_s\left(\left(\frac{u_0 \eta L}{KP_o}+1\right)^{0.5}+1\right)}u_o \qquad (15)$$

Equation (15) gives the variance per unit length of a GC column in terms of the outlet pressure (atmospheric); the outlet velocity; and physical and physicochemical properties of the column, packing, and phases and is independent of the inlet pressure. However, equation (13) is the recommended form for HETP measurements as the inlet pressure of a column is usually known, and is the less complex form of the HETP expression.

It is seen that equations (13) and (15) are very similar to equation (10) except that the velocity used is the *outlet* velocity and *not* the *average* velocity and that the diffusivity of the solute in the gas phase is taken as that measured at the outlet pressure of the column (atmospheric). It is also seen from equation (14) that the resistance to mass transfer in the stationary phase is now a function of the inlet-outlet pressure ratio (γ).

Lamentably, the *average* velocity is almost universally used in constructing HETP curves in both GC and LC largely because it is simple to calculate from the ratio of the column length to the dead time. Unfortunately, it provides very erroneous data and, for accurate column evaluation and column design, the *exit velocity* must be employed together with the inlet-outlet pressure ratio. An example of the kind of errors that can occur when using the *average* velocity as opposed to the *exit* velocity is demonstrated in Figure 4 from data obtained for a capillary column [10].

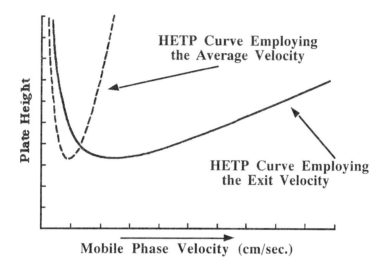

Figure 4. HETP Curves for the Same Column and Solute Using the Average Mobile Phase Velocity and the Exit Velocity

It is seen that the two curves are quite different and, if the results are fitted to the HETP equation, only the data obtained by using the *exit velocity* gives correct and realistic values for the individual dispersion processes. This point is emphasized by the graphs shown in Figure 5 where the HETP curve obtained by using *average velocity* data are deconvoluted into the individual contributions from the different dispersion processes.

Figure 5 shows that using *average velocity* data the extracted value for the multi-path term is *negative*, which is physically impossible, and, furthermore, for a capillary column should be zero or close to zero. In contrast, the extracted values for the different dispersion processes obtained from data involving the *exit velocity* give small positive, but realistic values for the multi-path term.

It is important to appreciate that, in all aspects of column evaluation and column design in GC, the compressibility of the mobile phase must be taken into account or serious errors will be incurred. Either equation (13) or (15) can be employed but, as already stated, equation (13) is recommended as the more simple to use.

Rate Theory Equations

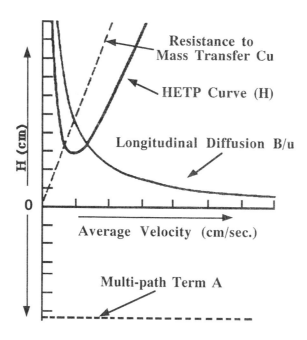

Figure 5. De-convolution of the HETP Curve Obtained Using the *Average* Mobile Phase Velocity

Mobile Phase Compressibility: Its Effect on the Interpretation of Chromatographic Data in LC

Liquids have relatively low compressibility compared with gases and, thus, the mobile phase velocity is sensibly constant throughout the column. As a consequence, elution volumes measured at the column *exit* can be used to obtain *retention volume* data and, unless extreme accuracy is required for special applications, there is no need for the retention volume to be corrected for pressure effects.

The compressibilities of solvents vary significantly from one solvent to another. The compressibility of cyclohexane is about 0.67% per thousand p.s.i. change in pressure [11] and, thus, for a column operated at 6,000 p.s.i. (mean pressure 3,000 p.s.i.), there will be an error in retention volume measurement of about 2%. In a similar manner, *n*-heptane has a compressibility of about 1.0% per 1,000 p.s.i. change in pressure [11] and, under similar circumstances, would give an error of about 3% in retention volume measurement. Fortunately, as already discussed in Part 1 of this book, there are other retention parameters that can be used for solute identification which are independent of column pressure and will not be susceptible to such errors.

Nevertheless, high pressures produce secondary effects in LC columns that can be even more serious than the effects of pressure alone. Such secondary effects can cause errors in all types of retention measurements and, unfortunately, are not easily circumvented. The energy used in forcing the mobile phase through the packed bed is evolved as heat throughout the whole length of the column. However, as the thermal conductivity of the packed bed is very low, most of the heat is carried away by the mobile phase. As a consequence, there is a negative temperature gradient from the center of the column to the column wall and a positive temperature gradient from the column inlet to the column outlet. This thermal effect was first reported by Halasz *et al.* [12], Lin and Horvath [13] and Poppe *et al.* [14] and temperature differences of 5°C or more were noted between the column inlet and outlet. Katz *et al.* [15] measured the rise in temperature of the eluent from a microbore column (1 mm I.D., 156 cm long) that was thermostatted with water. The results they obtained are shown in Figure 6.

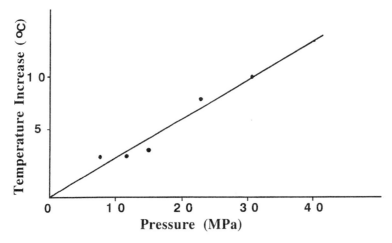

Figure 6. Graph of Change in Column Eluent Temperature against Inlet Pressure

It is seen that a very significant temperature increase occurs particularly at high pressures. It would appear that, due to the poor radial thermal transfer in the column, the thermostat had little effect on the temperature change. Katz *et al.* [15] concluded that increases in column temperature resulting from high inlet pressures could seriously affect the accuracy of retention measurements. The heat evolved on changing the flow rate from 4 to 18 ml/min., accompanied by a corresponding

Rate Theory Equations

pressure change from 5 to about 30 MPa, produced a rise of 18°C in effective column temperature. In the example taken, this temperature change furnished a corresponding reduction in retention volume or capacity ratio of about 6%. This effect can be particularly serious for short columns packed with small particles and operated at high flow rates to provide fast separations. For accurate retention measurements in LC, narrow bore columns should be used and operated at low flow rates (minimum pressure) and thermostatted with high thermal capacity liquids.

Effect of Mobile Phase Compressibility on the HETP Equation for a Packed LC Column

As the compressibility of liquids is small, the Van Deemter equation can be fitted to (H) versus (u) data using *average* velocity values. From such data, the multi-path dispersion, longitudinal diffusion and other dispersion processes can be accurately and reliably evaluated. However, solute diffusivity in liquids *is* significantly pressure dependent, the diffusivity falling as the pressure rises, and this places some limitations on the use of published diffusivity values in the Van Deemter equation for column design. In such calculations, the solute *diffusivity* at the *mean column pressure* must be employed and, unfortunately, most published data are determined at, or usually close to, atmospheric pressure. Regrettably, the pressure coefficients of solute diffusivity are rarely reported. The situation is further complicated by the heat generated in the column which tends to increase the solute diffusivity and, thus, partially compensate for the opposing effect of the pressure. In practice, when using columns 3 to 5 mm in diameter, the two effects tend to balance and, even without efficient thermostatting, there is usually little change in solute diffusivity [15]. Well thermostatted, small-diameter columns, however, will dissipate the heat more efficiently and, consequently, although retention measurements may then remain unaffected by pressure, the column efficiency will fall in the expected manner. This will be directly due to the decrease in solute diffusivity as there will be no compensating temperature effect.

Atwood and Goldstein [16] examined the effect of pressure on solute diffusivity and an example of some of their results is shown in Figure 7. It is seen that the diffusivity of the solutes appears to fall linearly with inlet pressure up to 40 MPa and the slopes of all the curves appear to be closely similar. This might mean that, in column design, diffusivities measured or calculated at atmospheric pressure might be used after they have been appropriately corrected for pressure using correction factors obtained from results such as those reported by Atwood and Goldstein [16] . It is also seen that the change in diffusivity with pressure is significant but not excessive. Nevertheless, the effect should probably be taken into account in column design.

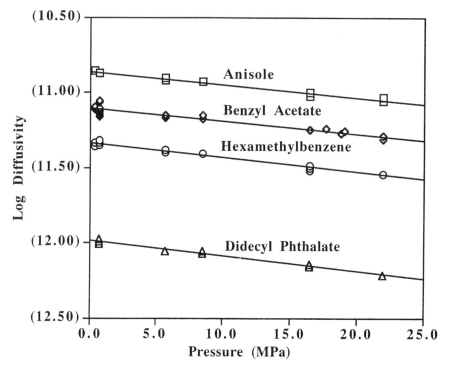

Figure 7. Graph of Solute Diffusivity against Pressure

Extensions of the HETP Equation

The HETP equation is not simply a mathematical concept of little practical use, but a tool by which the function of the column can be understood, the best operating conditions deduced and, if required, the optimum column to give the minimum analysis time calculated. Assuming that appropriate values of (u) and (D_m) and (D_s) are employed, equation (13) can be put into a simpler form which will be applicable to either GC or L C:

$$H = A + \frac{B}{u_o} + (C_m + C_s)u_o \qquad (16)$$

where $A = 2\lambda d_p$ $B = 2\gamma D_{m(o)}$ $C_m = \dfrac{f_1(k'')d_p^2}{D_{m(o)}}$ and $C_s = \dfrac{f_2(k'')d_f^2}{(\gamma+1)D_s}$

Equation (16) describes the 'HETP curve', or the curve that relates the variance per unit length or HETP of a column to the mobile phase linear velocity. A typical curve is shown in Figure 8.

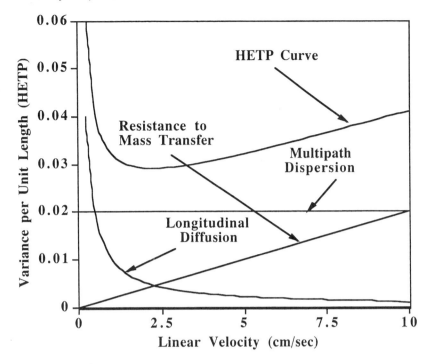

Figure 8. Graph of Variance per Unit Length against Linear Mobile Phase Velocity

It is seen that by a simple curve fitting process, the individual contributions to the total variance per unit length can be easily extracted. It is also seen that there is minimum value for the HETP at a particular velocity. Thus, the maximum number of theoretical plates obtainable from a given column (the maximum efficiency) can only be obtained by operating at the optimum mobile phase velocity.

Reiterating equation (16),

$$H = A + \frac{B}{u_o} + Cu_o \qquad (17)$$

where

$$C = C_m + C_s.$$

Differentiating (17),

$$\frac{dH}{dU} = -\frac{B}{u_o^2} + C$$

Equating to zero and solving for (u_{opt}),

$$u_{o(opt)} = \sqrt{\frac{B}{C}} \qquad (18)$$

Substituting for $(u_{o(opt)})$ from equation (18) in equation (17) gives a value for (H_{min}),

$$H_{min} = A + 2\sqrt{BC} \qquad (19)$$

Substituting for A, B and C in equations (24) and (25) expressions for (u_{opt}) and (H_{min}) can be obtained. Thus,

$$u_{opt} = \sqrt{\frac{2\gamma D_{m(o)}}{\dfrac{f_1(k'')d_p^2}{D_{m(o)}} + \dfrac{f_2(k'')d_f^2}{D_S(\gamma+1)}}} \qquad (20)$$

and

$$H_{min} = 2\lambda d_p + 2\sqrt{2\gamma D_{m(o)}\left(\frac{f_1(k'')d_p^2}{D_{m(o)}} + \frac{f_2(k'')d_f^2}{D_S(\gamma+1)}\right)} \qquad (21)$$

It is a common procedure to assume certain conditions for the chromatographic system and operating conditions and, as a result, simplify equations (20) and (21). However, in many cases the assumptions can easily be over-optimistic, to say the least. It is necessary, therefore, to carefully consider the conditions that may allow such simplifying procedures and take steps to ensure that such conditions are carefully met when such expressions are used in practice. Now, the relative magnitudes of the resistance to mass transfer terms will vary with the type of columns (packed or capillary), the type of chromatography (GC or LC) and the type of particle, *i.e.*, porous or microporous (diatomaceous support or silica gel).

The ratio of the resistance to mass transfer in the mobile phase to that in the stationary phase (R_{MS}) will indicate whether the expressions can be simplified or not. Now, (R_{MS}) will be given by,

$$R_{MS} = \frac{f_1(k'')d_p^2}{D_{m(o)}} \frac{D_S(\gamma+1)}{f_2(k'')d_f^2} = \frac{f_1(k'')(\gamma+1)}{f_2(k'')} \frac{D_S d_p^2}{D_{m(o)}d_f^2}$$

The ratio $\dfrac{f_1(k'')(\gamma+1)}{f_2(k')}$ will range within one order of magnitude and so any simplification of the expressions will largely depend on the magnitude of $\dfrac{D_S d_p^2}{D_{m(o)}d_f^2}$.

Packed GC Columns

Now, consider a packed GC column; due to the turbulence resulting from the rapid changes in direction of the gas velocity as it winds its tortuous path between the

Rate Theory Equations

particles of packing, the actual diffusivity of the solute between the particles is greatly increased, which significantly reduces the resistance to mass transfer in the gas between the particles. The mobile phase within the particles, however, is static and, thus, the solute diffusivity is not enhanced. Consequently, there will remain a resistance to mass transfer in the mobile phase within the particle but, as the mobile phase is not moving, it will arise as a resistance to mass transfer in a *stationary phase*, *i.e.*,

$$f_1(k'') = f_2(k'')$$

The diffusivity of the solute in the gas ($D_{m(o)}$) will, on average, be about 0.15 cm^2/s. whereas the diffusivity of the solute in the liquid phase will be about 1.5 x10^{-5} cm^2/s. In addition, the particle diameter (d_p) in GC is usually about 50 microns and the film thickness of the stationary phase will be about 0.5 micron. Thus,

$$R_{MS} = \frac{f_1(k'')d_p^2}{D_{m(o)}} \frac{D_S(\gamma+1)}{f_2(k'')d_f^2} = (\gamma+1)\frac{1.5\times10^{-5}\times(0.005)^2}{0.15\times(0.00005)^2} = (\gamma+1)$$

It is seen that, for GC packed columns operated under the conditions assumed, the two factors contributing to dispersion by resistance to mass transfer are of the same order of magnitude. Consequently, equations (20) and (21) cannot be simplified and must be used in their existing form for all optimization procedures using packed GC columns. If the conditions differ significantly from those assumed, then by using the same procedure the possibility of modifying expressions (20) and (21) can be re-examined.

Packed LC Columns

The mobile phase in LC is considered incompressible from the point of view of dispersion and, so, the equation will not contain the variable (γ). Thus,

$$u_{opt} = \sqrt{\frac{2\gamma D_{m(o)}}{\frac{f_1(k'')d_p^2}{D_{m(o)}} + \frac{f_2(k'')d_f^2}{D_S}}} \qquad (22)$$

$$H_{min} = 2\lambda d_p + 2\sqrt{2\gamma D_{m(o)}\left(\frac{f_1(k'')d_p^2}{D_{m(o)}} + \frac{f_2(k'')d_f^2}{D_S}\right)} \qquad (23)$$

$$\text{and} \quad R_{MS} = \frac{f_1(k'')d_p^2}{D_{m(o)}}\frac{D_S}{f_2(k'')d_f^2} = \frac{f_1(k'')}{f_2(k'')}\frac{D_S d_p^2}{D_{m(o)}d_f^2} \qquad (24)$$

Now, the diffusivity of the solute in the mobile phase ($D_{m(o)}$) will on average be about the same as that in the stationary phase, *viz.*, 1.5 x 10^{-5} cm^2/sec. The particle diameter (d_p) in LC is usually about 5 microns and the film thickness of the stationary phase about 0.1 micron. As in a GC column, the interparticle turbulence will reduce the actual diffusivity of the solute between the particles and, consequently, the resistance to mass transfer between the particles. The mobile phase within the particles, however, is static and, thus, the solute diffusivity is not enhanced. Therefore, there will remain a resistance to mass transfer in the mobile phase within the particle but, as the mobile phase is not moving, it again will appear as a resistance to mass transfer in the stationary phase. Thus,

$$f_1(k'') = f_2(k'')$$

and equation (24) becomes
$$R_{MS} = \frac{f_1(k'')d_p^2}{D_{m(o)}} \frac{D_S}{f_2(k'')d_f^2} = \frac{D_S d_p^2}{D_{m(o)} d_f^2}$$

Now, in LC (D_S) and ($D_{m(o)}$) will be closely similar, (d_f) about 0.25 micron and about 5 microns, thus,

$$R_{MS} = \frac{0.0005^2}{0.00001^2} = 2500$$

It is clear that, in LC, the resistance to mass transfer in the mobile phase (albeit within the pores of the particle) is much greater than the resistance to mass transfer in the stationary phase and, thus, simplifying equations (23) and (24), by ignoring the dispersion contribution from the resistance to mass transfer in the stationary phase, will be quite valid.

Thus, equations (23) and (24) become

$$u_{opt} = \sqrt{\frac{2\gamma D_{m(o)}}{\frac{f_1(k'')d_p^2}{D_{m(o)}}}} = \frac{D_{m(o)}}{d_p}\sqrt{\frac{2\gamma}{f_1(k'')}} \tag{25}$$

and
$$H_{min} = 2\lambda d_p + 2\sqrt{2\gamma D_{m(o)}\left(\frac{f_1(k'')d_p^2}{D_{m(o)}}\right)} = 2\lambda d_p + 2d_p\sqrt{2\gamma f_1(k'')}$$

$$= 2d_p\left(\lambda + \sqrt{2\gamma f_1(k'')}\right) \tag{26}$$

Rate Theory Equations

It is seen from equation (26) that the optimum velocity is determined by the magnitude of the diffusion coefficient and is inversely related to the particle diameter. Unfortunately, in LC (where the mobile phase is a liquid as opposed to a gas), the diffusivity is four to five orders of magnitude less than in GC. Thus, to achieve comparable performance, the particle diameter must also be reduced (*c.f.*, 3-5 µ) which, in turn, demands the use of high column pressures. However, the minimum plate height is seen to be controlled only by the particle diameter, the quality of the packing and the thermodynamic properties of the distribution system.

For a well-packed column, Giddings predicted that $\lambda = 0.5$ and $\gamma = 0.6$, thus,

$$H_{min} = d_p + 2d_p\sqrt{1.2\,f_1(k')} = d_p\left(1 + 2.2\sqrt{f_1(k')}\right) \tag{27}$$

Open Tubular GC Columns

As the open tubular column is geometrically simple, the respective functions of (k') have been explicitly developed and

$$H = \frac{2\,D_m}{u_o} + \frac{\left(1 + 6k'' + 11\,k''^2\right)r^2}{24(1+k'')^2 D_{m(o)}}u_o + \frac{2k''df^2}{3(1+k'')^2 D_s(\gamma+1)}u_o$$

Thus,

$$u_{opt} = \sqrt{\frac{2\gamma\,D_{m(o)}}{\dfrac{\left(1 + 6k'' + 11\,k''^2\right)r^2}{24(1+k'')^2 D_{m(o)}} + \dfrac{2k''df^2}{3(1+k')^2 D_s(\gamma+1)}}} \tag{28}$$

$$H_{min} = 2\sqrt{2\gamma\,D_{m(o)}\left(\frac{\left(1 + 6k'' + 11\,k''^2\right)r^2}{24(1+k'')^2 D_{m(o)}} + \frac{2k''df^2}{3(1+k'')^2 D_s(\gamma+1)}\right)} \tag{29}$$

and

$$R_{MS} = \frac{\left(1 + 6k'' + 11\,k''^2\right)r^2}{24(1+k'')^2 D_{m(o)}} \frac{3(1+k'')^2 D_s(\gamma+1)}{2k''df^2}$$

$$= \frac{\left(1 + 6k'' + 11\,k''^2\right)}{k''}\frac{D_s(\gamma+1)r^2}{16\,D_{m(o)}\,d_f^2} \tag{30}$$

Now, again, the diffusivity of the solute in a gas ($D_{m(o)}$) will, on average, be about 0.15 cm^2/sec., and the diffusivity of the solute in a liquid about 1.5 x 10^{-5} cm^2/sec.

In addition, the column diameter is usually about 360 (r=180 microns) and the film thickness of the stationary phase will be about 0.25 micron.

$$R_{MS} = \frac{(1 + 6k'' + 11 k''^2)}{k''} \frac{1.5 \times 10^{-5}(\gamma + 1)(180 \times 10^{-4})^2}{16 \times 0.15 \times (0.25 \times 10^{-4})^2}$$

$$= \frac{(1 + 6k'' + 11 k''^2)(\gamma + 1) 3.2}{k''}$$

Now, if $(k'') = 0$ and (γ) is about 39, (the inlet pressure is about 30 p.s.i.), then $R_{MS} = 12.8$ and, if (k'') is unity, then $R_{MS} = 230$.

It follows that, for all solutes eluted at a (k'') value of unity or more, the resistance to mass transfer in the stationary phase can be ignored and equations (28) and (29) reduce to

$$u_{opt} = \sqrt{\frac{2 \gamma D_{m(o)}}{\frac{(1 + 6k'' + 11 k''^2) r^2}{24(1 + k'')^2 D_{m(o)}}}} = \frac{4(1 + k'') D_{m(o)}}{r} \sqrt{\frac{3\gamma}{1 + 6k'' + 11 k''^2}} \quad (31)$$

and $H_{min} = 2\sqrt{2 \gamma D_{m(o)} \left(\frac{(1 + 6k'' + 11 k''^2) r^2}{24(1 + k'')^2 D_{m(o)}}\right)} = r \sqrt{\left(\frac{(1 + 6k'' + 11 k''^2)}{3(1 + k'')^2}\right)} \quad (32)$

Thus, for significant values of (k'') (unity or greater) the optimum mobile phase velocity is controlled primarily by the ratio of the solute diffusivity to the column radius and, secondly, by the thermodynamic properties of the distribution system. However, the minimum value of (H) (and, thus, the maximum column efficiency) is determined primarily by the column radius, secondly by the thermodynamic properties of the distribution system and is *independent* of solute diffusivity. It follows that for all types of columns, increasing the temperature increases the diffusivity of the solute in both phases and, thus, increases the optimum flow rate and reduces the analysis time. Temperature, however, will only affect (H_{min}) insomuch as it affects the magnitude of (k'').

The solute diffusivity will also depend on the nature of the mobile phase beitmay a gas or liquid. Very little work has been carried out on the effect of different carrier gases on column efficiency. Scott and Hazeldene [9] measured some HETP curves

Rate Theory Equations

on a Nylon capillary column 0.020 in. I.D. and showed that the lighter gases, hydrogen and hydrogen-nitrogen mixtures, gave much larger longitudinal diffusion dispersion than the heavier gases such as argon and carbon dioxide.

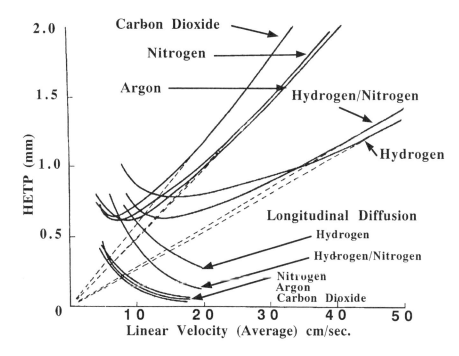

Figure 9. HETP Curves Obtained Employing Different Gases

The curves were obtained using *average* velocity data but, as the inlet-outlet pressure ratio was small, their curves, from the point of view of demonstrating the effect of gas diffusivity, will still be meaningful and are shown in Figure 15. It is also seen that, as theory predicts, the optimum velocity is higher for the less dense gases where solute diffusivity is greatest. In general, helium is the most commonly used carrier gas as not only does it provide faster separations but, when the katharometer is employed as the detector, it also provides the highest sensitivity.

Synopsis

A number of HETP equations were developed other than that of Van Deemter. Giddings developed an alternative form that eliminated the condition predicted by the Van Deemter equation that there was a finite dispersion at zero velocity. However, the Giddings equation reduced to the Van Deemter equation at velocities approaching the optimum velocity. Due to extra-column dispersion, the magnitude of which was originally unknown, experimental data were found *not* to fit the Van Deemter equation. As a consequence, other equations were developed to account for the difference. Huber *et al.* and Horvath and Lin introduced and expression for the resistance to mass transfer in the mobile phase that involved the mobile phase velocity to a power of less than unity. This produced an inflection in the HETP curve at higher velocities to match that which resulted from extra-column dispersion. Knox introduced an equation that employed the reduced plate height and the reduced velocity to demonstrate the effect of packing quality on column performance. His equation predicted that a well-packed column should have a plate height equivalent to about two particle diameters. The effect of mobile phase compressibility in gas chromatography requires the Van Deemter equation to be modified to accommodate the change in mobile phase velocity and diffusivity along the column. In a packed column gas compressibility need only be taken into account in the resistance to mass transfer term for the stationary phase, providing the solute diffusivity in the gas is measured at atmospheric pressure and the gas velocity is measured at the column outlet. The resistance to mass transfer term in the mobile phase requires modification by a simple function of the inlet-outlet pressure ratio. Due to the small compressibility of liquids, the Van Deemter equation can be used effectively as originally derived, as long as excessive inlet pressures are not needed. Increased pressure does reduce the diffusivity of a solute in a liquid but, more important, the energy lost in forcing the liquid through the packing is dissipated as heat. This causes a temperature gradient from the center of the column to the circumference and axially along its length. In turn, this can cause a significant change in the distribution coefficient of a solute and, thus, its retention. The effects of pressure and temperature on solute diffusivity tend to compensate and, so, the effect on the variance per unit length is small. However, the accurate measurement of retention data in LC is best carried out using a narrow bore column, thermostatted with a high heat capacity thermostatting medium. By differentiating the HETP equation and equating to zero, an expression for the optimum velocity can be obtained. By inserting the value for the optimum velocity into the original HETP equation, an expression for the minimum HETP can also be derived. Under certain operating conditions, column form and type of chromatography, the expressions for the optimum velocity and minimum HETP can be simplified for practical use in column design.

References

1. J. C. Giddings, *J. Chromatogr.* **5**(1961)46.
2. J. F. K. Huber and J. A. R. J. Hulsman, *Anal . Chim.. Acta* ,**38**(1967)305.

3. G. J. Kennedy and J. H. Knox, *J. Chromatogr. Sci.*, **10**(1972)606.
4. J. N. Done and J. H. Knox, *J. Chromatogr. Sci.,* **10**(1972)606.
5. J. N . Done, G. J. Kennedy and J. H. Knox, in "*Gas Chromatography 1972* ", (Ed. S. G. Perry), Applied Science Publishers, Barking, (1973)145.
6. J. C. Giddings, *Anal .Chem.,***35**(1963)1338.
7. J. C. Giddings,"*Dynamics of Chromatography*", Marcel Dekker, New York (1965)125.
8. Cs. Horvath and H. J. Lin, *J. Chromatogr.*, **149**(1976)401.
9. M. J. E. Golay, in "*Gas chromatography 1958"* (Ed. D. H. Desty), Butterworths, London, (1958)36.
10. K. Ogan and R. P. W. Scott, *J. High Res. Chromatogr.*, **7(July)** (1984)382.
11. "The Handbook of Chemistry and Physics" 51st Edition (Ed. R. C. Weast), The Chemical Rubber Co., Cleveland, Ohio (1970)F-12.
12. I. Haslasz, R. Endele and J. Asshauer, *J. Chromatogr.,* **112**(1975)37.
13. H. J. Lin and Cs. Horvath, *Chem. Eng. Sci.*, **36**(1981)47.
14. H. Poppe, J. C. Kraak, J. F. K. Huber and J. H. M. van der Berg, *Chromatographia,* **14**(1981)515.
15. E. Katz, K. Ogan and R. P. W. Scott, *J. Chromatogr.,* **260**(1983)277.
16. J. G. Atwood and J. Goldstein, *J. Phys. Chem.* **88**(1984)1875.
17. R. P. W. Scott, in "*Gas Chromatography 1960"*, (Ed. R. P. W. Scott), Butterworths and Co., London (1960)144.

Chapter 9

Extra-column Dispersion

To reiterate the definition of chromatographic resolution; *a separation is achieved in a chromatographic system by moving the peaks apart and by constraining the peak dispersion so that the individual peaks can be eluted discretely*. Thus, even if the column succeeds in meeting this criterion, the separation can still be destroyed if the peaks are dispersed in parts of the apparatus other than the column. It follows that extra-column dispersion must be controlled and minimized to ensure that the full performance of the column is realized.

There are four major sources of extra-column dispersion which can be theoretically examined and/or experimentally measured in terms of their variance contribution to the total extra-column variance. They are as follows:

1. Dispersion due to the sample volume (σ_S^2).
2. Dispersion in the valve–column and column–detector connecting tubing (σ_T^2).
3. Dispersion in the sensor volume resulting from Newtonian flow (σ_{CF}^2).
4. Dispersion in the sensor volume from band merging (σ_{CM}^2).

(Note: These sources do not include the response time of the detector sensor and detector electronics as, today, employing modern electronic technology, such sources of dispersion have been rendered virtually picayune.)

The sum of the variances will give the overall variance for the extra-column dispersion (σ_E^2). Thus,

$$\sigma_E^2 = \sigma_S^2 + \sigma_T^2 + \sigma_{CF}^2 + \sigma_{CM}^2 \tag{1}$$

Equation (1) shows how the extra-column dispersion is made up and according to Klinkenberg [1] must not exceed 10% of the column variance (σ_c^2) if the resolution of the column is to be maintained, *i.e.*,

$$\sigma_E^2 = \sigma_S^2 + \sigma_T^2 + \sigma_{CF}^2 + \sigma_{CM}^2 = 0.1\sigma_c^2$$

To realistically evaluate the effect of extra-column dispersion on column performance, it is necessary to evaluate the maximum extra-column dispersion that can be tolerated by different types of columns. Such data will indicate the level to which dispersion in the detector and its associated conduits must be constrained to avoid abrogating the chromatographic resolution.

From the Plate Theory (chapter 6) the column volume variance is given by $\left(\dfrac{V_r^2}{n}\right)$ and, for a peak eluted at the dead volume the volume variance will be $\left(\dfrac{V_o^2}{n}\right)$. Assuming that the peaks eluted close to the dead volume are given equal priority to all other peaks, then, for a capillary column of radius (r_t) and length (l_t),

$$\sigma_E^2 = \dfrac{V_o^2}{n} = \dfrac{0.1(\pi r_t^2 l_t)^2}{n}$$

Now, it will be shown later in this book, the following approximation is valid for most LC separations:

$$n = \dfrac{1}{0.6 r_t}$$

Thus,
$$\sigma_E^2 = 0.06\pi^2 r_t^5 l_t \qquad (2)$$

For a packed column of radius (r_p) and length (l_p), the permissible extra-column variance will be much larger. Again $\sigma_c^2 = \dfrac{V_r^2}{n}$ and, for the dead volume peak,

$$\sigma_c^2 = \dfrac{(\varepsilon \pi r^2 l)^2}{n}, \qquad \text{thus,} \qquad \sigma_E^2 = 0.1\dfrac{(\varepsilon \pi r^2 l)^2}{n} \qquad (3)$$

where (ε) is the fraction of the column volume that is occupied by the mobile phase, normally taken as 0.60 or 0.65.

Extra-Column Dispersion

Now, it will also be shown later in this book that the following approximation is valid for most LC separations:

$$\sigma_E^2 = 0.068\pi^2 r^4 l dp \quad (4)$$

Equations (2) and (4) allow the permissible extra-column dispersion to be calculated for a range of capillary and packed columns. To allow comparison, data was included for a GC column, in addition to LC columns. The results are shown in Table 1.

Table 1. The Permissible Extra-column Dispersion for a Range of Different Types of Column

Capillary Columns (GC)

Dimensions	Macrobore	Standard	High Efficiency
length (l)	10 m	100 m	400 m
radius (r)	0.0265 cm	0.0125 cm	0.005 cm
(σ_E)	2.78 µl	1.34 µl	0.27 µl

Packed Column

Dimensions	Packed LC	2 m GC	Microbore LC
length (l)	100 cm	200 cm	25 cm
radius (r)	0.05 cm	0.23 cm	0.05 cm
particle diam.	0.0020 cm	0.0080 cm	0.0005 cm
(σ_E)	0.91 µl	54.8 µl	0.23 µl

The standard deviation of the extra-column dispersion is given as opposed to the variance because, as it represents one-quarter of the peak width, it is easier to visualize from a practical point of view. It is seen the values vary widely with the type of column that is used. (σ_E) values for GC capillary columns range from about 12 µl for a relatively short, wide, macrobore column to 1.1 µl for a long, narrow, high efficiency column.

The packed GC column has a value for (σ_E) of about 55 µl, whereas the high efficiency microbore LC column only 0.23 µl. It is clear that problems of extra-column dispersion with packed GC columns are not very severe. However, shorter GC capillary columns with small diameters will have a very poor tolerance to extra-column dispersion. In the same way, short microbore LC columns packed with small particles will make very stringent demands on dispersion control in LC detecting systems.

Dispersion Generated by Different Parts of the Chromatographic System

The maximum allowable dispersion will include contributions from *all* the different dispersion sources. Furthermore, the analyst may frequently be required to place a large volume of sample on the column to accommodate the specific nature of the sample. The peak spreading resulting from the use of the maximum possible sample volume is likely to reach the permissible dispersion limit. It follows that the dispersion that takes place in the connecting tubes, sensor volume and other parts of the detector must be reduced to the absolute minimum and, if possible, kept to less than 10% of that permissible (*i.e.*, 1 % of the column variance) to allow large sample volumes to be used when necessary.

Dispersion Due to Sample Volume

The effect of sample volume on peak width has been considered and treated theoretically in Chapter 6; however, it is of interest to determine the maximum sample volume that can be tolerated with modern columns packed with small particles. The maximum sample volume is defined by the following equation,

(chapter 6),
$$V_i = \frac{1.1\, V_o (1+k')}{\sqrt{n}}$$

where the symbols have the meanings previously ascribed to them. Now, the limiting minimum sample volume will be determined by the peak eluted at the dead volume of the column (*i.e.*, k'=0), thus,

$$V_i = \frac{1.1\, V_o}{\sqrt{n}}$$

Thus, for a packed column length (L), radius (r), with a mobile phase volume equivalent to 60% of the column volume,

$$V_i = \frac{0.66\, L\pi r^2}{\sqrt{n}}$$

Assuming the minimum variance per unit length of a packed column is ($2d_p$), then

$$n = \frac{L}{2d_p} \quad \text{and} \quad \sqrt{n} = \left(\frac{L}{2d_p}\right) \quad \text{and} \quad V_i = 0.93\,\sqrt{Ld_p}\,\pi r^2 \qquad (5)$$

Extra-Column Dispersion

Equation (5) was used to calculate the maximum sample volume for a series of columns having different lengths and internal diameters and packed with particles 3 μ in diameter. The maximum sample volume will increase as the (k') of the solute increases and, for any solute that is retained, the maximum sample volume can be obtained by multiplying the results from equation (5) by (1+k'). The results obtained are shown as curves relating maximum sample volume to column radius for columns having lengths of 3, 5 and 10 cm, in Figure 1. The column length is limited to a maximum of 10 cm because columns of greater length, packed with particles 3 μ in diameter, would provide too great a flow impedance to be operated satisfactorily with many modern liquid chromatography pumps.

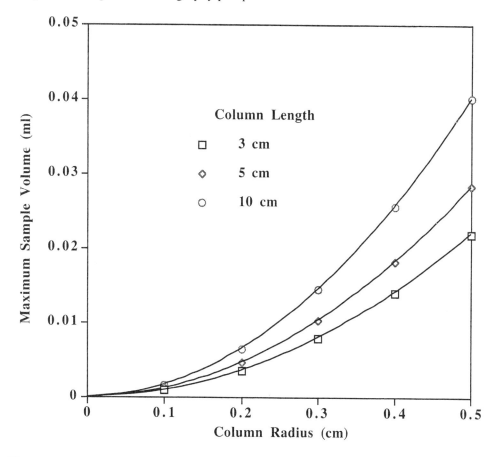

Figure 1. Curves Relating Maximum Sample Volume to Column Diameter for Columns of Different Length Packed with Particles 3 μ in Diameter

It is seen that columns having diameters less than 2 mm will only tolerate a maximum sample volume of a fraction of a microliter. Although larger volume valves can be used to inject sample volumes of this size, the dispersion from the valve is still likely used to inject sample volumes of this size, the dispersion from the valve is still likely to provide an unacceptable loss in column efficiency and, so, internal loop sample valves would appear to be preferable. It is also seen that, by increasing the column diameter, much larger sample volumes can be used without degrading the column performance. Although the use of large columns inevitably increases solvent consumption, it also helps in placing samples of very insoluble materials on the column. Similar curves relating maximum sample volume to column diameter. but for columns packed with 5 μm particles, are shown in Figure 2.

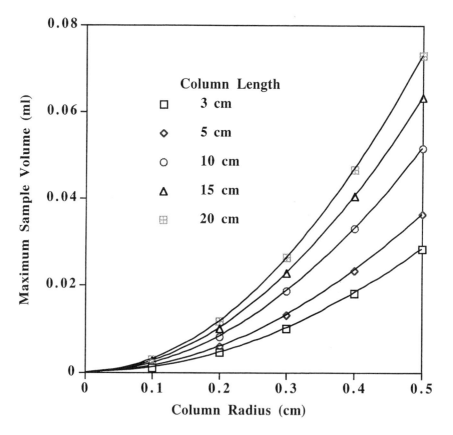

Figure 2. Curves Relating Maximum Sample Volume to Column Diameter for Columns of Different Length Packed with Particles 5 μ in Diameter

The use of 5 μm particles permits the use of much longer columns due to the increased permeability. This, in turn, permits the use of much larger sample volumes. In fact, for a column 20 cm long, eluting a solute at a (k') of 5, the maximum sample volume that can be used without increasing the peak variance by more than 10% will

Extra-Column Dispersion

be about 0.45 ml. Considering the column will have an efficiency of about 25,000 theoretical plates, a sample volume of 0.45 ml provides a great deal of useful flexibility in an analysis. However, there are also disadvantages to long columns of wide diameter that result from the large solvent composition, which often significantly increases the cost of an analysis and aggravates the problems of solvent disposal.

Dispersion in Sample Valves

There are two types of sample valve commonly used in LC, the *internal loop valve* and the *external loop valve*. In order to improve the seating and eliminate leaks, the valve faces are sometimes made from appropriate ceramics. The *internal loop valves* are largely used with small bore columns, that is to say columns having internal diameters of less than 1.5 mm. The *external loop valves* are used for larger diameter columns up to semi-preparative columns.

The sample volume of the *internal loop valve* is contained in the connecting slot of the valve rotor and can range in capacity from 0.1 µl to about 0.5 µl. The dispersion that takes place in internal loop valves was measured by Scott and Simpson [5] and the results they obtained are shown in Table 2.

Table 2. Variance of Solute Bands Resulting from Two Internal Loop Valves

Internal Bore of Valve	Variance μl^2
0.030 in.	0.667
0.010 in.	0.338

The *external loop valve* is the more commonly used sampling system and offers a wider choice of readily adjustable sample sizes. A modified form of the external loop sample valve has become very popular for quantitative LC analysis, a diagram of which is shown in Figure 3.

The basic difference between this type of valve and the normal external loop sample valve is the incorporation of an extra port at the front of the valve. This port allows the injection of a sample by a syringe directly into the front of the sample loop. Position (A) shows the load position. Injection in the front port causes the sample to flow into the sample loop. The tip of the needle passes through the rotor seal and, on injection, is in direct contact with the ceramic stator face. Note the needle is chosen so that its diameter is too great to enter the hole. After injection, the valve is rotated to position (B) and the mobile phase flushes the sample directly onto the column. The sample is actually *forced out of the beginning of the loop* so it does not have to flow through the entire length of the loop before entering the column. This type of injection system is ideally suited for quantitative LC and is probably, by far, the most popular injection system in use. Sample valves based on this design are available from a number of manufacturers *but the dispersion from this type of valve does not appear to be readily available.* Data on certain specific valves may be available on request from Rheodyne Inc.

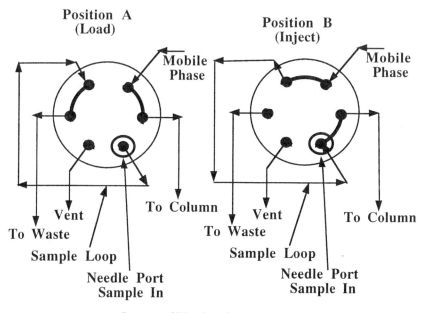

Courtesy of Rheodyne Instruments Inc.

Figure 3. The Modified External Loop Sample Valve

In addition, due to the likely variation between injections and the convoluted geometry of the internal parts of the valve, a useful theoretical treatment of the valve dispersion would be very difficult if not impossible to develop.

Dispersion in Unions and Stainless Steel Frits

Unions can also be a serious source of dispersion, depending on their design. Today, low dead volume unions are generally available which exhibit reduced dispersion characteristics but, again, actual data reporting their dispersion do not appear to be readily available. Scott and Simpson also measured the dispersion that occurred in normal unions and drilled-out unions. Drilled-out unions allow the ends of the

Extra-Column Dispersion

connecting tubes to butt against one another and thus reduce the union volume to virtually zero. Their results are shown in Table 3.

Table 3. Dispersion in Normal and Drilled-Out Unions

Union Type	Variance (μl^2)
Normal	1.464
Drilled-Out	0.113

It is seen that the normal union can cause significant dispersion and, by drilling out the union, the dispersion is greatly reduced. It must be assumed that the dispersion caused by the modern reduced volume union would fall somewhere between these two extremes. The same authors also measured the dispersion arising from stainless steel frits and their results are shown in Table 4.

Table 4. Dispersion Due to Stainless Steel Frits

No. of Frits	Variance (μl^2)
0	0.826
1	0.826
2	0.909
3	0.857
Average Contribution/Frit	0.017

It is seen that a frit contributes little or nothing to peak dispersion and this is probably due to eddy diffusion that takes place in the pore of the frit, much the same as that which occurs in the interstices of a column packing. The continual change in direction of the mobile phase, as it winds its way through the interstices of the frit, causes eddies that greatly increase the effective solute diffusivity and, thus, virtually eliminates diffusion related dispersion. It is true that the small dispersion that does occur will increase with the porosity of the frit but the net contribution would still be small unless the pores were exceedingly large. It follows that in general use, stainless steel frits need not be considered a significant source of dispersion in the chromatographic system.

Dispersion in Open Tubular Conduits

The dispersion that takes place in an open tube, as discussed in chapter 8, results from the parabolic velocity profile that occurs under conditions of Newtonian flow (*i.e.*, when the velocity is significantly below that which produces turbulence). Under condition of Newtonian flow, the distribution of fluid velocity across the tube is depicted diagramatically in Figure 6A. Due to the relatively high velocity at the center of the tube and the very low velocity at the walls, the center of the band of solute passing down the tube will move ahead of that situated at the walls. This dispersive effect is depicted in Figure 6B.

A. Newtonian Flow

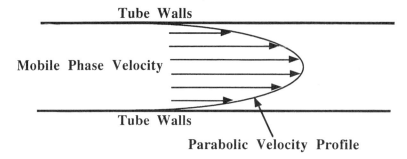

B. Band Dispersion Due to Newtonian Flow

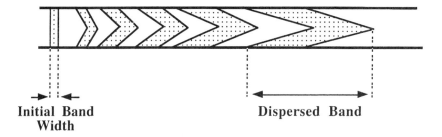

Figure 6. Dispersion Due to Newtonian Flow

The effect is much less when the fluid is a gas, where the solute diffusivity is high and the solute tends to diffuse rapidly across the tube and partially compensate for the nonlinear velocity profile. However, when the fluid is a liquid, the diffusivity is five orders of magnitude less and the dispersion proportionally larger.

Reiterating equation (8) from chapter 7,

$$H = \frac{2D_m}{u} + \frac{1}{24}\frac{r^2}{D_m}u \tag{6}$$

where the symbols have the meanings previously allotted to them.

If $u \gg D_m/r$, which will be true for all conditions except if open tubular columns are being used in LC, then equation (6) reduces to

Extra-Column Dispersion

$$\sigma_T^2 = H_T = \frac{r^2 u}{24 D_m} \quad (7)$$

Now, from the Plate Theory (Chapter 6) the peak variance in volume units (σ_V^2) is

$$\sigma_V^2 = \frac{(\text{Tube Volume})^2}{n}$$

$$= \frac{\left(\pi r^2 L\right)^2}{n} = \frac{\pi^2 r^4 L^2}{n} \quad (8)$$

where (n) is the number of theoretical plates in the tube.

Now, $\frac{L}{n} = H$, thus, replacing $\left(\frac{L}{n}\right)$ by (H), and substituting for (H) from (7),

$$\sigma_V^2 = \pi^2 r^4 L H = \frac{\pi^2 r^4 L r^2 u}{24 D_m} = \frac{\pi^2 r^6 L u}{24 D_m} \quad (9)$$

Now, the flow rate through the tube (Q) will be given by $Q = \pi r^2 u$, so

$$\sigma_T^2 = \frac{Q \pi r^4 L}{24 D_m} \quad (10)$$

The conditions required to minimize tube dispersion are clearly indicated by equation (10). Firstly, as the column should be operated at its optimum mobile phase velocity and the flow rate, (Q) is defined by column specifications it is not a variable that can be employed to control tube dispersion. Similarly, the diffusivity of the solute (D_m) is determined by the nature of the sample and the mobile phase that is chosen and so is also not a variable that can be employed to control tube dispersion. The major factors available to control dispersion are the tube radius and the tube length. Dispersion increases as the fourth power of the tube radius and linearly as the tube length; thus, a reduction in the tube radius by a factor of 2 will reduce the dispersion by a factor of sixteen and a similar reduction in the tube length by a factor of 2.

The radius of the column cannot be reduced indefinitely due to the pressure difference that will increase as (R) is reduced. From Poiseuille's equation the pressure drop across the tube is given by

$$\Delta P = \frac{8 L \eta u}{r^2}$$

Thus, as $Q = \pi r^2 u$ or $u = \dfrac{Q}{\pi r^2}$,

$$\Delta P = \frac{8 L \eta Q}{\pi r^4} \qquad (11)$$

Equation (11) shows that the pressure drop across the connecting tube increases inversely as the fourth power of the tube radius. It follows that, as it is impractical to dissipate a significant amount of the available pump pressure across a connecting tubing, there will be a limit to the reduction of (r) to minimize tube dispersion.

Reducing the length of the connecting tube has the same effect on both dispersion and pressure drop. Consequently, reducing (L) will linearly reduce dispersion and, at the same time, proportionally reduce the pressure drop across the connecting tube. Thus, reducing the length of the connecting tube is, by far, the best method of controlling dispersion and by making (L) as small as possible, both the dispersion and the pressure drop can be minimized. The diameter of the connecting tube should not be reduced to much less than 0.012 cm (0.005 in. I.D.), not only because of the high pressure drop that will occur across it, but for a more mundane but very important reason. If tubes of very small diameter are employed, they *will easily become blocked.* The volume variance and the volume standard deviation contribution from connecting tubes of different lengths were calculated employing equation (10) and the results are shown in Figure 7A and 7B, respectively. The tube radius was taken as 0.012 cm, the flow rate 1 ml per minute, and the diffusivity of the solute in the mobile phase 2.5×10^{-5} cm^2/sec.

Figure 1A shows that, as might be expected, the variance increases linearly with the tube length. The variance of the dispersion generated from all connecting tubes must be added to that of the column, and to other extra-column dispersion variances, to arrive at the ultimate peak variance. Theoretically, it might be possible to reduce the tube diameter to 0.008 cm (0.003 in.) I.D., and this would reduce the tube variance by a factor of eight. Unfortunately, however, this would, without doubt, increase the chance of tube blockage very significantly.

The curve in Figure 1B is probably more useful from a practical point of view. Although the standard deviations of any dispersion process are not additive, they do give a better impression of the actual dispersion that a connecting tube alone can cause. It is clear that a tube 10 cm long and 0.012 cm I.D. can cause dispersion resulting in a peak with a standard deviation of 4 µl. Now, a peak with a standard deviation of 4 µl would have a base width of 16 µl and, in practice, many short columns packed with particles 3 mm (or less) in diameter would produce peaks of commensurate size.

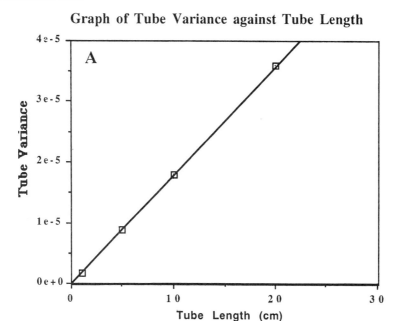

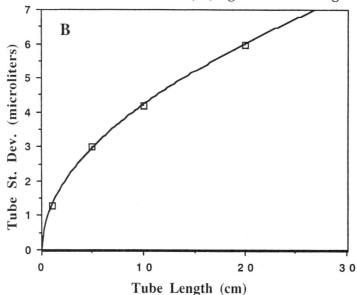

Figure 7. Graphs Relating the Variance and Standard Deviation of Peaks Passing Through an Open Tube against Tube Length

Thus, if high efficiency columns are to be used, with their inherent small plate heights, then connecting tubes must either be eliminated altogether, or reduced to the absolute minimum in length. Regrettably, it is often extremely difficult to use short lengths of a connecting tube, particularly for column detector connections. Unfortunately, manufacturers often design detectors that involve significant lengths of tubing to carry the column eluent to the sensor cell.

Low Dispersion Connecting Tubes

Although direct connection between sample valve and column, and column and detector sensor cell, is the ideal solution to the problem of extra-column dispersion; as already mentioned, this is not possible in practice with most contemporary instrument designs. Consequently, a low dispersion conduit would be highly desirable. In order to reduce the dispersion that occurs in an open tube, the parabolic velocity profile of the fluid flowing through the tube must be disrupted to allow rapid radial mixing. This can be achieved, and secondary flow introduced into the tube, by deforming its regular geometry. The dispersion that takes place in physically deformed tubes (squeezed, twisted and coiled) was studied by Halasz [7,8] and the effect of radial convection (secondary flow) on the dispersion in tightly coiled tubes has been examined both theoretically and experimentally by Tijssen (9). In addition, the effect of secondary flow produced in serpentine shaped tubes has also been examined by Katz and Scott [10]. To simplify the algebraic expressions developed by Tijssen, the equations for radial dispersion in coiled tube are given in conventional chromatographic terms. At relatively low linear velocities (but not low relative to the optimum velocity for the tube) Tijssen derived the equation

$$H = \frac{jr^2 u}{D_m} \qquad (12)$$

where (j) is a constant over a given velocity range and the other symbols have the meanings previously ascribed to them.

Equation (12) indicates that the band variance is directly proportional to the square of the tube radius, very similar to that for a straight tube. At high linear velocities, Tijssen deduced that

$$H = \frac{b D_m^{0.14}}{\phi u} \qquad (13)$$

where (b) is a constant for a given mobile phase and (ϕ) is the *coil aspect ratio* (the ratio of the tube radius to the coil radius), *i.e.*,

$$\phi = \frac{r_{tube}}{r_{coil}}$$

Extra-Column Dispersion

Equation (13) shows that, at the higher velocities, the value of (H) now depends on (D_m) to the power of 0.14 and inversely on the coil aspect ratio and velocity. It follows, from equations (12) and (13), that, at low velocities, the band dispersion *increases* with (u), whereas at high velocities, the band dispersion *decreases* with (u). Consequently, a plot of (H) against (u) should exhibit a maximum at a certain value of (H). Combining equations (8) and (9), an equation can be derived that allows the value of (u) at which (H) is a maximum to be calculated, *viz.*,

$$u_{(H_{Max.})} = \frac{c}{r\sqrt{\phi}}$$

where (c) is a constant for a given solute and mobile phase

The effect of tube radius and coil aspect ratio on the onset of radial mixing in coiled tubes was investigated using the above equations. In Table 5, the dimensions of the coiled tubes examined are given and the curves relating (H) and (u) in Figure 8.

Table 5.   Physical Dimensions of Coiled Tubes Examined

Tube	r (cm)	L (cm)	r(Coil) (cm)	y	L(Coil) (cm)
1	0.019	365.8	0.5	0.038	18.5
2	0.020	365.0	0.085	0.235	65.8
3	0.0127	998.0	0.0765	0.166	128.0
4	0.0127	337.5	0.0498	0.0255	73.7

It is seen from Figure 8, where there is little radial mixing, that is, at low linear velocities, (H) increases as (u) increases. In addition, in the larger radii coiled tubes 1 and 2, the dispersion is greater than that in the smaller radii tubes 3 and 4. It is also seen that radial mixing commences at the higher velocities and (H) *decreases* as (u) *increases*. As that range of velocities is approached where radial mixing dominates, dispersion becomes independent of the linear velocity (u). Figure 8 also shows that the maximum value of (H), for any particular coil, occurs at differing values of (u) which depend on the combined values of (r) and (ϕ). It would appear that, in general, the maximum value of (H) and the value of (u) at which it occurs, are reduced at high coil aspect ratios.

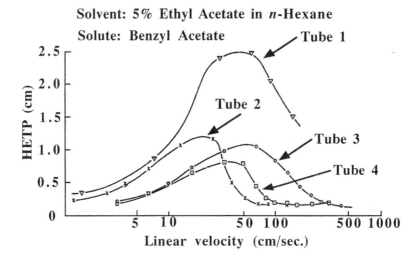

Figure 8. Graphs of (H) against (u) for Different Coiled Tubes

It is interesting to note that, although the straight tube theory of Golay is not applied to coiled tubes, his equation can be employed to qualitatively explain the shape of the curves given in Figure 8. At low velocities longitudinal diffusion dominates and controls the dispersion. As the velocity approaches the optimum, the resistance to mass transfer term becomes dominant and, consequently, (H) rapidly increases. However, as a result of induced radial flow that occurs at high velocities, the diffusivity of the solute dramatically increases, eventually reducing the resistance to mass transfer to virtually zero, causing a corresponding dramatic reduction in (H). Finally, at extremely high velocities, the only contribution to dispersion that remains is longitudinal diffusion, which is now barely significant and (H) is very small indeed.

Serpentine Tubes

The low dispersion serpentine tube was developed by Katz *et al.* [11] as an alternative to the coiled tube and was designed to increase secondary flow by actually *reversing* the direction of flow at each serpentine bend. A diagram of a serpentine tube is shown in Figure 9, which was originally designed as an interface between a liquid chromatograph and an atomic adsorption spectrometer. The serpentine tube is encased in an outer sheath to protect the tube and provide some mechanical strength. In a coiled tube, the flow of fluid is continually deflected in the same direction, which causes secondary radial flow and, thus, increases the solute transfer across the tube. In the serpentine tube, however, the reversal of the flow at each bend also induces

Extra-Column Dispersion

turbulence, augments the effect of radial flow and, thus, produces low dispersion at relatively low solvent velocities.

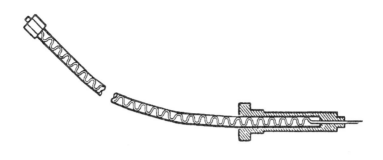

Internal radius, 0.010 in. (0.0127 cm), external radius 0.020 in. (0.025 cm), linear length 17 in., serpentine length 15 in., serpentine amplitude, 0.050 in.

Figure 9. The Low Dispersion Serpentine Tube

A graph relating the variance per unit length of the tube (H) against flow rate is shown in Figure 10, for a serpentine tube having the dimensions given in Figure 9.

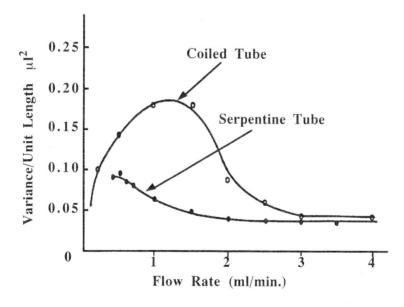

Figure 10. Graph of Variance against Flow Rate for Coiled and Serpentine Tubes

As the flow rate is defined by the column with which the low dispersion tubing is to be used, the flow rate is employed as the independent variable as an alternative to the more usual linear velocity. In fact, as already noted, the column flow rate is independently defined by the chromatographic characteristics of the column. A similar curve is obtained for the serpentine tube as that for the coiled tube, but the maximum value of (H) is reached at a much lower flow rate with the serpentine tube. More importantly, once the flow rate exceeds about 1.5 ml/min., the variance remains more or less constant over a wide flow rate range that encompasses those usually employed in normal LC separations.

Although the details of low dispersion tubing were published nearly 20 years ago, low dispersion connecting tubes are still not in common use in contemporary LC equipment. Low dispersion tubing has a another feature that, in fact, might be predicted from its operating principle. The secondary flow resulting from its serpentine form also greatly improves its thermal conducting properties. As a consequence, serpentine tubes can also be used as highly efficient heat exchangers. One manufacturer utilizes serpentine tubing as a heat exchanger to preheat the mobile phase before entering the thermostatted column. It was determined that a few centimeters of serpentine tubing were sufficient to achieve complete thermal equilibrium between the column and the mobile phase.

The different forms of dispersion profiles that are obtained from various types of connecting tubes used in LC are shown in Figure 10.

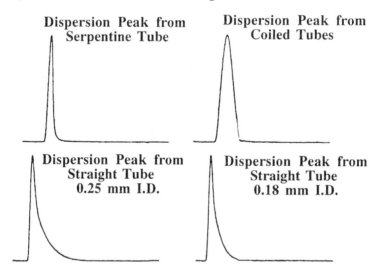

Figure 10. Dispersion Profiles from Different Types of Tube

The peaks shown were obtained using a low dispersion UV detector (cell volume, 1.4 µl) in conjunction with a sample valve with a 1 µl internal loop. All tubes were of the same length and a flow rate of 2 ml/min. was employed; the peaks were recorded on a high speed recorder. The peak from the serpentine tubing is seen to be

Extra-Column Dispersion

symmetrical and has the smallest width. The peak from the coiled tube, although still very symmetrical, is the widest at the points of inflection of all four peaks. The peak from the straight tube 0.25 mm I.D. is grossly asymmetrical and has an extremely wide base. The width and asymmetry are reduced using a tube with an I.D. of 0.18 mm but serious asymmetry remains. It is clear that under conditions where the tube length between the end of the column and the detector sensor can not be made sufficiently short to restrain dispersion, the use of low dispersion serpentine tubing may be a satisfactory alternative.

Dispersion in the Detector Sensor Volume

The sensor volume of a detector can cause dispersion and contribute to the peak variance in two ways. Firstly, there will be dispersion resulting from the viscous flow of fluid through the cell sensor volume, which will furnish a variance similar in form to that from cylindrical connecting tubes. Secondly, there will be a peak spreading which results from the finite volume of the sensor. If the sensor has a significant volume, it will not measure the instantaneous concentration at each point on the elution curve, but it will measure the average concentration of a slice of the peak equivalent to the sensor volume. Thus, the true profile of the peak will not be monitored. The net effect will first (at sensor volumes still significantly smaller than the peak volume) give the peak an apparent dispersion. Then (as the sensor volume becomes of the same order of magnitude as the peak volume) it will distort the profile and eventually seriously impair resolution. In the worst case, two peaks could coexist in the sensor at one time and only a single peak will be represented. This apparent peak spreading is not strictly a dispersion process but, in practice, will have a similar effect on chromatographic resolution. The effect of viscous flow on dispersion will first be considered.

Dispersion in Detector Sensors Resulting from Newtonian Flow

Most sensor volumes, whether in LC (*e.g.*, a UV absorption cell) or in GC (*e.g.*, a katharometer cell), are cylindrical in shape, are relatively short in length and have a small length-to-diameter ratio. The small length-to-diameter ratio is in conflict with the premises adopted in the development of the Golay equation for dispersion in an open tube and, consequently, its conclusions are not pertinent to detector sensors. Atwood and Golay [12] extended the theory of dispersion in open tubes to tubes of small length-to-diameter ratio. The theory developed is not pertinent here as it will be seen that, with correctly designed cells, that dispersion from viscous sources can be made negligible. Nevertheless, the effect of the cell on solute profiles is shown in Figure 11.

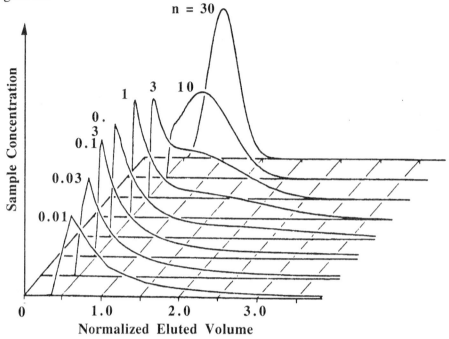

Figure 11. Elution Curves Presented as a Function of the Normalized Tube Length

Fortunately, this situation rarely arises in practice as the profile is further modified by the manner of entrance and exit of the mobile phase. The conduits are designed to produce secondary flow and break up the parabolic velocity profile, causing peak distortion as shown in Figure 12. The Newtonian flow through the detector is broken up by the conformation of the inlet and outlet conduits to and from the cell. Mobile phase enters the cell at an angle that is directed at the cell window. As a consequence, the flow has to virtually reverse its direction to pass through the cell, producing a strong radial flow and disrupting the Newtonian flow. The same situation is arranged to occur at the exit end of the cell. The flow along the axis of the cell must reverse its direction to pass out of the port that is also set at an angle to accomplish the same effect. Employing this type of cell geometry dispersion resulting from viscous flow is practically eliminated.

Apparent Dispersion from Detector Sensor Volume

The detector responds to an average value of the total amount of solute in the sensor cell. In the extreme, the sensor volume or cell could be large enough to hold two closely eluted peaks and, thus, give a response that would appear as though only a single solute had been eluted, albeit very distorted in shape.

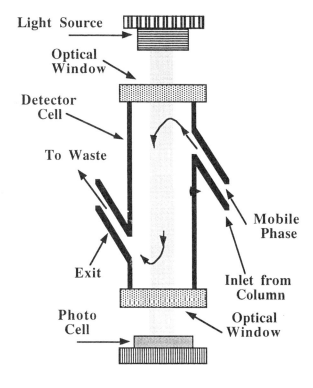

Figure 12. The Design of a Modern Absorption Cell

This extreme condition rarely happens but serious peak distortion and loss of resolution can still result. This is particularly so if the sensor volume is of the same order of magnitude as the peak volume. The problem can be particularly severe when open tubular columns and columns of small diameter are being used. Scott and Kucera measured the effective sensor cell volume on peak shape and their results are shown in Figure 13.

The column used in the upper chromatogram was 24 cm long, 4.6 mm I.D.; the solvent was tetrahydrofuran, the solute benzene and the flow rate 1 ml/min. The column used in the lower chromatogram was 1 m long, 1 mm I.D. using the same solvent and solute but at a mobile phase flow rate of 40 ml/min. It is seen that the reduction in cell volume has a dramatic effect on peak shape. The large 25 µl cell causes significant peak asymmetry as well as excessive peak dispersion which is predicted by the work of Atwood and Golay. Clearly, even cell volumes of 3 µl are too large for use with 1 mm I.D. columns and very few contemporary detectors have cell volumes less than 3 µl.

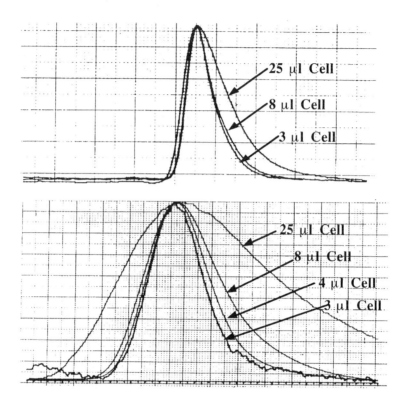

Figure 13. Peak Profiles from Detector Having Different Cell Volumes

Large cell volume can also have a very serious effect on solute resolution, but this can be examined theoretically. The situation is depicted in Figure 14. It is the elution profile of a peak eluted from a column 3 cm long, 3 mm I.D. packed with particles 3 µm in diameter.

The peak is considered to be eluted at a (k') of 2 and it is seen that the peak width at the base is about 14 µl wide. The sensor cell volume is 2.5 µl and the portion of the peak in the cell is included in the Figure (this cell volume is small compared with those of other contemporary detectors). It is clear the detector will respond to the mean concentration of the slice contained in the 2.5 µl sensor volume. It is also clear that as the sensor volume is increased, the greater part of the peak will be contained in the cell and the output will be an average value of an even larger portion of the peak resulting in serious peak distortion. The effect of a finite sensor volume can be easily simulated with a relatively simple computer program and the output from such a program is shown in Figure 15. The example given, although not the worst case

Extra-Column Dispersion

scenario, is a condition where the sensing volume of the detector can have a very serious effect on the peak profile and, consequently, the resolution.

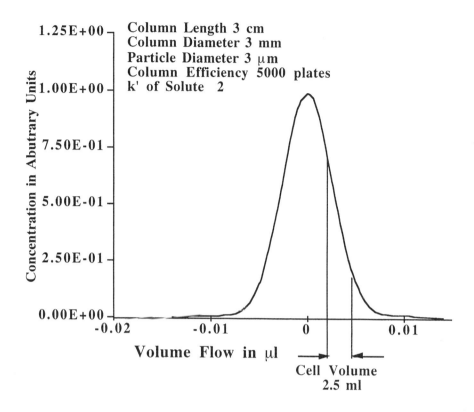

Figure 14. Effect of Sensor Volume on Detector Output

The column is a small bore column and, thus, the eluted peaks have a relatively small peak volume, which is commensurate with that of the sensing cell. It is seen that even a sensor volume of 1 µl has a significant effect on the peak width and it is clear that if the maximum resolution is to be obtained from the column, then the sensor cell volume should be certainly no greater than 2 µl. It should also be noted that the results from the use of a sensor cell having a volume of 5 µl are virtually useless and that many commercially available detectors do, indeed, have sensor volumes as great as, if not greater than, this. It follows that if small bore columns are to be employed, such sensor volumes must be studiously avoided. It should also be pointed out that small volume cells must be very carefully designed. In general, reducing the diameter of the cell reduces the volume and increases the noise thus reducing the usable sensitivity: reducing the length of the cell maintains the same noise, but reduces the response and, thus, also the sensitivity.

The Separation of Two Solutes 4σ Apart on a Microbore Column Employing Detectors Having Different Sensor Volumes

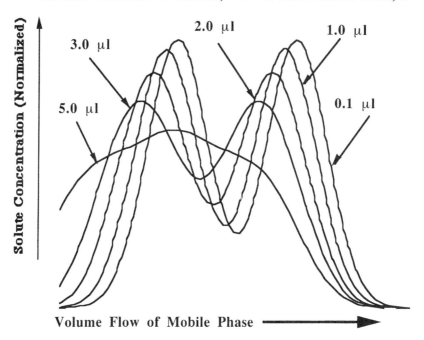

Figure 15. The Effect of Detector Sensor Volume on the Resolution of Two Solutes

Dispersion Resulting from the Overall Detector Time Constant

Although the response time of modern electronic systems is adequate for virtually all practical chromatographic separations, the effect of electronic response on peak shape will be briefly mentioned. Slow electronic response can cause the peak to appear to be significantly dispersed and if significant,can produce peak asymmetry. The term *appear* is used as the solvent profile itself is not actually changed; only the profile as presented on the recorder or printer. The effect of the detector time constant, and that of its electronics, can be calculated by employing the elution equation from the plate theory and multiplying by an exponential delay function. The results from such a calculation are shown in Figure 16. The undistorted peak, that would be monitored by a detector with a zero time constant, is about 4 seconds wide. Thus, for a GC

Extra-Column Dispersion

packed column operating at 20 ml/min. this would represent a peak having a volume of about 1.3 ml.

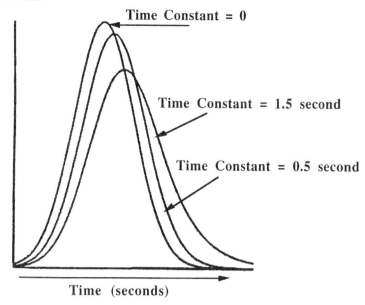

Figure 16. Peak Profiles Demonstrating Distortion Resulting from Detector Time Constant

For an LC column operating at a flow rate of 1 ml/min., a peak with a base width of 4 seconds would represent a peak volume of about 67 µl. Consequently, the peaks depicted would represent those eluted fairly late in the chromatogram. It is seen that, despite the late elution, the distortion is still quite severe. In order to elute early peaks without distortion, the time constant would need to be at least an order of magnitude less.

In the next chapter, experimental data supporting the Van Deemter equation will be described and discussed. Only data that has been acquired with equipment that has been specifically designed to eliminate, or reduce to an insignificant level, the dispersion sources described in this chapter can be used reliably for such a purpose.

Synopsis

Extra-column dispersion can arise in the sample valve, unions, frits, connecting tubing, and the sensor cell of the detector. The maximum sample volume, *i.e.*, that volume that contributes less than 10% to the column variance, is determined by the type of column, dimensions of the column and the chromatographic characteristics of the solute. In practice, the majority of the permitted extra-column dispersion should be allotted to the sample volume, as this may be demanded by the sample type for successful analysis. Consequently, the dispersion from other parts of the system must be kept to the absolute minimum. Dispersion in sample valves can be minimized by mechanical design. Internal loop valves tend to provide the minimum dispersion. Dispersion from unions can be minimized by using drilled-out unions or low dead volume unions. Stainless steel frits cause very little dispersion and can be employed without great concern for their contribution to the overall dispersion of the system. Connecting tubes are one of the major sources of extra-column dispersion and should be kept as short as possible and the radius reduced to a minimum commensurate with the tube not becoming blocked. Tubes 0.005 in. in diameter are a useful compromise. If it is inevitable that the detector sensor must be a significant distance from the column exit and, thus, a relatively long length of connecting tube is necessary, then low dispersion tubing, such as serpentine tubing, may be the solution. Dispersion in the detector sensor is the second largest source of extra-column dispersion and can only be reduced to a satisfactory level by reducing the volume of the sensor cell to 2 μl or less. Reducing the cell length will reduce its response but maintain the same or similar noise level, thus, reducing the sensitivity (minimum detectable concentration). Maintaining the same length, but reducing the cell radius, will maintain the same response, but increase the noise level, also reducing the overall sensitivity. It is clear a compromise is necessary and this compromise will depend on the type of detector, UV absorption, fluorescence, etc., that is employed. Nevertheless, to realize the maximum column efficiency and, thus, the maximum resolution, very small detector cell volumes will be essential.

References

1. R. P. W. Scott and P. Kucera, *J. Chromatogr. Sci.*, **9**(1971)641.
2. M. Martin, C. Eon and G. Guiochon, *J. Chromatogr.*, **108**(1975)229.
3. J. H. Knox and M. T. Gilbert, *J. Chromatogr.*, **186**(1979)405.
4. C. E. Reese and R. P. W. Scott, *J. Chromatogr. Sci.*, **18**(1980)479.
5. R. P. W. Scott and C. F. Simpson, *J. Chromatogr. Sci.*, **20**(1982)62.
6. A. Klinkenberg,*"Gas Chromatography 1960"* (Ed. R. P. W. Scott), Butterworths, London, (1960)194.
7. I. Halasz, H. O. Gerlach, K. F. Gutlich and P. Walking, *U.S. patent*, 3,820,660 (1974).
8. K. Hofmann and I. Halasz, *J. Chromatogr.*, **173**(1979)211.
9. K. Hofmann and I. Halasz, *J. Chromatogr.*, **199**(1980)3.

10. R. Tijssen, *Separ. Sci. Technol.*, **13**(1978)681.
11. E. D. Katz and R. P. W. Scott, *J. Chromatogr.*, **268**(1983)169.
12. J. G. Atwood and M. J. E. Golay, *J. Chromatogr.*, **218**(10981)97.
13. R. P. W. Scott and P. Kucera, *J. Chromatogr.*, **169**(1979)51.

Chapter 10

Experimental Validation of the Van Deemter Equation

The chromatography literature contains a vast amount of dispersion data for all types of chromatography and, in particular, much of the data pertains directly to GC and LC. Unfortunately, almost all the data is unsuitable for validating one particular dispersion equation as opposed to another. There are a number of reasons for this; firstly, the necessary supporting data (*e.g.*, diffusivity data for the solutes in the solvents employed as the mobile phase, accurate distribution and/or capacity factor constants (k")) are not available; secondly, the accuracy and precision of much of the data are inadequate, largely due to the use of inappropriate apparatus with high extra-column dispersion.

The need for extra physical chemical data to test the equations arises from their close similarity. It might appear adequate to apply each equation to a number of experimental data sets of (H) and (u) and to identify the equation that provides the best fit. Unfortunately, due to the basically similar form of the dispersion functions, all would provide an excellent fit to any given experimentally derived data set. Consequently, a mere fit to experimental data is insufficient to identify the true form of the dispersion equation. However, each term in a particular dispersion equation *purports to describe a specific dispersive effect*. That being so, if the dispersion effect described is to be physically significant over the mobile phase velocity range examined, all the constants for the proposed equations derived from a curve fitting procedure *must be positive and real*. Those equations that do not consistently provide positive and real values for all the constants would obviously not be valid expressions for peak dispersion.

Unfortunately, any equation that *does provide a good fit* to a series of experimentally determined data sets, and meets the requirement that all constants were positive and real, *would still not uniquely identify* the correct expression for peak dispersion. After a satisfactory fit of the experimental data to a particular equation is obtained, the constants, (A), (B), (C) etc. must then be replaced by the explicit expressions derived from the respective theory. These expressions will contain constants that define certain physical properties of the solute, solvent and stationary phase. Consequently, if the pertinent physical properties of solute, solvent and stationary phase are varied in a systematic manner to change the magnitude of the constants (A), (B), (C) etc., the changes as predicted by the equation under examination must then be compared with those obtained experimentally. The equation that satisfies both requirements can then be considered to be the true equation that describes band dispersion in a packed column.

To identify the pertinent HETP equation that describes dispersion in a packed bed, the following logical procedure will require to be carried out.

1. All the equations must be fitted to a series of (H) and (u) data sets and those equations that give positive and real values for the constants of the equations identified.

2. The explicit form of those equations that satisfy the preliminary data criteria, must then be tested against a series of data sets that have been obtained from different chromatographic systems. As an example, such systems might involve columns packed with different size particles, employed mobile phases or solutes having different but known physical properties such as diffusivity or capacity ratios (k").

The equation, or equations, that satisfied both criteria would be identified as that which was pertinent to the system examined.

The Accurate Measurement of Column Dispersion

Katz *et al.*[1] searched the literature for data that could be used to identify the pertinent dispersion equation for a packed column in liquid chromatography. As a result of the search, no data was found that had been measured with the necessary accuracy and precision and under the sufficiently diverse solute/mobile phase conditions required to meet the second criteria given above. It became obvious that a

Experimental Validation of Dispersion Equations

special chromatograph was required that must be designed specifically to provide dispersion data with the essential accuracy and precision. There are two ways of reducing the *effect* of extra-column dispersion; the first is to make the peak volume so large that the contribution of extra-column dispersion to the peak dispersion is negligible; the second is to reduce the extra-column dispersion by careful instrument design to render its contribution to the column dispersion insignificant. Both methods were used by Katz *et al*, to eliminate any effect from extra-column dispersion.

To minimize the effect of sample volume on dispersion, and ensure that there was minimum dispersion from the valve and valve connections, a 0.2 µl Valco internal loop valve was employed to place the sample on the column. In addition, the sample valve was used with an intra-column injection system [2] that ensured the sample was placed in the center of the column about 5 mm below the packing surface. This device also determined that, even if the packing settled, there would be no void above the point of injection and the effective length of the column was not changed. All connecting tubes were 0.007 in. I.D. and their lengths were kept to a minimum (*i.e.*, <5 cm). The detector employed was the Perkin Elmer LC-85B fitted with a cell 1.4 µl volume and was used in conjunction with an electronic amplifier that had an effective time constant of 24 millisec. The columns employed had internal diameters of 8-9 mm to ensure the peak volumes were very large compared to any dispersion introduced by extra-column dispersion. As a result, the extra-column dispersion was maintained at a level of less than 2% of the peak volume for all measurements and, in most instances, was maintained at a level of less than 1%.

The initial work was carried out using a silica column 25 cm long and 9 mm I.D. packed with Partisil 10 silica (actual mean particle diameter 8.5 µm) thermostatted at 25°C. In their first series of experiments, six data sets were obtained for (H) and (u), employing six solvent mixtures, each exhibiting different diffusivities for the two solutes. This served two purposes as not only were there six different data sets with which the dispersion equations could be tested, but the coefficients in those equations supported by the data sets could be subsequently correlated with solute diffusivity. The solvents employed were approximately 5%v/v ethyl acetate in *n*-pentane, *n*-hexane, *n*-heptane, *n*-octane, *n*-nonane and *n*-decane. The solutes used were benzyl acetate and hexamethylbenzene. The diffusivity of each solute in each solvent mixture was determined in the manner of Katz *et al*. [3] and the values obtained are included in Table 1. The solvent mixtures were adjusted slightly to produce a (k') value of 2 for benzyl acetate in all solvent mixtures.

Table 1. Physical Chemical Data for the Different Mobile Phases and Solutes

Solvent Et. Acetate in n-Alkane %v/v	Benzyl Acetate			Hexamethyl-benzene
	k'	k"	D_m $10^{-5} cm^2 s^{-1}$	D_m $10^{-5} cm^2 s^{-1}$
4.58 in C5	2.05	4.07	3.61	3.51
4.86 in C6	1.97	3.94	3.06	2.73
4.32 in C7	2.04	4.06	2.45	2.23
4.50 in C8	2.01	4.01	2.01	1.71
4.41 in C9	2.12	4.20	1.65	1.35
4.82 in C10	2.01	4.01	1.46	1.17

Table 1 includes the values of (k') and (k") determined in the manner discussed in earlier chapters, *i.e.*,

$$k' = \frac{V_r - V_o}{V_o} \quad \text{and} \quad k'' = \frac{V_r - V_e}{V_e}$$

where the symbols have the meanings previously ascribed to them.

The excluded volume (V_e) was taken as the retention volume of the fully excluded solute polystyrene (mean molecular weight 83,000). The variance per unit length (H) for each column was measured at twelve different linear velocities, over a range that extended from 0.02 cm/sec. to 0.6 cm/sec. (H) was taken as the ratio of the effective length of the column to the column efficiency (n), and (n) was taken as four times the square of the ratio of the retention distance to the peak width measured at 0.06065 of the peak height. Each measurement was made in triplicate (three chromatograms) and if any individual measurement differed from the others by more than 3%, then further replicate measurements were made. The velocity was measured in two ways; firstly, as the ratio of the column length to the fully permeating, but unretained, solute (hexamethylbenzene) (u); secondly, as the ratio of the column length to the fully excluded solute (polystyrene) (u_e). Thus,

$$u_e = \frac{u V_o}{V_e}$$

Experimental Validation of Dispersion Equations

The results obtained were probably as accurate and precise as any available and, consequently, were unique at the time of publication and probably unique even today. Data were reported for different columns, different mobile phases, packings of different particle size and for different solutes. Consequently, such data can be used in many ways to evaluate existing equations and also any developed in the future. For this reason, the full data are reproduced in Tables 1 and 2 in Appendix 1. It should be noted that in the curve fitting procedure, the true linear velocity calculated using the retention time of the totally excluded solute was employed. An example of an HETP curve obtained for benzyl acetate using 4.86%v/v ethyl acetate in hexane as the mobile phase and fitted to the Van Deemter equation is shown in Figure 1.

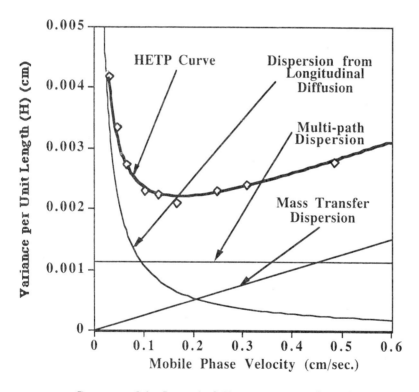

Courtesy of the Journal of Chromatography [ref. 1]

Figure 1. The (H) versus (u) Curve for Benzyl Acetate Using 4.86%v/v Ethyl Acetate in *n*-Hexane as the Mobile Phase

(H) was determined for each solute, in each solvent mixture, for 10 different linear velocities ranging from 0.02 cm/sec. to 0.6 cm/sec. Three measurements were taken at each velocity, which produced a minimum of 180 values of (H) for each solute. Each data set was then fitted to each dispersion equation and the respective constants

(A), (B), (C), etc. calculated. The data were sufficient in both quantity, and quality to allow the most appropriate dispersion equation to be identified. The results obtained are shown in Table 2.

Table 2. Experimental Values for the Dispersion Equation Coefficients Obtained by a Curve Fitting Procedure

	n-pentane 4.7%w/w Et. Acet.	n-hexane 4.9%w/w Et. Acet.	n-heptane 4.3%w/w Et. Acet.	n-octane 4.3%w/w Et. Acet.	n-nonane 4.5w/w Et. Acet.	n-decane 4.4%w/w Et. Acet.

The Van Deemter equation $H = A + \dfrac{B}{u} + Cu$

A	0.001189	0.001144	0.001144	0.001210	0.001208	0.001237
B	0.000108	0.000091	0.000081	0.000065	0.000049	0.000045
C	0.002525	0.003008	0.003362	0.003661	0.004298	0.004786

The Giddings equation $H = \dfrac{A}{1+\dfrac{E}{u}} + \dfrac{B}{u} + Cu$

A	0.001189	0.001144	0.001123	0.001210	0.001407	0.001257
B	0.000108	0.000091	0.000086	0.000065	0.000067	0.000053
C	0.002525	0.003008	0.003348	0.003661	0.004001	0.004754
D	0	0	0.005243	0	0.0337	0.008100

The Huber equation $H = \dfrac{A}{1+\dfrac{E}{u^{0.5}}} + \dfrac{B}{u} + Cu + Du^{0.5}$

A	0.001455	0.001408	0.000986	0.001196	0.000702	0.1612
B	0.000104	0.000092	0.000084	0.000065	0.000057	0.000056
C	0.003302	0.002728	0.002979	0.003622	0.002769	0.002310
D	-0.00092	0.000331	0.000447	0.000046	0.001775	-0.06804
E	0	0	0	0	0	2.13100

The Knox equation $h = \dfrac{B}{\nu} + A\nu^{1/3} + C\nu$

A	0.002509	0.002422	0.002390	0.002545	0.002608	0.002626
B	0.000123	0.000105	0.000096	0.000080	0.000064	0.000061
C	0.008720	0.001407	0.001754	0.002003	0.002518	0.002304

The Horvath equation $H = \dfrac{A}{1+\dfrac{E}{u^{1/3}}} + \dfrac{B}{u} + Cu + Du^{2/3}$

A	0.001366	0.001507	0.001013	0.001197	0.000825	0.005583
B	0.000104	0.000092	0.000083	0.000065	0.000056	0.000051
C	0.003572	0.002495	0.002744	0.003585	0.001948	0.009169
D	-0.00110	0.000541	0.000647	0.000080	0.002454	-0.008577
E	0	0	0	0	0	97.30

Examination of the data given in Table 2 shows a rational fit of the experimental data to the equations of Van Deemter, Giddings and Knox. Both the Huber and Horvath equations, however, gave alternating positive and negative values for the constant (D)

Experimental Validation of Dispersion Equations

(the coefficient of the term involving a fractional power of (u)). In addition, the value of coefficient (E) for the Huber equation is consistently zero and, similarly, for the Horvath equation, the value of (E) is zero for four solvent mixtures out of six, with an extreme value of 97.3 for one solvent. Katz *et al* also reported that the data for hexamethylbenzene produced the same poor fit with the Huber and Horvath equations. Consequently, due to the irrational fits of the data to the Huber and Horvath equations, these expressions were not considered to describe the relationship between (H) and (u) in a satisfactory manner.

From the results shown in Table 2, it is seen that, although a good fit is obtained to the Giddings equation, the value of (E) is numerically equal to zero. This would indicate that the Van Deemter equation is a special case of the Giddings equation, where, at the linear velocities employed (*i.e.*, those normally employed in practical LC) the coupling effect is complete and the multi-path term is independent of the velocity (u). It is possible, however, that at velocities outside the range studied (*i.e.*, at very low velocities where coupling is not complete), the Giddings equation might be more appropriate. To date, sufficiently precise data have not become available to test this possibility. Considering the practical range of velocities, however, it is the Van Deemter and Knox equations that are the two that must be examined further to decide which more accurately describes dispersion in a packed bed.

The Multi-path, or (A), Term

Stating the Van Deemter and Knox equations in the explicit form, the Van Deemter equation is

$$H = 2\lambda d_p + \frac{2\gamma D_m}{u_e} + \frac{f_1(k)d_p^2}{D_m}u_e + \frac{f_1(k)d_f^2}{D_s}u_e \qquad (1)$$

where the symbols have the meanings previously ascribed to them (as defined in chapter 7), and that of Knox,

$$h = \frac{B'}{v} + A'v^{1/3} + C'v$$

To compare the two equations, the *reduced plate height* (h) and *reduced velocity* (v) must be put in terms of (H) and (u_e). The reduced plate height is given by $h = \frac{H}{d_p}$ and the reduced velocity by $v = \frac{u_e d_p}{D_m}$, thus, the Knox equation becomes

$$\frac{H}{d_p} = \frac{BD_m}{u_e d_p} + A\left(\frac{u_e d_p}{D_m}\right)^{1/3} + C\frac{d_p}{D_m}u_e$$

or
$$H = \frac{BD_m}{u_e} + A\left(\frac{u_e d_p^2}{D_m}\right)^{1/3} + C\frac{d_p^2}{D_m}u_e \qquad (2)$$

where $A' = A\left(\frac{d_p^2}{D_m}\right)^{1/3}$, $B' = BD_m$ and $C' = C\frac{d_p^2}{D_m}$.

It is interesting to note that the function from the Knox equation $\left(\frac{u_e d_p^2}{D_m}\right)$ is the inverse of the coupling term (E) in the Giddings equation (chapter 8). It is also seen that the C term in the Knox equation is very similar to that in the equations of both Giddings and Van Deemter but at no value of (u_e) does any term tend to a constant value as it does in both the Giddings and the Van Deemter equations. Obviously, there is a very significant difference between equations (1) and (2). First, only the Van Deemter equation contains a term (A) that is independent of both the linear mobile phase velocity and the solute diffusivity.

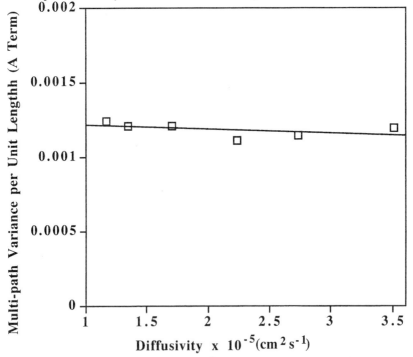

Figure 2. Graph of Multi-path Variance against Solute Diffusivity for Benzyl Acetate

Experimental Validation of Dispersion Equations

The linear curve in Figure 2 indicates that the magnitude of the (A) term appears, within experimental error, to be independent of the diffusivity of the solute in the mobile phase. Close examination of the graph, however, discloses a possible slight residual dependence of the (A) term on diffusivity. This might imply that the velocity range examined was not quite high enough to ensure that the first term of the Giddings equation was reduced to a constant as in the simple Van Deemter equation. Consider the detailed expression for the first term of the Giddings equation, *viz.*,

$$\frac{1}{\dfrac{1}{2\lambda d_p} + \dfrac{D_m}{\omega u_e d_p^2}}$$

As already discussed, when (u_e) is sufficiently large, the right-hand function in the denominator becomes negligible and the term simplifies to $2\lambda d_p$. However, if (u) takes values where the right-hand side of the denominator was not quite zero, then the value of the (A) term would show a slight decrease with increasing values of D_m. This might explain the slight slope of the straight line in Figure 2.

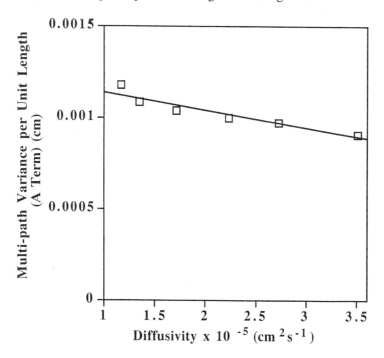

Figure 3. Graph of Multi-path Variance against Solute Diffusivity for Hexamethylbenzene

Figure 3, A similar graph of the (A) term against solute diffusivity for hexamethylbenzene, provides even stronger support for this possibility. The somewhat greater slope for the solute hexamethylbenzene, relative to benzyl acetate, is probably due to the smaller change in diffusivity between the different solvents. In any event, the magnitude of (A) changes little with change in diffusivity and the small change can be accounted for by the coupling term of Giddings. Thus, although the data shows some slight dependence of the (A) term on D_m, as a result of the curve fitting procedure to the equation $H = A + \frac{B}{u} + Cu$, it is shown not to be dependent on (u). This supports the Van Deemter equation as opposed to the Knox equation. It does, however, also support the idea that the Van Deemter equation is indeed a special case of the Giddings equation.

The Longitudinal Diffusion, or the (B), Term

Now, equations (1) and (2) indicate that both the Knox equation and the Van Deemter equation predict a linear relationship between the value of the (B) term (the longitudinal diffusion term) and solute diffusivity.

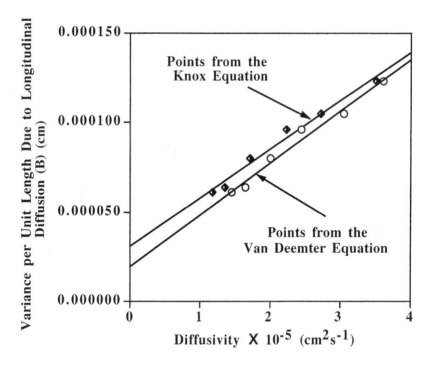

Figure 4. Graph of Variance Due to Longitudinal Diffusion against Solute Diffusivity for Benzyl Acetate and Hexamethylbenzene

A plot of the (B) term against diffusivity for benzyl acetate and hexamethylbenzene is shown in Figure 4. The predicted linear relationship is clearly confirmed. However,

Experimental Validation of Dispersion Equations

the values for the (B) term from the Knox equation curve fit give as good a linear relationship with solute diffusivity as those from the Van Deemter curve fit. It follows that the linear curves shown in Figure 4 do not *exclusively* support either equation.

The Optimum Velocity and the Minimum Plate Height

The *static* phase, for a silica gel column, would consist almost entirely of that portion of the mobile phase that is trapped in the pores of the silica gel. Consequently, it would appear reasonable to assume that, in equation (1),

$$D_m = D_s$$

However, there might be exceptions; if the mobile phase consists of a binary mixture of solvents, then a layer of the more polar solvent would be adsorbed on the surface of the silica gel and the mean composition of the solvent in the pores of the silica gel would differ from that of the mobile phase exterior to the pores. Nevertheless, it would still be reasonable to assume that

$$D_s = eD_m$$

where (e) is a constant, probably close to unity.

It should be pointed out, however, that the diffusivity of the solute in the mobile phase can be changed in two ways. The solute that is chromatographed can be changed, in which case the above assumptions are clearly valid, as (D_s) is likely to change linearly with (D_m). However, the solute diffusivity can also be changed by the employing a *different mobile phase*. In this case, (D_m) will be changed but (D_s) will remain the same. In the second case, the above assumptions are not likely to be *precisely* correct. Nevertheless, if the resistance to mass transfer in the *stationary phase* makes only a small contribution to the overall value of (H) (*i.e.*, because $d_f \ll d_p$ (see equation (1)), then the assumption $D_m = eD_s$ will still be approximately true. Thus, assuming that the diffusivity is varied by changing the nature of the solute and not the mobile phase, then D_s can be replaced by eD_s and equation (1) can be simplified to

$$H = 2\lambda d_p + \frac{2\gamma D_m}{u_e} + \frac{bu_e}{D_m} \tag{3}$$

where
$$b = f_1(k)d_p^2 + \frac{f_2(k)d_f^2}{e}$$

Differentiating equation (3) with respect to (u_e),

$$\frac{dH}{du_e} = -\frac{2\gamma D_m}{u_e^2} + \frac{b}{D_m}$$

Equating to zero and solving for $u_{opt.}$, it can be seen that

$$u_{e(opt)}^2 = \frac{2\gamma D_m^2}{b} \quad \text{and, thus,} \quad u_{e(opt)} = \sqrt{\frac{2\gamma}{b}} D_m \qquad (4)$$

Further, by substituting for ($u_{opt.}$) in equation (3), it can be seen that the minimum plate height ($H_{min.}$) is given by,

$$H_{min} = 2\lambda d_p + 2\sqrt{2\gamma b} \qquad (5)$$

Equations (4) and (5) predict that the optimum linear velocity should be linearly related to the diffusivity of the solute in the mobile phase, whereas the minimum value of the HETP should be constant and independent of the solute diffusivity. This, of course, will only be true for solutes eluted at the same (k'). It is seen, from Table 1, that (by appropriate adjustment of the concentration of ethyl acetate) the values of both (k') and (k'_e) have been kept approximately constant for all the mobile phase mixtures employed. Consequently, assuming the resistance to mass transfer in the stationary phase contribution to the total variance (H) is relatively small, then equations (4) and (5) can be tested against experimental data.

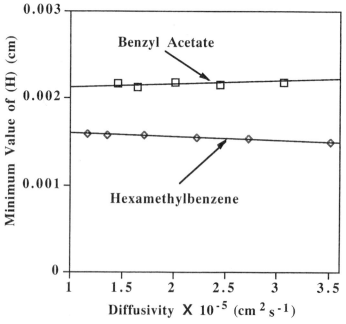

Figure 5. Graph of H_{min} against Solute Diffusivity

Experimental Validation of Dispersion Equations

The values of (H_{min}), plotted against solute diffusivity, are shown in Figure 5 and it is seen that the independence of (H_{min}) on diffusivity is largely confirmed. Close examination, however, shows that neither of the lines for the two solutes are completely horizontal with the baseline; nevertheless the dependence of (H_{min}) on diffusivity is extremely small for the solute benzyl acetate. The slight slope of the line for the solute hexamethylbenzene might well have two explanations; the slight slope is due to the (A) term not being completely independent of the diffusivity (D_m), as shown by the results in Figure 3 (the residual effect of the coupling term); the resistance to mass transfer in the stationary phase *does* make a small, but significant, contribution to the value of (H).

Curves relating the optimum velocity to the solute diffusivity are shown in Figure 6. It is seen that the straight lines predicted by the Van Deemter equation are realized for both solutes.

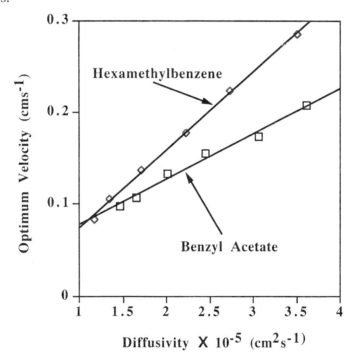

Figure 6. Graph of Optimum Velocity against Solute Diffusivity

Similar treatment of the Knox equation does not predict that values of H(min) should be independent of the solute diffusivity; neither does it predict that (u_{opt}) should vary linearly with solute diffusivity. Consequently, the relationships shown in Figures 5 and 6 are strong evidence supporting the validity of the Van Deemter equation, as opposed to that of Knox.

Dispersion Due to Resistance to the Mass Transfer of the Solute between the Two Phases

Reiterating equation (1), $$H = 2\lambda d_p + \frac{2\gamma D_m}{u_e} + \frac{f_1(k)d_p^2}{D_m}u_e + \frac{f_1(k)d_f^2}{D_s}u_e$$

i.e., $$C = \frac{f_1(k)d_p^2}{D_m} + \frac{f_1(k)d_f^2}{D_s}.$$

It is seen that the Van Deemter equation predicts that the total resistance to mass transfer term must also be linearly related to the reciprocal of the solute diffusivity, either in the mobile phase or the stationary phase. Furthermore, it is seen that if the value of (C) is plotted against $1/D_m$, the result will be a straight line and if there is a significant contribution from the resistance to mass transfer in the stationary phase, the curves will show a positive intercept. Such curves are shown in Figure 7

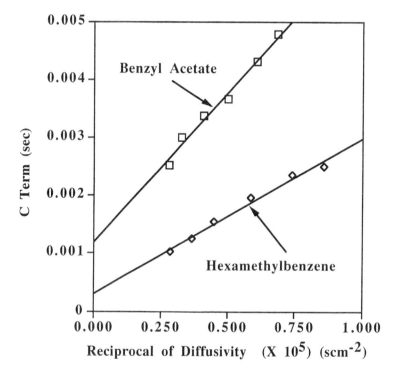

Figure 7. Graph of Resistance to Mass Transfer Term against the Reciprocal of the Solute Diffusivity in the Mobile Phase

Experimental Validation of Dispersion Equations

In Figure 7, the resistance to mass transfer term (the (C) term from the Van Deemter curve fit) is plotted against the reciprocal of the diffusivity for both solutes. It is seen that the expected linear curves are realized and there is a small, but significant, intercept for both solutes. This shows that there is a small but, nevertheless, significant contribution from the resistance to mass transfer in the stationary phase for these two particular solvent/stationary phase/solute systems. Overall, however, all the results in Figures 5, 6 and 7 support the Van Deemter equation extremely well.

The Effect of Particle size on the Magnitude of the Van Deemter C Term

Katz et al. [1] also examined the effect of particle diameter on resistance to mass transfer constant (C). They employed columns packed with 3.2 μm, 4.4 μm, 7.8 μm, and 17.5 μm, and obtained HETP curves for the solute benzyl acetate in 4.3%w/w of ethyl acetate in *n*-heptane on each column. The data were curve fitted to the Van Deemter equation and the values for the A, B and C terms for all four columns extracted. A graph relating the value of the (C) term with the square of the particle diameter is shown in Figure 8.

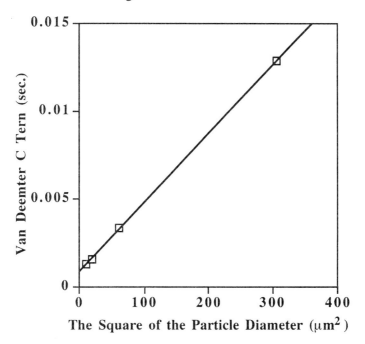

Figure 8. Graph of C Term against the Square of the Particle Diameter

The predicted linear relationship between the resistance to mass transfer term and the square of the particle diameter is clearly demonstrated in Figure 8. The linear correlation is extremely good and it is seen that there is an intercept on the (C) term axis at zero particle diameter. The intercept confirms that a small but significant contribution from the resistance to mass transfer in the stationary phase is present in the system. It should be pointed out, however, that the two systems differ and the intercept cannot be directly compared with that obtained with a series of different solvents.

The Effect of the Function of (k') on Peak Dispersion

If it is assumed that the column is operated at relatively high velocities, such that the contribution from longitudinal diffusion is no longer significant, then to a first approximation,

$$H = 2\lambda d_p + \frac{f_1(k)d_p^2}{D_m}u_e + \frac{f_1(k)d_f^2}{D_S}u_e \qquad (6)$$

Assuming that the diffusivity of the solute in the stationary phase (D_S) is simply related to the diffusivity in the mobile phase (D_m), i.e., $D_S = eD_m$

Substituting for (D_S) in equation (7)

$$H = 2\lambda d_p + \left(\frac{f_1(k)d_p^2 + \frac{f_1(k)d_f^2}{e}}{D_m}\right)u_e \qquad (7)$$

Inserting the expression for $f_1(k')$ recommended by Purnell [4] $\frac{1+6k+11k^2}{24(1+k)^2}$ and the expression for $f_1(k')$ as derived by Van Deemter [5] $\frac{k}{(1+k)^2}$ and rearranging,

$$H - 2\lambda d_p = \frac{1}{D_m}\left(\frac{1+6k+11k^2}{24(1+k)^2}d_p^2 + \frac{8k}{\pi^2(1+k)^2}\frac{d_f^2}{e}\right)u_e \qquad (8)$$

Dividing throughout by $\frac{1+6k+11k^2}{24(1+k)^2}$, and rearranging,

$$(H - 2\lambda d_p)\frac{D_m}{u_e}\left(\frac{24(1+k)^2}{1+6k+11k^2}\right) = d_p^2 + \frac{8k}{\pi^2(1+6k+11k^2)}\frac{d_f^2}{e} \qquad (9)$$

It follows, that if the Van Deemter equation is correct, a graph relating,

Experimental Validation of Dispersion Equations

$(H - 2\lambda d_p)\dfrac{D_m}{u_e}\left(\dfrac{24(1+k)^2}{1+6k+11k^2}\right)$ to $\dfrac{k}{(1+6k+11k^2)}$, should provide a straight line.

Katz and Scott [6] developed a method for measuring the molecular weight of a solute from chromatographic data (the particulars of which will be discussed in detail in a later chapter). However, their work generated sufficient data to test the relationship given in equation (9). Furthermore, the equation could be tested against the two alternative values for the capacity factor, either (k') calculated by employing the fully permeating dead volume (the thermodynamic dead volume), or (k") derived by employing the excluded dead volume (the dynamic dead volume). The correct value for (k) would provide a linear relationship between

$$(H - 2\lambda d_p)\dfrac{D_m}{u_e}\left(\dfrac{24(1+k)^2}{1+6k+11k^2}\right) \text{ and } \dfrac{k}{(1+6k+11k^2)}$$

Katz and Scott measured (k') and (k"), the diffusivities, and the total HETP curves (identifying the magnitude of $2\lambda d_p$) for 69 different solutes. This data were inserted in the functions and values of $(H - 2\lambda d_p)\dfrac{D_m}{u_e}\left(\dfrac{24(1+k)^2}{1+6k+11k^2}\right)$ were plotted against $\dfrac{k}{(1+6k+11k^2)}$ and the results are shown in Figure 9.

It is seen that a linear curve is not obtained with the use of (k') values derived from the fully permeating dead volume and, thus, (k') can not be used in the kinetic studies of columns. In contrast, the linear curve shown when using (k"), obtained from the use of the dynamic dead volume, confirms that (k'_e) values based on the excluded dead volume must be employed in *all column kinetic* studies. Thus, the excluded dead volume must be used for measuring mobile phase velocities and solute capacity ratios in all kinetic studies of columns and also in column design.

In summary, it can be said that all the dispersion equations that have been developed will give a good fit to experimental data, but only the Van Deemter equation, the Giddings equation and the Knox equation give positive and real values for the constants in the respective equations.

The basically correct equation appears to be that of Giddings but, over the range of mobile phase velocities normally employed (*i.e.,* velocities in the neighborhood of the optimum velocity), the Van Deemter equation is the simplest and most appropriate to use.

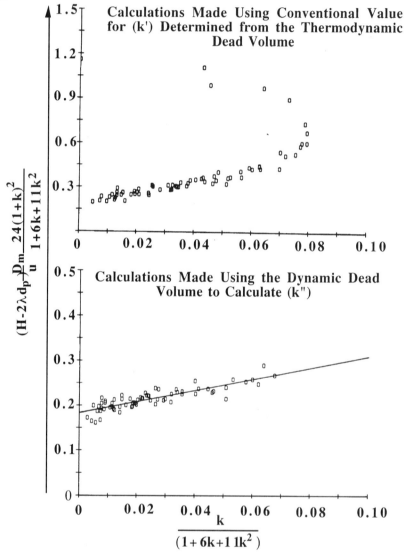

Figure 9. Graph of $\left(H-2\lambda d_p\right)\dfrac{D_m}{u_e}\left(\dfrac{24(1+k)^2}{1+6k+11k^2}\right)$ **against** $\dfrac{k}{\left(1+k+11k^2\right)}$

The Van Deemter equation appears to be a special case of the Giddings equation. The form of the Van Deemter equation and, in particular, the individual functions contained in it are well substantiated by experiment. The Knox equation is obtained from an empirical fit to experimental data and the individual functions contained in the equation are not all substantiated by experiment.

It would appear, from the data available at this time, that the Van Deemter equation would be the most appropriate to use in column design.

Synopsis

To confirm the pertinence of a particular dispersion equation, it is necessary to use extremely precise and accurate data. Such data can only be obtained from carefully designed apparatus that provides minimum extra-column dispersion. In addition, it is necessary to employ columns that have intrinsically large peak volumes so that any residual extra-column dispersion that will contribute to the overall variance is not significant. Merely obtaining a good fit between experimental data and a particular dispersion equation will not identify the pertinent equation that accurately describes column dispersion. First, the constants for the individual expressions in the equation that describe specific forms of dispersion must all be positive and real. Second, each dispersion function must relate in the predicted manner to the physical properties of the system, such as solute diffusivity and solute capacity ratio. Employing published data, the Giddings, Knox and Van Deemter equations are all shown to be pertinent, but the Knox equation is not correct in detail and is not suitable for column design. The Giddings equation appears to be the most pertinent, but the Van Deemter equation is a special case of the Giddings equation and can be used confidently in column design, providing the mobile phase velocity is close to, or significantly greater than, the optimum velocity. In all kinetic studies, the mobile phase velocity must be taken as the ratio of the column length to the dead time *measured by a fully excluded solute*. In addition, capacity ratios used in dispersion equations must be calculated using the dynamic dead volume and not the thermodynamic dead volume.

References

1. E. D. Katz, K. L. Ogan and R. P. W. Scott, *J. Chromatogr.*, **270**(1983)51.
2. E. D. Katz and R. P. W. Scott, *J. Chromatogr.* **270**(1983)29.
3. E. D. Katz, K. L. Ogan and R. P. W. Scott, *J. Chromatogr.*, **260**(1983)277.
4. J. H. Purnell, *Nature (London)* **Suppl. 26**(1959)184.
5. J. J. Van Deemter, F. Zuiderweg and A. Klinkenberg, *Chem. Eng. Sci.,* **5**(1956)271.
6. E. D. Katz and R. P. W. Scott, *J. Chromatogr.*, **270**(1983)29.

Chapter 11

The Measurement of Solute Diffusivity and Molecular Weight

Analytical information taken from a chromatogram has almost exclusively involved either retention data (retention times, capacity factors, etc.) for peak identification or peak heights and peak areas for quantitative assessment. The width of the peak has been rarely used for analytical purposes, except occasionally to obtain approximate values for peak areas. Nevertheless, as seen from the Rate Theory, the peak width is inversely proportional to the solute diffusivity which, in turn, is a function of the solute molecular weight. It follows that for high molecular weight materials, particularly those that cannot be volatalized in the ionization source of a mass spectrometer, peak width measurement offers an approximate source of molecular weight data for very intractable solutes.

The major problem in this procedure is not in obtaining an accurate value for the solute diffusivity from peak width measurements (which is relatively straightforward) but in identifying the best relationship between diffusivity and molecular weight to employ in the subsequent data processing.

Unfortunately, there are many expressions in the literature that give molecular weight as a function of diffusivity, and the most appropriate expression must be identified in order to permit a reasonably accurate value for the molecular weight to be calculated. Thus, the diffusivities of a large number of solutes of known molecular weight need to be measured in a solvent that is commonly used in the liquid chromatography, so that a practical relationship between diffusivity and molecular weight can be identified.

Katz and Scott [1] measured the diffusivity of 69 different solutes having molecular weights ranging from 78 to 446. The technique they employed was to measure the dispersion of a given solute band during passage through an open tube.

The dispersion in an open tube has been previously discussed (chapter 9 page 287). Reiterating equation (8) from Chapter 8 page 266

$$H = \frac{2D_m}{u} + \frac{1}{24}\frac{r^2}{D_m}u \tag{1}$$

where the symbols have the meaning previously ascribed to them.

If the linear velocity is high and $u \gg D_m/r$, then equation (1) reduces to,

$$\sigma_x^2 = H = \frac{r^2 u}{24 D_m} \tag{2}$$

Now, from the Plate Theory (Chapter 6, page 182) the peak variance in volume units (σ_v^2) is given by,

$$\sigma_v^2 = \frac{(\text{Tube Volume})^2}{n}$$

$$= \frac{(\pi r^2 L)^2}{n} = \frac{\pi^2 r^4 L^2}{n} \tag{3}$$

where (n) is the number of theoretical plates in the tube.

Now, $\dfrac{L}{n} = H$, thus, replacing $\left(\dfrac{L}{n}\right)$ by (H), and substituting for (H) from (2),

$$\sigma_v^2 = \pi^2 r^4 L H = \frac{\pi^2 r^4 L r^2 u}{24 D_m} = \frac{\pi^2 r^6 L u}{24 D_m} \tag{4}$$

Now, the flow rate through the tube (Q) will be given by $Q = \pi r^2 u$, so,

$$\sigma_v^2 = \frac{Q \pi r^4 L}{24 D_m} \tag{5}$$

Thus, $$D_m = \frac{Q \pi r^4 L}{24 \sigma_v^2}$$

Now, if the peak width at 0.6065 of the peak height is (x) and the chart speed (c), then the time variance of the peak (σ_t^2) is

$$\sigma_t^2 = \left(\frac{x}{2c}\right)^2$$

and thus the volume variance (σ_v^2) will be

$$\sigma_v^2 = (Q\sigma_t)^2 = \left(\frac{Qx}{2c}\right)^2 \qquad (6)$$

Substituting for (σ_v^2) from (6) in (5),

$$D_m = \frac{Q\pi r^4 L}{24\left(\frac{Qx}{2c}\right)^2} = \frac{\pi r^4 L c^2}{6Qx^2} \qquad (7)$$

It is seen from equation (7) that the diffusivity of a solute can be calculated from length and radius of the tube through which it passes, the flow rate, chart speed and the peak width at 0.6065 of the peak height.

Katz and Scott used equation (7) to calculate diffusivity data from measurements made on a specially arranged open tube. The equation that explicitly relates dispersion in an open tube to diffusivity (the Golay function) is only valid under condition of perfect Newtonian flow. That is, there must be *no radial flow induced in the tube* to enhance diffusion and, thus, the tube must be *perfectly straight*. This necessity, from a practical point of view, limits the length of tube that can be employed.

A low volume (0.2 µl) Valco sample valve was employed with one end of the open tube connected directly to the valve and the other connected directly to the sensor cell of the detector. The UV detector was the LC 85B manufactured by Perkin Elmer, and specially designed to provide low dispersion with a sensor volume of about 1.4 µl. The total variance due to extra-column dispersion was maintained at all times less than 2% of that of the elution curve of the solute. The open tube was 365.4 cm long with a radius of 0.0184 cm: it was enclosed in a 1 cm I.D. plastic tube through which thermostatting fluid passed to maintain the temperature 25°C ± 0.2°C. The detector output was passed to a high speed Bascomb Turner recorder that acquired data at a rate of five data points per second.

The solvent used was 5 %v/v ethyl acetate in *n*-hexane at a flow rate of 0.5 ml/min. Each solute was dissolved in the mobile phase at a concentration appropriate to its extinction coefficient. Each determination was carried out in triplicate and, if any individual measurement differed by more than 3% from either or both replicates, then further replicate samples were injected. All peaks were symmetrical (*i.e.*, the asymmetry ratio was less than 1.1). The efficiency of each solute peak was taken as four times the square of the ratio of the retention time in seconds to the peak width in seconds measured at 0.6065 of the peak height. The diffusivities obtained for 69 different solutes are included with other physical and chromatographic properties in table 1. The diffusivity values are included here as they can be useful in many theoretical studies and there is a dearth of such data available in the literature (particularly for the type of solutes and solvents commonly used in LC separations).

There are many equations that relate diffusivity to various physical and molecular properties of both solute and solvent [2-5], but the one that appeared to fit the data best was that of Arnold [2] that gave an expression for (D_m) of the following form,

$$D_m = \frac{A\left(\frac{1}{M_1} + \frac{1}{M_2}\right)^{\frac{1}{2}}}{\left(V_1^{\frac{1}{3}} + V_2^{\frac{1}{3}}\right)^2} \qquad (8)$$

where (A) is a constant,
(M$_1$) is the molecular weight of the solute,
(M$_2$) is the molecular weight of the solvent,
(V$_1$) is the molar volume of the solute,
and (V$_2$) is the molar volume of the solvent.

By trial and error it was found, however, that the best fit was obtained with the following arbitrary function,

$$\frac{1}{D_m} = A + BV_1^{\frac{1}{3}}M^{\frac{1}{2}} \qquad (9)$$

Table 1. Physical and Chromatographic Data for 70 Solutes in 5 %v/v Ethyl Acetate in *n*-Hexane

Compound	Mol. Wt.	d gml^{-1}	Dx10^5 cm^{2}s^{-1}	H cm	k'	k"
1.Benzene	78	0.874	4.196	0.02528	0.17	0.94
2.Benzonitrile	103	1.001	3.772	0.03302	2.21	4.30
3.*p*-Xylene	106	0.861	3.755	0.02627	0.12	0.87
4.Benzaldehyde	106	1.046	3.839	0.03281	1.98	3.91
5.Anisole	108	0.989	3.767	0.02945	0.56	1.60
6.Antranil	119	1.183	3.524	0.03517	3.92	7.14
7.Acetophenone	120	1.028	3.550	0.03628	2.95	5.59
8.Nitrobenzene	123	1.207	3.732	0.03318	1.75	3.57
9.Benzyl Chloride	126	1.103	3.423	0.02927	0.42	1.35
10.Naphthalene	128	1.145	3.694	0.02767	0.30	1.17
11.Phenyl-2-propanone	134	1.028	3.138	0.03883	5.64	10.05
12. *p*-Methylacetophenone	134	1.005	3.416	0.03805	3.00	5.61
13.2-Benzothiazole	135	1.248	3.294	0.03962	6.37	11.28
14.*o*-Nitrotoluene	137	1.168	3.557	0.03414	1.26	2.73
15.*p*-Dimethoxybenzene	138	1.053	3.487	0.03465	1.50	3.14
16.α,α,α-Trifluorotoluene	146	1.199	3.631	0.02834	0.40	1.37
17.4-Phenyl-3-buten-2-one	146	1.020	3.047	0.04106	7.38	12.95
18. Anethole	148	0.991	3.281	0.03331	0.60	1.67
19.Benzylacetone	148	0.989	3.020	0.04149	5.41	9.64
20.2-Methylbenzothiazole	149	1.203*	3.321	0.04078	6.30	11.12
21.Benzyl Acetate	150	1.057	3.127	0.03820	2.35	4.58
22.Ethyl Benzoate	150	1.051	3.055	0.03483	0.98	2.27
23.*p*-Tolyl Acetate	150	1.049	3.059	0.03716	2.11	4.17
24.Biphenyl	154	1.041	3.123	0.02843	0.31	1.17
25.1-Chloro-3-nitrobenzene	157	1.534	3.349	0.03510	1.53	3.19
26.2-Methoxynaphthalene	158	1.013*	3.102	0.03382	0.79	1.97
27.2,4-Dichlorotoluene	161	1.280*	-	0.02865	0.23	1.06
28.1,3,5-Triethylbenzene	162	0.863	2.585	0.03085	0.06	0.76
29.*n*-Propyl Benzoate	164	1.021	2.859	0.03557	0.79	1.97
30.Ethylphenyl Acetate	164	1.031	2.839	0.03906	1.80	3.62
31.*p*-Diethoxybenzene	164	1.008	2.983	0.03575	0.89	2.12
32.2,3Dimethoxybenzaldehyde	166	1.019*	2.942	0.04106	6.39	11.27
33.Carbazole	167	1.10	2.482	-	-	-
34.*m*-Dinitrobenzene	168	1.575	3.136	0.03687	11.99	20.49
35.2-Acetonnaphthone	170	1.147*	2.966	0.04079	4.15	7.50
36.1-Nitronaphthalene	173	1.331	2.920	0.03750	2.04	4.01
37.Ethyl Cinnamate	176	1.049	2.830	0.04004	1.64	3.40
38.4-Biphenylcarbonitrile	179	1.041*	2.812	0.03893	2.08	4.09
39.2',5-Dimethoxyacetophenone	180	1.126*	2.810	0.04204	8.03	14.02
40.Benzophenone	182	1.080	2.746	0.03830	1.47	3.11
41.2,4-Dinitrotoluene	182	1.521	2.812	0.03843	8.88	15.45
42.Bibenzyl	182	0.978	2.756	0.03272	0.25	1.08
43.Azobenzene	182	1.203	2.694	0.04080	2.64	5.05
44.1,2-Dimethoxy-4-nitrobenzene	183	1.348*	2.819	0.04120	18.56	31.43
45.*o*-Nitro-α,α,α-trifluorotoluene	191	1.281*	2.941	0.03760	6.84	11.94
46.Dibenzyl Ether	198	1.036	2.617	0.03796	0.86	2.09
47.Phenyl Benzoate	198	1.235	2.646	0.03764	1.10	2.47
48.*p*-Bromoacetophenone	199	1.647	3.143	0.03998	3.80	6.98
49.9-Cyanoanthracene	203	1.097*	2.510	0.04123	2.27	4.42
50.Benzil	210	1.23	2.497	0.04148	2.23	4.36
51.Benzyl Benzoate	212	1.112	2.587	0.03811	1.13	2.54

Table 1 Continued Compound	Mol. Wt.	d gml^{-1}	Dx10^5 cm^{2}s^{-1}	H cm	k'	k"
52.1,2--Diphenoxyethane	214	1.098*	2.475	0.04020	1.29	2.78
53.2,5-Diphenyloxazole	221	1.152#	2.505	0.04757	5.40	9.55
54.Triphenylene	228	1.302	2.498	0.03800	0.78	1.97
55.p-Terphenyl	230	1.221#	2.422	0.03657	0.47	1.44
56.7H-Benz[de]anthracene-7-one	230	1.249*	2.484	0.04328	3.77	6.93
57.Diethyl Phenylmalonate	236	1.095	2.250	0.04796	4.45	8.00
58.2-Naphthyl Benzoate	248	1.160*	2.363	0.04030	1.34	2.87
59.Dipropyl Phthalate	250	1.059	2.220	0.04842	4.09	7.47
60.Perylene	252	1.35	2.351	0.03892	0.97	2.28
61.Bis(2-phenoxyethyl) Ether	258	1.125*	2.168	0.04958	6.66	11.74
62.Tridecylbenzene	260	0.881	2.055	0.03359	0	0.66
63.Dibutyl Phthalate	278	1.043	2.024	0.05002	3.18	5.91
64.Hexachlorobenzene	285	2.044	2.604	0.03020	0.06	0.81
65.O,O-Diethyl O-p-Nitrophenyl Phosphorothioate	291	1.286*	2.149	0.04582	3.15	5.90
66.1,2,4,5-Dibenzopyrene	302	1.288*	2.180	0.04749	1.32	2.87
67.O,O-Diethyl O-[2-isopropyl-4-methyl-6-pyrimidyl] phosphorothioate	304	1.107*	1.922	0.05840	10.77	17.96
68. m-Quaterphenyl	306	1.206	1.890	0.04347	0.61	1.66
69.Dioctyl Phthalate	390	0.981	1.635	0.05803	1.83	3.71
70.Didecyl Phthalate	446	0.965	1.462	0.06208	1.55	3.24

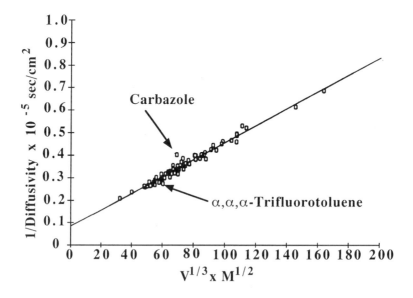

Figure 1. Graph of the Reciprocal of the Diffusivity against the Product of the Cube Root of the Molar volume and the Square Root of the Molecular Weight

Now, bearing in mind that $V_1 = \dfrac{M_1}{d_1}$ where (d_1) is the density of the solute,

Measurement of Diffusivity and Molecular Weight 341

$$\frac{1}{D_m} = A + B \frac{M_1^{0.833}}{d_1^{\frac{1}{3}}} \tag{10}$$

It is seen that if the diffusivity is to be correlated with the molecular weight, then a knowledge of the density of the solute is also necessary. The result of the correlation of the reciprocal of the diffusivity of the 69 different compounds to the product of the cube root of the molecular volume and the square root of the molecular weight is shown in Figure 1. A summary of the errors involved is shown in Figures 2 and 3

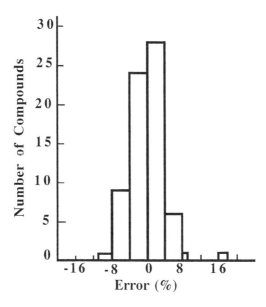

Figure 2. Graph of Number of Compounds against the Distribution of the Percentage Error

Figure 2 is a histogram that reveals the number of compounds associated with a particular error, whereas in Figure 3, the percent of the compounds is plotted against their respective error. It is seen that the linear relationship holds for 90% of all compounds within an error of less than 7%, and 95% of the compounds gave values that exhibited errors of less than 8%. In fact, the correlation held so well that 98% of the compounds exhibited a maximum error of 11%. The two substances with the greatest error were carbazole and α,α,α-trifluorotoluene which gave errors of 18% and 10%, respectively. 1,3,5-Triethylbenzene also gave a significant error of 9% but a possible explanation for this could be the presence of isomers causing greater peak dispersion and thus a false diffusivity value.

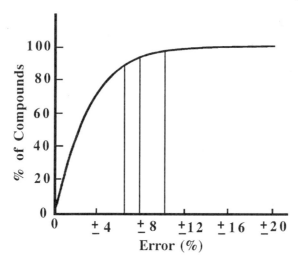

Figure 3. Graph of Percent of Compounds against Percent Error

It follows, that the peak width of a solute could give an indication of its molecular weight and, although the data may not be precise, approximate values could be extremely valuable when dealing with very large molecular weight substances such as polypeptides and proteins. In particular, the technique would be very useful for those substances that are extremely difficult, or impossible, to vaporize in the ion source of a mass spectrometer to provide mass data.

However, other factors that might effect the accuracy and precision of the diffusivity measurement need to be taken into account. In particular, due to the need for high pressures in LC, the effect of pressure on diffusivity must be considered. In general, diffusivity falls linearly with pressure as shown by the graph of diffusivity against pressure depicted in figure 4. It is seen that, over the pressure range examined, the diffusivity falls by 0.69 % per megaPascal. In fact, this agrees well with the previous work of Katz *et al.* [6], who predicted the change to be about 0.80 % per megaPascal. Unfortunately, the effect of pressure on diffusivity in LC is confused by the accompanying change in mobile phase temperature. The work done in forcing the mobile phase through the column against viscous resistance is dissipated as heat which raises the temperature of the mobile phase [6]. Thus, as the pressure falls along the column, the temperature rises which increases the diffusivity. The thermal increase in diffusivity, to some extent, counteracts the reduced diffusivity resulting from the higher pressure. It should be noted that the average column pressure can be taken as half the inlet pressure and, consequently, the change in diffusivity would only be 0.35 % per MPa at the column inlet.

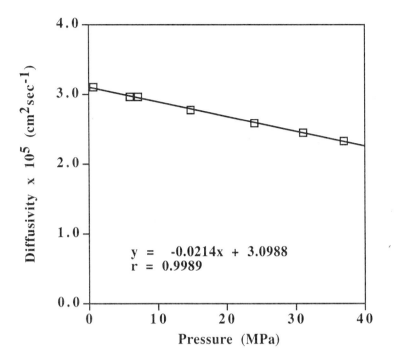

Figure 4. Graph of the Diffusivity of Benzyl Acetate against Absolute Pressure

The Relationship between Dispersion in a Packed Column to Solute Molecular Weight

Reiterating the Van Deemter equation,

$$H = 2\lambda d_p + \frac{2\gamma D_m}{u} + \frac{f_m(k)d_p^2 u}{D_m} + \frac{f_s(k)d_f^2 u}{D_s}$$

where all the symbols have the meanings previously ascribed to them.

Now, at high linear velocities, the longitudinal diffusion term will become insignificant and, equally important, the resistance to mass transfer term that incorporates the inverse function of diffusivity will become large, thus improving the precision of measurement.

Therefore,
$$H = 2\lambda d_p + \frac{f_m(k)d_p^2 u}{D_m} + \frac{f_s(k)d_f^2 u}{D_s}$$

or,
$$H - 2\lambda d_p = \frac{f_m(k)d_p^2 u}{D_m} + \frac{f_s(k)d_f^2 u}{D_s}$$

Inserting the previously established expressions for $f_m(k)$ and $f_s(k)$,

$$H - 2\lambda d_p = \frac{\left(1+6k''+11k''^2\right)d_p^2 u}{24(1+k'')^2 D_m} + \frac{8k'' d_f^2 u}{\pi^2 (1+k'')^2 D_S} \quad (11)$$

From equation (11), it is seen that, by measuring values of (H) over a series of linear velocities that are well above the optimum, a value for the diffusivity of the solute in the mobile phase can be obtained and hence an estimate of the molecular weight. A similar apparatus was used to that employed for the basic diffusivity measurements. However, due to the evolution of heat in packed columns at the high flow rates (necessary to reduce the significance of the longitudinal diffusion term), it was imperative to employ a microbore column (1 mm I.D. and 100 cm long) appropriately thermostatted at a temperature of 25°C ± 0.2°C. Microbore columns generate less heat than conventional columns as they operate at much lower flow rates and, due to their small dimensions, lose heat more rapidly and are, thus, more easily thermostatted. To enhance, still further, the resistance to mass transfer contribution to dispersion, large particles were use for packing (*viz.* nominally 20 μ and actually 17.5 μ). As the resistance to mass transfer increases as the square of the particle diameter, the use of such particles ensured a significant resistance to mass transfer contribution. The mobile phase was 5 %v/v ethyl acetate in *n*-heptane and the same solutes employed so their diffusivities were known from previous measurements. Efficiencies and the values for (H) were determined in the manner described previously at a mobile flow rate of 363 μl per minute, equivalent to a linear velocity of 0.98 cm per second. The values obtained for (H), (k') and (k") for each solute are included in table 1.

In order to relate the value of (H) to the solute diffusivity and, consequently, to the molecular weight according to equation (11), certain preliminary calculations are necessary. It has already been demonstrated in the previous chapter (page 303) that the dynamic dead volume and capacity ratio must be used in dispersion studies but, for equation (11) to be utilized, the value of the multipath term ($2\lambda d_p$) must also be evaluated.

Interactions in the stationary phase employing a porous stationary phase or support must also involve the mobile phase trapped in a static form inside the pores. It follows that the diffusivity of the solute in the stationary phase (D_S) will be similar to that in the mobile phase (D_m). Thus, to a first approximation, it can be assumed that $D_S = \omega D_m$, where (ω) is a constant probably close to unity. Thus, equation

Measurement of Diffusivity and Molecular Weight

(11) becomes,

$$H - 2\lambda d_p = \frac{(1+6k''+11k''^2)d_p^2 u}{24(1+k'')^2 D_m} + \frac{8k''d_f^2 u}{\pi^2(1+k'')^2 \omega D_m} \quad (12)$$

If a series of solutes is chosen from the group determined and all have approximately the same capacity ratio (k") but significantly different values for the diffusivity (D_m) then equation (11) reduces to

$$H = 2\lambda d_p + \left[\frac{(1+6k''+11k''^2)d_p^2 u}{24(1+k'')^2} + \frac{8k''d_f^2 u}{\pi^2(1+k'')^2 \omega}\right]\frac{1}{D_m} = 2\lambda d_p + \frac{\xi}{D_m} \quad (13)$$

$$\text{where } \xi = \frac{(1+6k''+11k''^2)d_p^2 u}{24(1+k'')^2} + \frac{8k''d_f^2 u}{\pi^2(1+k'')^2 \omega}$$

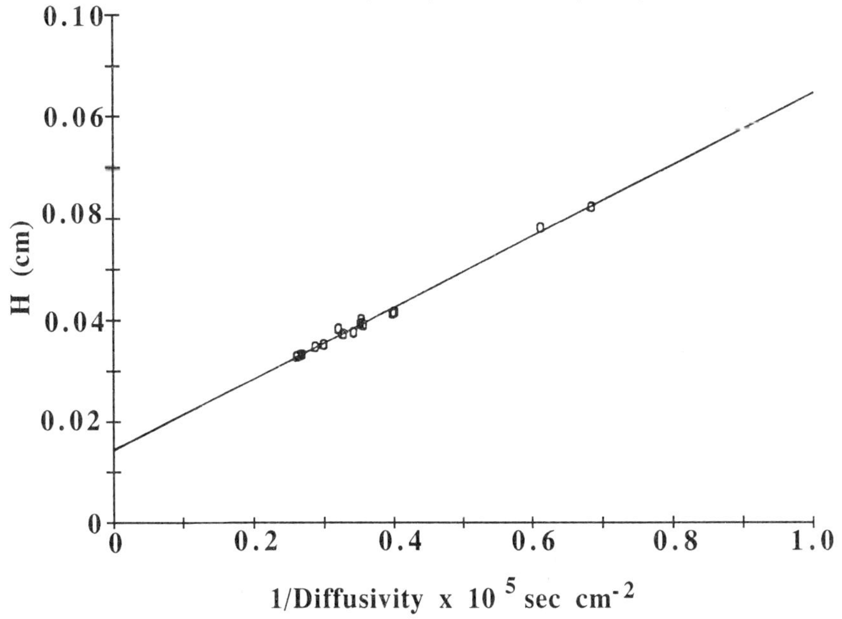

Figure 5. Graph of the Variance per Unit Length of the Column (H). against the Reciprocal of the Diffusivity

Thus, from equation (13) a value for ($2\lambda d_p$) can obtained by plotting (H) against ($1/D_m$) for data that has been obtained at a constant linear mobile phase velocity (u). Such a curve is shown in figure 5. It is seen that, from the intercept of the linear curve with the (H) axis, the value for ($2\lambda d_p$) for the particular column can be educed and can be used in further calculations providing the same column is employed.

It is now necessary to identify the correct functions of the capacity factor (k") to be used in equation (12) to evaluate the diffusivity. Knox [7] suggested the following approach which required a polynomial curve fitting procedure to identify the necessary constants. Rearranging equation (12),

$$(H - 2\lambda d_p)D_m(1+k")^2 = \frac{(1+6k"+11k"^2)d_p^2 u}{24} + \frac{8k"d_f^2 u}{\pi^2 \omega} \quad (14)$$

For a given column, operated at a constant linear velocity $(d_p^2 u)$ and $\left(\frac{d_f^2 u}{\omega}\right)$ are constants and, so, equation (14) can be put in the form,

$$(H - 2\lambda d_p)D_m(1+k")^2 = a + bk" + ck"^2 \quad (15)$$

Now, $\left((H - 2\lambda d_p)D_m(1+k")^2\right)$ can be calculated from the data given in table 1 and, thus, the constants (a), (b) and (c) can be found. However, the determination of the constants (a), (b) and (c) must be carried out with some circumspection. If a curve fitting routine to a second order polynomial is attempted, the precise value of (a) that is obtained is strongly influenced by small errors in the high values of (k") due to the second order term (ck"²). As a consequence, (a), (b) and (c) should first be obtained by a curve fitting procedure of $\left((H - 2\lambda d_p)D_m(1+k")^2\right)$ to a second order polynomial in (k") *for values of* (k" < 2). This will give precise values for (a) and (b) but not for (c): a precise value for (c) can only be obtained when large values of (k") are employed. Rearranging equation (15)

$$\frac{\left((H-2\lambda d_p)D_m(1+k")^2 - a\right)}{k"} = b + ck" \quad (16)$$

A linear fit of $\left(\frac{(H-2\lambda d_p)D_m(1+k")^2 - a}{k"}\right)$ to (k") will give another value for (b) (close in value to the previous) and a more accurate value for (c).

A graph relating $\left[\frac{(H-A)(1+k")^2}{a+bk"+ck"^2}\right]$ to the reciprocal of the diffusivities (employing the previously established values for (2λd_p) (a), (b) and (c)) is shown in figure 6.

Measurement of Diffusivity and Molecular Weight

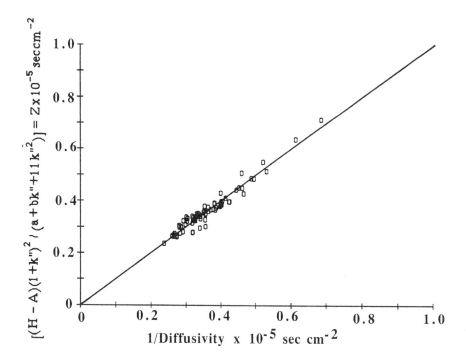

Figure 6. Graph of Corrected Peak Dispersion $\left[\dfrac{(H-A)(1+k'')^2}{a+bk''+ck''^2}\right]$ **against the Reciprocal of the Diffusivity**

It is seen that good linear relationship is obtained, substantiating the dependence of the resistance to mass transfer dispersion on the reciprocal of the diffusivity. The error expressed as a histogram relating the number of solutes to the percentage error is given Figure 7 and the percentage of samples plotted against the percentage error is shown in Figure 8. It is seen that for 90% of the samples, the error is less than 9%. The relationship between diffusivity and molecular weight also includes the solute density and, unfortunately, this information will not be available for an unknown sample. However, a large proportion of the solutes of interest and which are separated today by liquid chromatography techniques (particularly those of biological origin) will have densities close to unity. Those substances that have densities lying between 0.85 and 1.25 g/ml (where the cube root of the density approaches unity) were selected from the 68 solutes (totaling 56 substances) and values for $\left[\dfrac{(H-2\lambda d_p)(1+k'')^2}{a+bk''+ck''^2}=Z\right]$ were plotted against the molecular weight to

348 Chromatography Theory

the power of 0.833. The curve obtained is shown in figure 9.

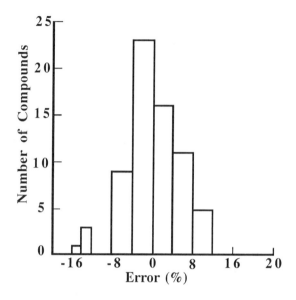

Figure 7. Graph of Number of Compounds against Error % obtained from the Linear Regression of the Graph of the Corrected Peak Dispersion against the Reciprocal of the Diffusivity

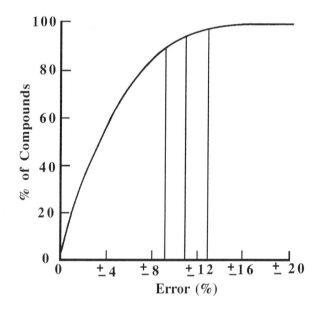

Figure 8. Graph of Percentage of Compounds against Error % obtained from the Linear Regression of the Graph of the Corrected Peak Dispersion against the Reciprocal of the Diffusivity

Measurement of Diffusivity and Molecular Weight 349

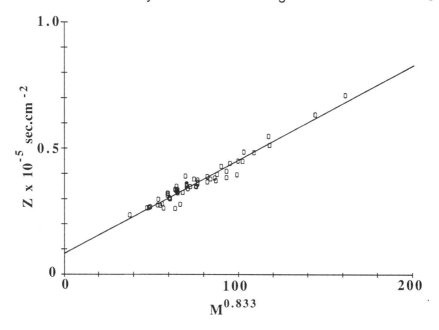

Figure 9. Graph of Corrected Peak Dispersion (Z) against the Molecular Weight to the Power of 0.833

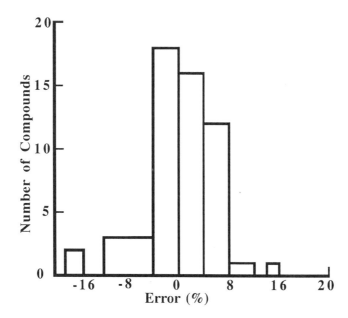

Figure 10. Graph of Number of Compounds against Error % Obtained from the Linear Regression of the Graph of the Corrected Peak Dispersion (Z) against the Reciprocal of the Diffusivity Molecular Weight to the Power of 0.833

It is seen that, again, a very good linear relationship is obtained between the function (Z) and the molecular weight to the power of 0.833. In figure 10, the number of samples is plotted against the percentage error, and in figure 11, the percent of the samples is plotted against the percentage error. It is seen from figure 11 that over 90 % of the samples involve an error of less than 9 %, the greatest error being 16 % . This level of error may appear gross in general analytical terms, but when examining samples of biological origin and, in particular, samples involving biopolymers accuracy's of this magnitude can be extremely valuable. In many instances, even employing the modern sophisticated ionization techniques, such materials are impossible to examine by mass spectrometry due to the difficulty in producing molecular ions. Under such circumstances molecular weight data, accurate to within ± 16 % can be a very valuable asset.

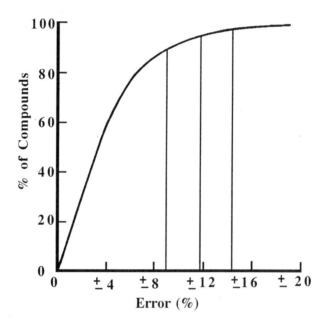

Figure 11. Graph of Percentage of Compounds against Error % obtained from the Linear Regression of the Graph of the Corrected Peak Dispersion (Z) against the Reciprocal of the Diffusivity Molecular Weight to the Power of 0.833

The values for the molecular weight obtained experimentally are shown plotted against the actual molecular weights in figure 12. The number of samples plotted against percentage error is shown in figure 13 and the percentage of all the samples plotted against percentage error in figure 14.

Measurement of Diffusivity and Molecular Weight

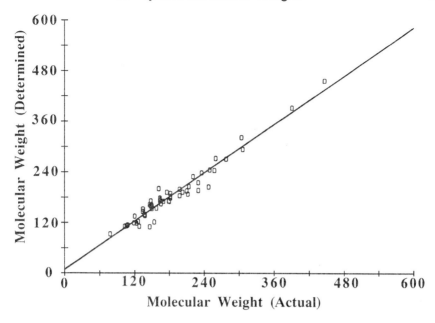

Figure 12. Graph of Molecular Weight Experimentally Determined against the Actual Molecular Weight

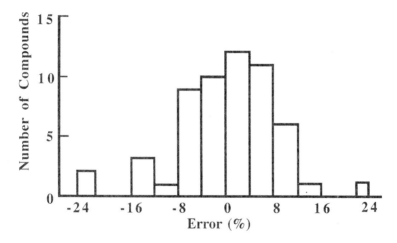

Figure 13. Graph of Number of Compounds against Error % Obtained from the Linear Regression of the Graph of the Determined Molecular Weight against the Actual Molecular Weight

It is seen that the molecular weight of a solute having a density between 0.85 and 1.25 can be estimated experimentally from peak width measurements for 90% of the compounds within an error of 13% (80% of the samples gave an error of less than 10%). Again, depending on the field of application, it should be noted that such data for a given substance can be obtained in addition to its separation from contaminating materials together with a quantitative estimate of its content in the mixture.

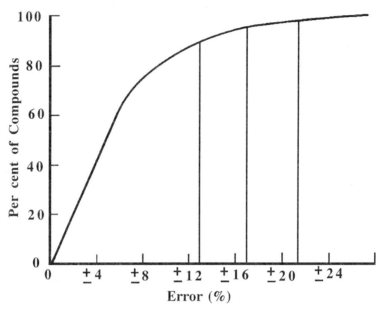

Figure 14. Graph of Number of Compounds against Error % obtained from the Linear Regression of the Determined Molecular Weight against the Actual Molecular Weight

The use of chromatographic procedures to measure solute diffusivity and molecular weight has been carefully, and extensively, examined by Katz in her Ph.D. thesis. (7). Those interested in the use of chromatographic techniques for this purpose are strongly recommended to refer to her thesis. Katz was particularly interested in using chromatography to provide approximate values for the molecular weight of proteins/peptides. She determined the diffusivities of 12 protein/peptides employing the open tube procedure and then determined their dispersion in a packed column using the same mobile phase (an acetonitrile/water mixture). She developed an extremely simple relationship between solute dispersion in a packed column and the molecular weights of the protein/peptides. Restating Arnold's equation (equation 8),

$$D_m = \frac{A\left(\frac{1}{M_1} + \frac{1}{M_2}\right)^{\frac{1}{2}}}{\left(v_1^{\frac{1}{3}} + v_2^{\frac{1}{3}}\right)^2} \qquad (17)$$

where all the symbols have the meanings previously ascribed to them.

Measurement of Diffusivity and Molecular Weight

Assuming the effective molecular weight of the solvent is significantly less than that of the protein/peptides, then equation (17) can be reduced to

$$D_m = \frac{A'\left(\frac{1}{M_2}\right)^{\frac{1}{2}}}{\left(v^{\frac{1}{3}}\right)^2} = \frac{A'\left(\frac{1}{M_2}\right)^{\frac{1}{2}}}{\left(\left(\frac{M_2}{d_2}\right)^{\frac{1}{3}}\right)^2}$$

where (d_2) is the density of the protein/peptide and (A') is a constant.

The vast majority of protein/peptides will have densities lying between 0.85 an 1.25 and, thus, the cube root of their density will tend to unity. Consequently, the following approximation can be made,

$$D_m = \frac{A'\left(\frac{1}{M_2}\right)^{\frac{1}{2}}}{\left((M_2)^{\frac{1}{3}}\right)^2} = \frac{A''}{(M_2)^{\frac{1}{3}}} \qquad (18)$$

where (A") is another constant.

Katz then tested the assumption that the dispersion of a high molecular weight solute such as a protein/peptide was largely independent of its capacity ratio, but strongly dependent on its relatively low diffusivity. That is,

$$H = E + \frac{F}{D_m} \qquad (19)$$

The dispersions of 12 solutes were then measured in a packed reverse phase column using the same mobile phase as that used in the determination of their diffusivities (*i.e.*, acetonitrile/water). The column dispersion (H) was then plotted against $1/D_m$ and the results are shown in Figure 15. It is seen that a fairly close linear relationship does, indeed, still exist between (H) and ($1/D_m$) which validates the approximations assumed by Katz.

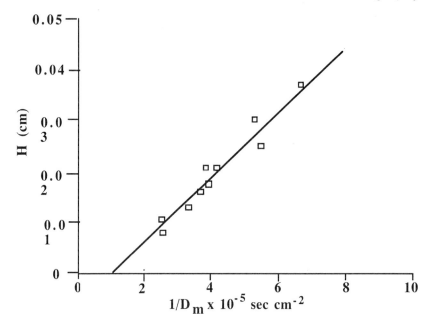

Figure 15 Graph of HETP against $1/D_m$ for 12 Proteins and Peptides

Katz then combined equations (18) and (19) arriving at the following simple expression,

$$H = E + F' M^{\frac{1}{3}} \qquad (20)$$

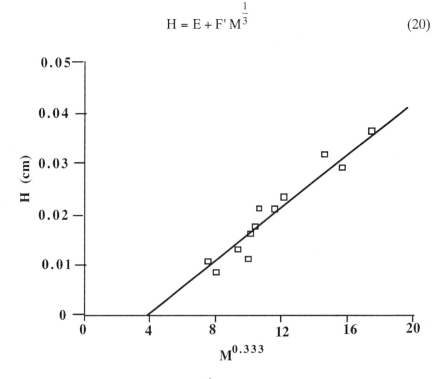

Figure 16. Graph of (H) against $M^{0.333}$ for 11 Peptides

Curves relating (H) and ($M^{1/3}$) are shown in figure 16 and the molecular weights determined together with their actual molecular weights are shown in Table 2.

Table 2 The Actual and Experimental Values of the Molecular Weight of 11 Proteins/Peptides

	Peptides	M Actual	M Det.	% Error
1	Insulin B (22-25)	525	568	8.2
2	Cholecystokin Tetrapeptide (30-33)	598	458	-3.4
3	Insulin B (21-26)	758	706	-6.8
4	Angiotensin III (human)	931	928	-.02
5	Cholecystokin Heptapeptide (30-33)	948	-	-
6	Angiotensin II (human)	1046	1074	2.7
7	Angiotensin I (human)	1297	1633	25.9
8	Gastrin Releasing Peptide (1-16) (Porcine)	1547	164	6.6
9	Neurotensin	1673	1923	14.9
10	Insulin Chain A, Oxidized	2530	3596	42.1
11	Insulin Chain B, Oxidized	3496	2518	-28.0
12	Insulin (Bovine)	6000	4905	-18.3

It is seen from figure 12 that a good linear relationship was obtained for all solutes (maximum error 30%) except for insulin chain A which exhibited an error of 40%. This error was probably due to closely eluting impurities but this was not confirmed experimentally. It follows, that the relationship between peak dispersion (uncorrected for its capacity ratio) and the cube root of the molecular weight can be employed to provide a reasonable estimate of the molecular weight. In figure 17, the experimental values obtained for the molecular weights for all solutes (except insulin chain A) are shown plotted against their true molecular weights. It is seen that considering the simple procedure that was employed, there is good agreement between the experimental values for the molecular weight of the protein/peptides and their true values, the curve having a slope of 0.759 and an intercept of 280.4.

Thus, a practical procedure would be as follows. Initially the HETP of a series of peptides of known molecular weight must be measured at a high mobile phase velocity to ensure a strong dependence of peak dispersion on solute diffusivity.

Such data can provide a calibration curve and allow the constants (E) and (F') in equation (20) to be determined. The value of the molecular weight of an unknown solute can then be obtained from its (H) value by reading the value directly from the curve or by calculation using the predetermined constants (E) and (F') in equation (20). It should be pointed out that an error of up to 30% may not appear to be very useful but, in fact, such precision can be extremely valuable in the preliminary examination of many biochemical substances where only very small quantities of material are available. It is also an ideal method for molecular weight determination before more accurate, labor-intensive and time-consuming methods are considered.

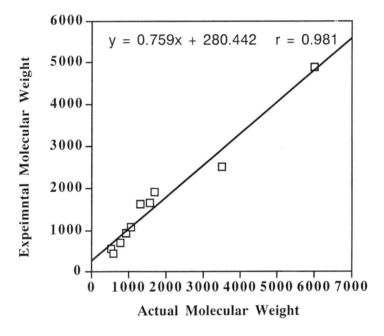

Figure 17 Graph of Experimentally Determined Values o Molecular Weight against the Actual Values for 10 Protein/Peptides

Synopsis

To day peak widths are rarely used in chromatographic analysis except for the purpose of calculating peak areas. Peak widths, however, can provide a means of measuring the diffusivity of a solute which is a function of the molecular weight. Consequently, if a reliable relationship between diffusivity and molecular weight can be identified, then the molecular weight of the solute can be assessed. Peak widths of solutes eluted from an open tube can give very precise values of diffusivity. There are a number of equations that purport to relate diffusivity to molecular weight and one of the more reliable forms is that of Arnold that expresses diffusivity as a function of molecular weight and molar volume of solute and solvent, respectively. An empirical relationship based on experimental data relates the reciprocal of the solute diffusivity to the product of the cube root of the solute molar volume and the square root of the molecular weight. The diffusivities and molecular weights of 69 different solutes are reported, and correlation between experimental values of the molecular weight and the true values confirmed that the molecular weight of 98% of all the solutes could be determined with an error of less than 11%. The measurement of diffusivity requires certain precautions to be taken. Extra-column (tube) dispersion must be reduced to a minimum and any radial flow that make take pace in the tube must also be minimized by employing a straight tube with very smooth walls. As diffusivity is a function of pressure, the minimum pressure must be used to control the flow of solvent. Diffusivities can also be determined from the dispersion in a packed column, but the column must be operated at high mobile phase velocities to strongly reduce the significance of dispersion from longitudinal diffusion. As a consequence, the peak dispersion is simply the sum of the multi-path dispersion and the two resistance to mass transfer effects. The actual value of the multi-path term can be taken as the intercept of the linear curve relating the total dispersion (H) to the reciprocal of the diffusivity for a series of compounds eluted at approximately the same (k") value. The appropriate function of (k") can be identified from a given set of dispersion data by a curve fitting procedure employing a second-order polynomial in (k"). In this manner the diffusivity of a solute can be determined from its dispersion in a packed column and then related to the molecular weight. It was found the procedure gave molecular weight values within 9% of the true value for over 90% of the solutes examined. Although such errors may appear large this precision can be extremely valuable when assessing the molecular weight of large biomolecules for which there is no other method available. The procedure can be simplified by making the assumption that the effect of changes in diffusivity on the plate height is much greater than changes in capacity factor (*i.e.*, changes in retention). In addition, it can also be assumed that for solute densities ranging from 0.85 and 1.25 (which will include many polymers of biological origin) the cube root of the density can be taken as unity. As a result, the variance per unit length of the column (H) for a particular solute can be shown to be a simple linear function of the cube root of the molecular weight. This relationship was shown to be true for a range of protein/peptide solutes, providing molecular weight values with a maximum error lying between

Chapter 12

Chromatography Column Design

The Design of Packed Columns for GC and LC

All chromatographic analyses involve a large number of interacting variables, not all of which, are under the control of the chromatographer. The nature of the sample presented for analysis, the productivity goals of the analytical service and the economic constraints of the laboratory will all have their own individual influence on the chromatographic system and, in particular, its design. These exigencies (which, unfortunately, are sometimes in conflict with each other) result from the various and different needs of the overall system and, for the sake of clarity, need to be incorporated into a chromatography *design protocol*. From such a protocol a procedure can be developed that can identify the basic characteristics of the fully optimized chromatographic system.

The *Design Protocol* must contain three different data sources which, for convenience, will be termed the *Design Data Bases*. The first data base will contain the exigencies of the analyst and will be given the title *Performance Criteria*. The performance criteria must contain explicit statements, given, wherever possible, in numerical terms that define the quality of the separation that is required in order to achieve an accurate analysis.

The apparatus employed for any given analysis will have operating specifications that are unique to the particular instrument that is selected or that is available. These specifications have been determined by the design and method of manufacture of the instrument and can differ significantly from one chromatograph to another. Some will control and limit the ultimate performance achieved by any column with which the instrument is used. However, it is likely that, as a result of careful design by the manufacturer, the important instrument characteristics affecting column design will remain sensibly constant over the lifetime of the instrument. This will allow any column that is designed for optimum use with the instrument to also have a reasonable life-span. The instrument specifications provide the second data base necessary for the design of the optimum column and this data base is given the term *Instrument Constraints*. It is important to realize, and it will become increasingly apparent during the development of the design procedure, that it is the instrument constraints that ultimately control and limit the optimum performance of the column.

Finally, the analyst is left with some choice in the strategy that can be used in the analysis by way of the chromatographic media selected, and in the level of some operating variables that may be considered appropriate or necessary. The range of variables left to the choice of the analyst constitutes the third data base necessary for optimum column design and this will be termed the *Elective Variables*. However, as most of the conditions that need to be specified will be defined under performance criteria and designated under instrument constraints, the analyst is not left with a very wide choice of variables from which to choose. This might be considered advantageous, however, as the fewer the decisions that are left in the hands of the operator, the less skill and experience will be required and fewer mistakes will be made.

The information contained in the three data bases provides the necessary information required to design the optimum column. In addition, once the column has been designed, and its properties defined, a complementary set of *Analytical Specifications* can also be calculated. Thus, the design protocol contains three data bases, *Performance Criteria*, *Elective Variables* and *Instrument Constraints*.

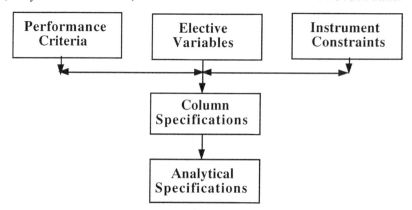

Figure 1. The Design Protocol

Design of Packed Columns

Performance Criteria

These data bases will provide, first, the *column specifications* and, second, the *analytical specifications*. A diagram representing the overall design protocol is shown diagramatically in Figure 1.

A satisfactory chromatographic analysis demands, *a priori*, on an adequate separation of the constituents of the sample that will permit the accurate quantitative evaluation of each component of interest. To achieve this, an appropriate phase system must be chosen so that the individual components of the mixture will be moved apart from one another in the column. In addition, their dispersion must be constrained sufficiently to allow all the solutes of interest to be eluted discretely. At this stage it is necessary to introduce the concept of the *Reduced Chromatogram*.

The Reduced Chromatogram

Any chromatogram that represents the separation of a complex mixture of solutes can be reduced to a relatively simple separation that will concisely and accurately represent the limits and extent of the separation problem. The simple separation can be depicted in the form of a *reduced chromatogram*, an example of which is given in Figure 2.

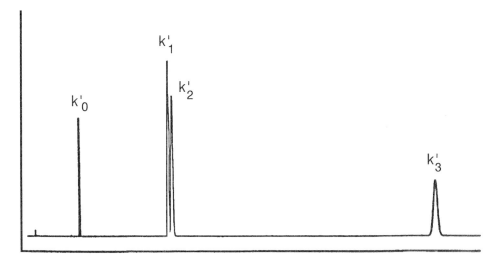

Figure 2. The Reduced Chromatogram

The reduced chromatogram contains four peaks; the first will be the dead volume peak (which, as has been shown previously, must be the fully excluded peak determined from the retention volume of a salt or solute of large molecular weight).

The second and third peaks will be the pair of peaks in the mixture that are eluted closest together and, thus, the most difficult to separate (usually given the term the *critical pair* as they define the severity of the separation). Finally, the fourth peak will be that which is eluted last from the mixture and will determine when the analysis is complete and establishes the total analysis time. The chromatographic system must be designed to separate the critical pair and, as this is the pair that is eluted closest together, all other peaks should also be resolved

However, any given column operated at a specific flow rate will exhibit a range of efficiencies depending on the nature and capacity ratio of the solute that is chosen for efficiency measurement. Consequently, under *exceptional* circumstances, the predicted conditions for the separation of the critical pair may *not* be suitable for another pair, and the complete resolution of all solutes may still not be obtained. This could occur if the separation ratio of another solute pair, although larger, was very close to that of the critical pair but contained solutes, for example, of widely different molecular weight (and, consequently, very different diffusivities). Fortunately, the possibility of this situation arising is remote in practice, and will not be considered in this discussion. It follows that the efficiency required to separate the critical pair, numerically defined, is the first performance criterion.

Laboratory economy will also require the maximum sample throughput from the equipment and, thus, the second criterion will require the analysis to be completed in the minimum time. It should be pointed out,that just a *rapid separation* is *not* the criterion. The separation must be achieved in the *minimum time.* In practice, the column must be designed so that, when employed with the chosen phase system and the specific apparatus, *no other column will separate the mixture in less time.*

Another aspect of cost reduction would be solvent economy. The need to preferentially select inexpensive solvents and employ the minimum amount of solvent per analysis would be the third performance criteria. Finally, to conserve sample and to have the capability of determining trace contaminants, the fourth criterion would be that the combination of column and detector should provide the maximum possible mass sensitivity and, thus, the minimum amount of sample. The performance criteria are summarized in Table 1. Certain operating limits are inherent in any analytical instrument and these limits will vary with the purpose for which the instrument was designed. For example, the preparative chromatograph will have very different operating characteristics from those of the analytical chromatograph.

Design of Packed Columns

Table 1. Performance Criteria

1. A defined resolution must be obtained.

2. The analysis must be completed in the minimum time.

3. The analysis must be completed with the minimum solvent consumption.

4. The apparatus must provide the maximum mass sensitivity.

Instrument Constraints

The first, and obvious operating limit, will be the maximum column inlet pressure that the solvent pump will provide. It will be seen that the maximum inlet pressure that is available (or can be tolerated) will determine the optimum column length, the optimum particle diameter of the packing material and, as a consequence, the minimum analysis time. It should also be noted that the pressure limit is not usually defined by the pump design, but by the specifications of the sample valve. Most pumps can provide pressures of at least 6,000 p.s.i. but the sample valve operating pressure will often limit the column inlet pressure to as little as 3,000 p.s.i. Although it is claimed by many manufacturers, that their sample valves will operate at 6,000 p.s.i. or even 10,000 p.s.i., the life time of the valves, when operated at these pressures, is often relatively short. As a consequence, for successful and continuous operation over an extended period of time, the operating pressure of the chromatographic system may well be limited by the long-term pressure tolerance of the sample valve and not by the available pressure from the solvent pump.

The maximum and minimum flow rate available from the solvent pump may also, under certain circumstances, determine the minimum or maximum column diameter that can be employed. As a consequence, limits will be placed on the mass sensitivity of the chromatographic system as well as the solvent consumption. Almost all commercially available LC solvent pumps, however, have a flow rate range that will include all optimum flow rates that are likely to be required in analytical chromatography

Another critical instrument specification is the total extra-column dispersion. The subject of extra-column dispersion has already been discussed in chapter 9. It has been shown that the extra-column dispersion determines the minimum column radius and, thus, both the solvent consumption per analysis and the mass sensitivity of the overall chromatographic system. The overall extra-column variance, therefore, must be known and quantitatively specified.

Finally, the speed of response of the detector sensor and the associated electronics once played an important part in optimum column design. The speed of response, or the overall time constant of the detector and associated electronics, would be particularly important in the analysis of simple mixtures where the analysis time can be extremely short and the elution of each peak extremely rapid. Fortunately, modern LC detector sensors have a very fast response and the associated electronic circuits very small time constants and, thus, the overall time constant of the detector system does not significantly influence column design in contemporary instruments. The instrument constraints are summarized in Table 2

Table 2. Instrument Constraints

1. The maximum operating pressure.

2. The extra-column dispersion.

3. The minimum flow rate.

4. The maximum flow rate.

5. The response time of the detecting system.

Elective Variables

The choice of variables remaining with the operator, as stated before, is restricted and is usually confined to the selection of the phase system. Preliminary experiments must be carried out to identify the best phase system to be used for the particular analysis under consideration. The best phase system will be that which provides the greatest separation ratio for the critical pair of solutes and, at the same time, ensures a minimum value for the capacity factor of the last eluted solute. Unfortunately, at this time, theories that predict the optimum solvent system that will effect a particular separation are largely empirical and those that are available can be very approximate, to say the least. Nevertheless, there are commercially available experimental routines that help in the selection of the best phase system for LC analyses, the results from which can be evaluated by supporting computer software. The program may then suggest further routines based on the initial results and, by an iterative procedure, eventually provides an optimum phase system as defined by the computer software.

Whether the optimum phase system is arrived at by a computer system, or by trial and error experiments (which are often carried out, even after computer optimization), the basic chromatographic data needed in column design will be

Design of Packed Columns

identified. The phase system will define the separation ratio of the critical pair, the capacity ratio of the first eluted peak of the critical pair and the capacity ratio of the last eluted peak. It will also define the viscosity of the mobile phase and the diffusivity of the solute in the mobile phase.

There remains little more for the operator to decide. Sometimes, alternative but similar solvent mixtures that have a lower viscosity and higher solute diffusivity can be preferentially selected. For example, an *n*-hexane/methanol mixture might be chosen as an alternative to the more viscous *n*-heptane/isopropyl alcohol mixture, as it has similar elution properties. It will be shown later that if a fully optimized column is employed, the viscosity of the mobile phase does not seem to affect the column performance as it is taken into account in the optimization procedure. The difference in diffusivity, which varies approximately as the inverse of the viscosity, however, does have a very significant impact on column performance. The operator may, under some circumstances, be free to choose a less toxic or a less costly solvent. For example, in reverse phase chromatography the operator might be able to select a methanol/water solvent mixture as opposed to acetonitrile/water mixture on the basis of lower cost or 'less toxicity'. However, it must be remembered that the elution characteristics of methanol and acetonitrile, although similar, are certainly not identical. Furthermore, it should be emphasized that the solvent optimization procedure must take place *after* the individual solvents have been chosen and, if subsequently changed again, then the optimization procedure must be repeated. The elective variables are summarized in Table 3.

Table 3. Elective Variables

The selection of the mobile phase and the conditions of development. Having chosen the solvent(s) the following are defined from the reduced chromatogram,

1. The separation ratio of the critical pair.

2. The capacity ratio of the first solute of the critical pair.

3. The capacity ratio of the last eluted solute.

4. The viscosity of the solvent.

5. The diffusivity of the first solute of the critical pair in the mobile phase.

Column Specifications and Operating Conditions

Employing the conditions defined in the three data bases and the appropriate equations derived from the Plate and Rate Theories, the optimum physical properties of the column and operating conditions can be defined. The precise column length and particle diameter (or open tube radius) that will achieve the necessary resolution and provide the *minimum analysis time* can then be calculated. It should again be emphasized that, the specifications will be such, that for the specific separation carried out, on the phase system selected and the equipment available, the minimum analysis time will be *absolute. No other column will allow the analysis to be carried out in **less** time.*

The optimum *mobile phase velocity* will also be determined in the above calculations together with the minimum radius to achieve minimum solvent consumption and maximum mass sensitivity. The column specifications and operating conditions are summarized in Table 4.

Table 4. Column Specifications and Operating Conditions

1. The minimum column length.
2. The optimum column radius.
3. The optimum particle diameter.
4. The optimum mobile phase linear velocity.
5. Mobile phase flow rate.

Analytical Specifications

The analytical specifications must prescribe the ultimate performance of the total chromatographic system, in appropriate numerical values, to demonstrate the performance that has been achieved.

Table 5. Analytical Specifications.

1. Column efficiency in theoretical plates.
2. Analysis time.
3. Maximum volume of charge.
4. Solvent consumption per analysis.
5. Overall mass sensitivity.
6. Total peak capacity.

The separation of the critical pair would require a minimum column efficiency and, so, the number of theoretical plated produced by the column must be reported. The

Design of Packed Columns

second most important requisite is that the analysis is achieved in the minimum time and thus is appropriately reported. The analyst will also want to know the maximum volume of charge that can be placed on the column, the solvent consumption per analysis, the mass sensitivity and finally the total peak capacity of the chromatogram.

The analytical specifications are summarized in Table 5. It is obvious that such a protocol would not be employed to design a chromatographic system for a single analysis or even for a few dozen analyses. It will be seen that the optimization procedure entails a considerable amount of work and, therefore, would only be justified for a routine analysis that was repetitive and would be carried out regularly many times a day. Under such circumstances, the use of the protocol to construct the fully optimized column could be economically advantageous, increasing the sample throughput dramatically and significantly reducing operating costs.

Unfortunately, some of the data that are required to calculate the specifications and operating conditions of the optimum column involve instrument specifications which are often not available from the instrument manufacturer. In particular, the total dispersion of the detector and its internal connecting tubes is rarely given. In a similar manner, a value for the dispersion that takes place in a sample valve is rarely provided by the manufactures. The valve, as discussed in a previous chapter, can make a significant contribution to the extra-column dispersion of the chromatographic system, which, as has also been shown, will determine the magnitude of the column radius. Sadly, it is often left to the analyst to experimentally determine these data.

The Column Design Process for Packed Columns

Column design involves the application of a number of specific equations (most of which have been previously derived and/or discussed) to determine the column parameters and operating conditions that will provide the analytical specifications necessary to achieve a specific separation. The characteristics of the separation will be defined by the *reduced chromatogram* of the particular sample of interest. First, it is necessary to calculate the efficiency required to separate the critical pair of the reduced chromatogram of the sample. This requires a knowledge of the capacity ratio of the first eluted peak of the critical pair and their separation ratio. Employing the Purnell equation (chapter 6, equation (16)).

$$n = \left(\frac{16(1+k'_A)^2}{k'^2_A(\alpha-1)^2}\right) \quad (1)$$

where $\alpha_{(AB)}$ is the separation ratio of the critical pair of solutes (A) and (B) and (k'_A) is the capacity ratio of the first of the critical pair.

It should be noted that Purnell's equation utilizes the *thermodynamic capacity ratio* calculated using the *thermodynamic dead volume*.

Knowing the number of theoretical plates that are required, the length of the column (L) is defined as the product of the number of plates and the variance per unit length of the column (H), *i.e.*,

$$L = nH \quad (2)$$

From the Van Deemter equation, $\quad H = A + \dfrac{B}{u} + Cu \quad (3)$

For convenience the Van Deemter equation for the LC column will be generally used in the following argument. However, when GC columns are considered the equations will be appropriately modified, where

$$A = 2\lambda d_p \quad (4)$$

$$B = 2\gamma_p D_{o(m)} \quad (5)$$

$$C = \left(\frac{(1+6k''+11k''^2)d_p^2}{24(1+k'')^2 D_m} + \frac{8k'' d_f^2}{\pi^2(1+k'')^2 D_S}\right) \quad (6)$$

and for GC, $\quad C = \left(\dfrac{(1+6k''+11k''^2)d_p^2}{24(1+k'')^2 D_{o(m)}} + \dfrac{8k'' d_f^2}{\pi^2(1+k'')^2 (\gamma+1)D_S}\right) \quad (7)$

It should be noted that the velocity employed in equation (7) will be the exit velocity (u_o) and not the mean velocity (u_{mean}). Differentiating equation (3) and equating to zero, to obtain an expression for the optimum velocity (u_{opt}),

$$u_{opt} = \sqrt{\frac{B}{C}} \quad (8)$$

Design of Packed Columns

Substituting for (B) and (C) from equations (5) and (6) in equation (8) and simplifying,

$$u_{opt} = \left(\frac{2\gamma_p D_m}{\left(\frac{(1+6k''+11k''^2)d_p^2}{24(1+k'')^2 D_m} + \frac{8k'' d_f^2}{\pi^2 (1+k'')^2 D_S} \right)} \right)^{\frac{1}{2}} \qquad (9)$$

For GC, substituting for (B) and (C) from (5) and (7) in (8) and simplifying,

$$u_{o(opt)} = \left(\frac{2\gamma_p D_{o(m)}}{\left(\frac{(1+6k''+11k''^2)d_p^2}{24(1+k'')^2 D_{o(m)}} + \frac{8k'' d_f^2}{\pi^2 (1+k'')^2 (\gamma+1) D_S} \right)} \right)^{\frac{1}{2}} \qquad (10)$$

By substituting for (u_{opt}) from equation (8) in equation (3) and simplifying an expression for (H_{min}) can be obtained,

$$H_{min} = A + 2\sqrt{BC} \qquad (11)$$

Again substituting for (A), (B) and (C) from equations (4), (5) and (6),

$$H_{min} = 2\lambda d_p + 2\left(2\gamma_p D_m \left(\frac{(1+6k''+11k''^2)d_p^2}{24(1+k'')^2 D_m} + \frac{8k'' d_f^2}{\pi^2 (1+k'')^2 D_S} \right) \right)^{\frac{1}{2}} \qquad (12)$$

and, for GC columns,

$$H_{min} = 2\lambda d_p + 2\left(\frac{2\gamma_p D_{o(m)}}{(1+k'')^2} \left(\frac{(1+6k''+11k''^2)d_p^2}{24 D_{o(m)}} + \frac{8k'' d_f^2}{\pi^2 (\gamma+1) D_S} \right) \right)^{\frac{1}{2}} \qquad (13)$$

Substituting for (H_{min}) and (n) from equations (12) and (1), respectively, in equation (2), an expression for the column length (l) can be obtained,

$$L = \left(\frac{16(1+k'_A)^2}{k'^2_A(\alpha-1)^2}\right)\left\{2\lambda d_p + 2\left(2\gamma_p D_m\left(\frac{(1+6k''+11k''^2)d_p^2}{24(1+k'')^2 D_m} + \frac{8k''d_f^2}{\pi^2(1+k'')^2 D_s}\right)\right)\right\}^{\frac{1}{2}}$$

(14)

And for a GC column,

$$L = \left(\frac{16(1+k'_A)^2}{k'^2_A(\alpha-1)^2}\right) \times$$

$$\left\{2\lambda d_p + \left(\frac{\gamma_p(1+6k''+11k''^2)d_p^2}{6(1+k'')^2} + \frac{32\gamma_p D_{o(m)} k''d_f^2}{\pi^2(1+k'')^2(\gamma+1)D_s}\right)\right\}^{\frac{1}{2}} \quad (15)$$

It should be pointed out that equations (14) and (15) do not give an expression for the *minimum* column lengths, as the optimum particle diameter has yet to be identified.

The Optimum Particle Diameter

A given number of theoretical plates can be obtained from a range of columns by choosing different column lengths packed with particles of different size. As you reduce the particle size, you reduce the variance of the eluted peak and, thus, increase the number of theoretical plates. Consequently, the column can be shorter and the separation faster. The process of reducing the particle size and column length can be continued until it requires the maximum pressure available from the chromatographic system to obtain the optimum velocity. This will be the optimum particle size and minimum column length. Thus, the optimum particle size will permit the use of the shortest column, which, in turn, will give the minimum analysis time.

The column length, as well as providing the required efficiency, is also defined by the D'Arcy equation. The D'Arcy equation describes the flow of a liquid through a packed bed in terms of the particle diameter, the pressure applied across the bed, the viscosity of the fluid and the linear velocity of the fluid. The D'Arcy equation for an incompressible fluid is given as follows,

Design of Packed Columns

$$L = \frac{D'_A P d_p^2}{\eta u} \tag{16}$$

where (P) is the inlet pressure to the column,

(η) is the viscosity of the mobile phase,

(D'_A) is the D'Arcy's constant, which for a well-packed LC column takes a value of about 35 when the pressure is measured in p.s.i.

Substituting for (L) from equation (2) in equation (16),

$$\frac{D'_A P d_p^2}{\eta u_{opt}} = n H_{min}$$

Substituting for (u_{opt}) and (H_{min}) from equations (8) and (11) respectively,

$$\frac{D'_A P d_p^2}{\eta \sqrt{\frac{B}{C}}} = n\left(A + 2\sqrt{BC}\right)$$

Rearranging,

$$\frac{D'_A P d_p^2}{\eta} = n\left(A\sqrt{\frac{B}{C}} + 2B\right) \tag{17}$$

Substituting for (A), (B) and (C) from equations (4), (5) and (6), respectively, in equation (17),

$$\frac{D'_A P d_p^2}{\eta} = \left(\frac{16(1+k'_A)^2}{k'^2_A(\alpha-1)^2}\right) \times \left(2\lambda d_p \sqrt{\frac{2\gamma_p D_{o(m)}}{\frac{(1+6k''+11k''^2)d_p^2}{24(1+k'')^2 D_{o(m)}} + \frac{8k'' d_f^2}{\pi^2(1+k'')^2 D_S}}} + 4\lambda D_{o(m)}\right) \tag{18}$$

It is seen that there will be a unique value for (d_p), the *optimum* particle diameter, ($d_{p(opt)}$), that will meet the equality defined in equation (14) and allow the minimum HETP to be realized when operating at a maximum column inlet pressure

(P). Note the expression for (C) is also a function of the particle diameter (dp) and includes known thermodynamic and physical properties of the chromatographic system. Consequently, with the aid of a computer, the optimum particle diameter (d$_{p(opt)}$) can be calculated as that value that will meet the equality defined in equation (18). However, it will be seen in due course that these equations can be simplified. The equation for a flow of liquid though a packed bed will, however, differ for a compressible fluid, *i.e.*, a gas. Due to the compressibility of a gas, the flow rate can not be described by the simple D'Arcy law for liquids. From chapter 2, it is seen that

$$P_o Q_o L = a_f \frac{K}{\eta}(P_i^2 - P_o^2)$$

where the symbols have the meanings previously described.

Dividing through by (a$_f$) and rearranging,

$$L = \frac{K}{\eta u_o P_o}(P_i^2 - P_o^2) = \frac{D_{A(g)} d_p^2 P_o}{\eta u_o}(\gamma^2 - 1) \qquad (19)$$

where $(D_{A(g)})$ is D'Arcy's constant for a gas,

(K) is $D_{A(g)} d_p^2$,

and (γ) is the inlet/outlet pressure ratio.

Thus, equation (17) must be modified to the form

$$\frac{D'_{A(g)} P_o d_p^2}{\eta}(\gamma^2 + 1) = n\left(A\sqrt{\frac{B}{C}} + 2B\right)$$

Substituting for (A), (B) and (C),

$$\frac{D'_{A(g)} P_o d_p^2}{\eta}(\gamma^2 - 1) = \left(\frac{16(1+k'_A)^2}{k'^2_A(\alpha-1)^2}\right) \times$$

$$\left(2\lambda d_p \sqrt{\frac{2\gamma_p D_m}{\left(\frac{(1+6k''+11k''^2)d_p^2}{24(1+k'')^2 D_m} + \frac{8k'' d_f^2}{\pi^2(1+k'')^2(\gamma+1)D_S}\right)}} + 4\gamma_p D_m\right) \qquad (20)$$

It follows that knowing the optimum particle diameter, the optimum column length can also be identified. It must be emphasized that this optimizing procedure

Design of Packed Columns

assumes that a consistent, and efficient, packing procedure is available which is effective for a wide range of particle diameters. In general, most column manufacturers have well-established packing techniques that meet these requirements. The above theory will now be applied to both packed GC columns and packed LC columns and the equations will be simplified in the light of practical limitations.

Packed GC Columns

In GC, (D_m) and (D_S) differ by about 4 orders of magnitude (from the literature D_m is usually about $1\text{-}2 \times 10^{-1}$ cm^{2}sec.$^{-1}$ and $D_S = 1\text{-}2 \times 10^{-5}$ cm^{2}sec.$^{-1}$). Chromosorb W is the most commonly used support which is usually screened to a particle diameter range of 127 μm–149 μm, with a mean value of 138 μm (100-120 mesh). Such packing has a surface area of about 2 m^{2}g^{-1}. The stationary phase can be loaded on the support at levels ranging from 2.5%w/w to 15%w/w but by far the most common loading is 10%w/w. If the density of the stationary phase is taken as approximately unity, then the film thickness (d_f) is given by

$$d_f = \frac{0.1}{2 \times 10^4} = 5 \times 10^{-6} \text{ cm} = 0.05 \ \mu m$$

Reiterating the equation for the variance per unit length for a packed GC column,

$$H = 2\lambda d_p + \frac{2\gamma_p D_{m(o)}}{u_o} + \frac{(1+6k''+11k''^2)d_p^2}{24(1+k'')^2 D_{m(o)}} u_o + \frac{8k'' d_f^2}{\pi^2(1+k'')^2 D_S(\gamma+1)} u_o \quad (21)$$

Now, it is of interest to determine if either the resistance to mass transfer term for the mobile phase or, the resistance to mass transfer term in the stationary phase dominate in the equation for the variance per unit length of a GC packed column. Consequently, taking the ratio of the two resistance to mass transfer terms (G)

$$G = \left(\frac{(1+6k''+11k''^2)d_p^2}{24(1+k'')^2 D_{m(o)}} \quad \frac{\pi^2(1+k'')^2 D_S(\gamma+1)}{8k'' d_f^2} \right)$$

$$= \left(\frac{(1+6k''+11k''^2)}{19.5k''} \quad \frac{D_S(\gamma+1)d_p^2}{D_{m(o)}d_f^2} \right)$$

Thus,
$$G = \left(\frac{(1+6k''+11k''^2)((\gamma+)}{19.5k''} \frac{1.5\times10^{-5} \times 1.9\times10^{-4}}{1.5\times10^{-1} \times 2.5\times10^{-11}}\right)$$

$$= \left(\frac{39(1+6k''+11k''^2)(\gamma+1)}{k''}\right)$$

Thus as (γ) will always be greater than unity, the resistance to mass transfer term in the mobile phase will be, at a minimum, about forty times greater than that in the stationary phase. Consequently, the contribution from the resistance to mass transfer in the stationary phase to the overall variance per unit length of the column, relative to that in the mobile phase, can be ignored. It is now possible to obtain a new expression for the optimum particle diameter ($d_{p(opt)}$) by eliminating the resistance to mass transfer function for the liquid phase from equation (14).

Therefore, equation (20) can now be reduced to,

$$\frac{D_{A(g)}d_p^2 P_o}{\eta_o}(\gamma^2 - 1) = \left(\frac{16(1+k'_A)^2}{k'^2_A(\alpha-1)^2}\right)\left(2\lambda d_p \left(\left(\frac{48(1+k'')^2 \gamma_p D_m^2}{(1+6k''+11k''^2)d_p^2}\right)\right)^{\frac{1}{2}} + 4\gamma_p D_m\right)$$

Simplifying,

$$d_{p(opt)}^2 = \left(\frac{64\eta_o D_m(1+k'_A)^2}{P_o D'_{A(g)} k'^2_A (\alpha-1)^2 (\gamma^2-1)}\right)\left(\lambda \sqrt{\frac{12\gamma_p(1+k'')^2}{1+6k''+11k''^2}} + \gamma_p\right)$$

$$d_{p(opt)} = \frac{8(1+k'_A)}{k'_A(\alpha-1)}\sqrt{\frac{\eta_o D_m}{P_o D'_{A(g)}(\gamma^2-1)}}\left(\lambda \sqrt{\frac{12\gamma_p(1+k'')^2}{1+6k''+11k''^2}} + \gamma_p\right) \quad (22)$$

Employing the new equations for (C), from equation (8) the optimum velocity (u_{opt}) is given by

$$u_{o(opt)}' = \left(\frac{2\gamma_p D_{m(o)}}{\left(\frac{(1+6k''+11k''^2)d_{p(opt)}^2}{24(1+k'')^2 D_{m(o)}}\right)}\right)^{\frac{1}{2}} = \left(\frac{48(1+k'')^2 \gamma_p D_{m(o)}^2}{(1+6k''+11k''^2)d_{p(opt)}^2}\right)^{0.5}$$

Design of Packed Columns

$$= \frac{4D_{m(o)}}{d_{p(opt)}} \sqrt{\frac{3(1+k'')^2 \gamma_p}{(1+6k''+11k''^2)}} \tag{23}$$

The equation for the minimum value of (H) becomes

$$H_{min} = 2\lambda d_{p(opt)} + 2d_{p(opt)} \left(\gamma_p \frac{(1+6k''+11k''^2)}{12(1+k'')^2} \right)^{\frac{1}{2}}$$

$$= 2d_{p(opt)} \left(\lambda + \left(\gamma_p \frac{(1+6k''+11k''^2)}{12(1+k'')^2} \right)^{\frac{1}{2}} \right) \tag{24}$$

Thus, the minimum column length (L) will be given by

$$L_{(min)} = nH_{min} = \frac{32(1+k'_A)^2 d_{p(opt)}}{k'^2_A (\alpha-1)^2} \left(\lambda + \left(\gamma_p \frac{(1+6k''+11k''^2)}{12(1+k'')^2} \right)^{\frac{1}{2}} \right) \tag{25}$$

and, consequently, the minimum analysis time (t_{min}) will be given by

$$\frac{L_{min}}{u_{mean}}((1+k'_2)) = \frac{\frac{32(1+k'_2)(1+k'_A)^2 d_{p(opt)}}{k'^2_A (\alpha-1)^2} \left(\lambda + \left(\gamma_p \frac{(1+6k''+11k''^2)}{12(1+k'')^2} \right)^{\frac{1}{2}} \right)}{\frac{4D_{m(o)}}{d_{p(opt)}} \sqrt{\frac{3(1+k'')^2 \gamma_p}{(1+6k''+11k''^2)}}}$$

Now, applying the pressure correction factor from chapter 2,

$$u_{mean(opt)} = \frac{3(\gamma^2-1)}{2(\gamma^3-1)} u_{o(opt)}$$

Thus,

$$t_{min} = \frac{L_{min}}{\frac{3(\gamma^2-1)}{2(\gamma^3-1)} u_{o(opt)}} (1+k'_2) = \frac{\frac{32(1+k'_2)(1+k'_A)^2 d_{p(opt)}}{k'^2_A (\alpha-1)^2} \left(\lambda + \left(\gamma_p \frac{(1+6k''+11k''^2)}{12(1+k'')^2} \right)^{\frac{1}{2}} \right)}{\frac{4D_{m(o)}}{d_{p(opt)}} \sqrt{\frac{3(1+k'')^2 \gamma_p}{(1+6k''+11k''^2)}}}$$

$$= \frac{\dfrac{32(1+k'_2)(1+k'_A)^2 d_{p(opt)}}{k'^2_A(\alpha-1)^2}\left(\lambda + \left(\gamma_p \dfrac{(1+6k''+11k''^2)}{12(1+k'')^2}\right)^{\frac{1}{2}}\right)}{\dfrac{3(\gamma^2-1)}{2(\gamma^3-1)}\dfrac{4D_{m(o)}}{d_{p(opt)}}\sqrt{\dfrac{3(1+k'')^2 \gamma_p}{(1+6k''+11k''^2)}}}$$

$$= \frac{64(\gamma^3-1)(1+k'_2)(1+k'_A)^2 d^2_{p(opt)}}{(\gamma^2-1)3k'^2_A(\alpha-1)^2 4D_{m(o)}\sqrt{\dfrac{3(1+k'')^2 \gamma_p}{(1+6k''+11k''^2)}}}\left(\lambda + \left(\gamma_p \dfrac{(1+6k''+11k''^2)}{12(1+k'')^2}\right)^{\frac{1}{2}}\right) \quad (26)$$

Packed LC Columns

Consider first the equation for the optimum particle diameter. Reiterating equation (18),

$$\frac{D'_A P d^2_p}{\eta} = \left(\frac{16(1+k'_A)^2}{k'^2_A(\alpha-1)^2}\right)\left(2\lambda d_p \sqrt{\frac{2\gamma_p D_m}{\left(\dfrac{(1+6k''+11k''^2)d^2_p}{24(1+k'')^2 D_m} + \dfrac{8k'' d^2_f}{\pi^2(1+k'')^2 D_S}\right)}} + 4\gamma D_m\right)$$

(18)

It is first necessary to identify the relative magnitudes of the resistance to mass transfer in the mobile phase,

$$\frac{(1+6k''+11k''^2)d^2_p}{24(1+k'')^2 D_m}$$

and in the stationary phase,

$$\frac{8k'' d^2_f}{\pi^2(1+k'')^2 D_S}$$

A liquid mobile phase is far denser than a gas and, therefore, carries more momentum. Thus, in its progress through the interstices of the packing, violent eddies are formed in the inter-particular spaces which provides rapid solute transfer and, in effect, greatly increases the effective diffusivity. Thus, the resistance to mass transfer in that mobile phase which *is situated in the interstices of the column* is virtually zero. However, assuming the particles of packing are porous (*i.e.*, silica based) the particles of packing will be filled with the mobile phase and so there will

Design of Packed Columns

still be a resistance to transfer in the mobile phase, but it will occur in the static solvent contained in the particle pores. There will, therefore, be another resistance to mass transfer term for the stationary phase, but it will result from *static* mobile phase. So, now, only the resistance to mass transfer in the stationary and static phase remain. It can be established that, for a silica gel surface, or for the layer of organic material on the surface of a bonded phase, that the effective thickness of the film will be less than 0.05 μm. It follows that as the resistance to mass transfer in the stationary phase is a function of (d_f^2), then the contribution from the stationary phase to the overall variance per unit length of the column will be minimal. There remains, however, the contribution from the static mobile phase in the pores. The pores of a silica gel based stationary phase are homogeneous throughout each particle and therefore the effective thickness of the mobile phase in the particle will be equivalent to the particle diameter. Thus, the function for the resistance to mass transfer in the static mobile phase will be the same as that for the stationary phase, but the diffusivity will be that of the mobile phase (D_m) and the film thickness equivalent to the particle diameter (d_p).

Thus, the resistance to mass transfer term for the static mobile phase will be

$$\frac{8k''d_p^2}{\pi^2(1+k'')^2 D_m}$$

and equation (18) becomes

$$\frac{D_A' P d_p^2}{\eta} = \left(\frac{16(1+k_A')^2}{k_A'^2(\alpha-1)^2}\right)\left(2\lambda d_p \sqrt{\frac{2\gamma_p D_m}{\left(\frac{8k''d_p^2}{\pi^2(1+k'')^2 D_m}\right)}} + 4\gamma D_m\right)$$

$$\frac{D_A' P d_p^2}{\eta} = \left(\frac{16(1+k_A')^2 D_m}{k_A'^2(\alpha-1)^2}\right)\left(\lambda\sqrt{\frac{\gamma_p \pi^2(1+k'')^2}{k''}} + 4\gamma\right)$$

or

$$d_{p(opt)} = \frac{4(1+k_A')}{k_A'(\alpha-1)}\sqrt{\left(\frac{\eta D_m}{D_A' P}\right)\left(\lambda\sqrt{\frac{\gamma_p \pi^2(1+k'')^2}{k''}} + 4\gamma\right)} \qquad (27)$$

and, from equation (9),

$$u_{opt} = \left(\frac{2\gamma_p D_m}{\left(\frac{8k''d_{p(opt)}^2}{\pi^2(1+k'')^2 D_m} \right)} \right)^{\frac{1}{2}}$$

$$u_{opt} = \pi(1+k'')\frac{D_m}{2d_{p(opt)}}\left(\frac{\gamma_p}{k''}\right)^{\frac{1}{2}} \qquad (28)$$

and, from equation (10),

$$H_{min} = 2\lambda d_{p(opt)} + 2\left(2\gamma_p D_m \left(\frac{8k''d_{p(opt)}^2}{\pi^2(1+k'')^2 D_m}\right)\right)^{\frac{1}{2}}$$

$$H_{min} = 2\lambda d_{p(opt)} + \frac{8d_{p(opt)}}{\pi(1+k'')}\sqrt{k''\gamma_p}$$

$$= 2d_{p(opt)}\left(\lambda + \frac{4}{\pi(1+k'')}\sqrt{k''\gamma_p}\right) \qquad (29)$$

and, from equation (14),

$$L_{min} = \left(\frac{16(1+k'_A)^2}{k'^2_A(\alpha-1)^2}\right)\left(2\lambda d_{p(opt)} + 2\left(2\gamma_p D_m\left(\frac{8k''d_{p(opt)}^2}{\pi^2(1+k'')^2 D_m}\right)\right)^{\frac{1}{2}}\right)$$

$$L_{min} = \left(\frac{16(1+k'_A)^2}{k'^2_A(\alpha-1)^2}\right)\left(2d_{p(opt)}\left(\lambda + \frac{4}{\pi(1+k'')}\sqrt{k''\gamma_p}\right)\right) \qquad (30)$$

and, consequently,

$$t_{min} \frac{L_{min}(1+k'_2)}{u_{opt}} = \frac{\left(\frac{16(1+k'_A)^2}{k'^2_A(\alpha-1)^2}\right)(1+k'_2)\left(2d_{p(opt)}\left(\lambda + \frac{4}{\pi(1+k'')}\sqrt{k''\gamma_p}\right)\right)}{\pi(1+k'')\frac{D_m}{d_{p(opt)}}\left(\frac{\gamma_p}{4k''}\right)^{\frac{1}{2}}}$$

Design of Packed Columns

$$t_{min} = \left(\frac{32(1+k'_A)^2}{k'^2_A(\alpha-1)^2}\right)(1+k'_2)\left[\frac{d^2_{p(opt)}}{\pi(1+k'')D_m\sqrt{\frac{\gamma_p}{4k''}}}\left(\lambda+\frac{4}{\pi(1+k'')}\sqrt{k''\gamma_p}\right)\right] \quad (31)$$

There remains the need to obtain expressions for the optimum column radius $(r_{(opt)})$, the optimum flow rate $(Q_{(opt)})$, the maximum solvent consumption $(S_{(sol)})$ and the maximum sample volume $(v_{(sam)})$.

The Optimum Column Radius

The maximum value for the extra-column dispersion that is acceptable, (σ_E), is given by (chapter 9)

$$\sigma_E = 0.32\sigma_C$$

where (σ_E) is the total *extra-column* dispersion
and (σ_C) is the *column* dispersion.

As has been previously discussed, the limitation of (σ_E) to $(0.32\sigma_C)$, allows the variance of the peak eluted from the column to be increased by a maximum of 10% by any extra-column dispersion and, consequently, the width of the peak by a maximum of 5%.

Now, from the Plate Theory, $\sigma_C = \sqrt{n}(v_o + Kv_s)$ and $(v_o + Kv_s) = V_r/n$

>   where (n) is the efficiency of the column,
> (v_o) is the volume of mobile phase per plate,
> (v_s) is the volume of stationary phase per plate,
> and   (K) is the distribution coefficient of the solute between the two phases,

Thus, $\sigma_E = 0.32\dfrac{V_r}{\sqrt{n}}$ and as $V_r = V_o + KV_s = V_o(1+k')$

where (V_o) is the total volume of mobile phase in the column
and (V_S) is the total volume of stationary phase in the column.

Thus, $$\sigma_E = 0.32\frac{V_o(1+k')}{\sqrt{n}} \quad (32)$$

Now,
$$V_o = e\pi r^2 L_{min} \quad (33)$$

where (r) is the optimum column radius,
(L_min) is the minimum column length,
and (e) is the fraction of the column occupied by the mobile phase.

In addition, for an optimized column,
$$L_{min} = nH_{min} \quad (34)$$

Thus, substituting for (L_{min}) from equation (34) in (33)
$$V_o = e\pi r^2 n H_{min} \quad (35)$$

Substituting for (V_o) from equation (35) in equation (32),
$$\sigma_E = 0.32 e\pi r^2 \sqrt{n}\, H_{min}(1+k') \quad (36)$$

Thus for a GC column, substituting for (H_{min}) from equation (24) in equation (36),

$$\sigma_E = 0.32 e\pi r^2 \sqrt{n}\, 2 d_{p(opt)} \left(\lambda + \left(\gamma_p \frac{(1+6k''+11k''^2)}{12(1+k'')^2} \right)^{\frac{1}{2}} \right) (1+k')$$

Substituting for ($\sqrt{n}$) from equation (1)

$$\sigma_E = 0.32 e\pi r^2 \frac{8(1+k')^2}{k'(\alpha-1)} d_{p(opt)} \left(\lambda + \left(\gamma_p \frac{(1+6k''+11k''^2)}{12(1+k'')^2} \right)^{\frac{1}{2}} \right) \quad (37)$$

Rearranging and solving for (r_{opt}),

$$r_{opt} = \left(\frac{0.391 k'(\alpha-1)\sigma_E}{e\pi(1+k')^2 d_{p(opt)} \left(\lambda + \left(\gamma_p \frac{(1+6k''+11k''^2)}{12(1+k'')^2} \right)^{\frac{1}{2}} \right)} \right)^{0.5} \quad (38)$$

For an LC column, substituting for (H_{min}) from equation (29) in equation (36),

Design of Packed Columns

$$\sigma_E = 0.32 e\pi r^2 \sqrt{n}\, 2d_{p(opt)}\left(\lambda + \frac{4}{\pi(1+k'')}\sqrt{k''\gamma_p}\right)(1+k')$$

Substituting for ($\sqrt{n}$) from equation (1)

$$\sigma_E = 0.32 e\pi r^2 \frac{4(1+k')}{k'(\alpha-1)} 2d_{p(opt)}\left(\lambda + \frac{4}{\pi(1+k'')}\sqrt{k''\gamma_p}\right)(1+k')$$

Rearranging and solving for (r_{opt})

$$r_{opt} = \left(\frac{0.391 k'(\alpha-1)\sigma_E}{e\pi(1+k')^2 d_{p(opt)}\left(\lambda + \frac{4}{\pi(1+k'')}\sqrt{k''\gamma_p}\right)} \right)^{0.5} \tag{39}$$

The Optimum Flow Rate

The optimum flow rate is obviously the product of the fraction of the cross-sectional area occupied by the mobile phase and the optimum mobile phase velocity, i.e.,

$$Q_{opt} = e\pi r_{opt}^2 u_{opt} \tag{40}$$

Thus, for a GC column, by substituting for (r_{opt}) and (u_{opt}) from equations (38) and (23), respectively, and simplifying,

$$Q_{(opt)} = \frac{e\pi 0.391 k'(\alpha-1)\sigma_E}{e\pi(1+k')^2 d_{p(opt)}\left(\lambda + \frac{4}{\pi(1+k'')}\sqrt{k''\gamma_p}\right)} \frac{4D_{m(o)}}{d_{p(opt)}}\sqrt{\frac{3(1+k'')^2 \gamma_p}{(1+6k''+11k''^2)}}$$

$$Q_{(opt)} = \frac{1.561 k'(\alpha-1)\sigma_E D_{m(o)}}{(1+k')d_{p(opt)}^2\left(\lambda + \frac{4}{\pi(1+k'')}\sqrt{k''\gamma_p}\right)}\sqrt{\frac{3\gamma_p}{(1+6k''+11k''^2)}} \tag{41}$$

For an LC column, substituting for (r_{opt}) and (u_{opt}) in (40) from equations (39) and (28) respectively, and simplifying,

$$Q_{(opt)} = e\pi \frac{0.391 k'(\alpha-1)\sigma_E}{e\pi(1+k')^2 d_{p(opt)}\left(\lambda + \frac{4}{\pi(1+k'')}\sqrt{k''\gamma_p}\right)} \pi(1+k'')\frac{D_m}{2 d_{p(opt)}}\left(\frac{\gamma_p}{k''}\right)^{\frac{1}{2}}$$

$$Q_{(opt)} = \frac{0.195 k'(\alpha-1)\sigma_E \pi D_m}{(1+k') d_{p(opt)}^2 \left(\lambda + \frac{4}{\pi(1+k'')}\sqrt{k''\gamma_p}\right)}\left(\frac{\gamma_p}{k''}\right)^{\frac{1}{2}} \qquad (42)$$

The Minimum Solvent Consumption

The minimum solvent consumption will be obtained from the product of the optimum flow rate and the analysis time,

$$V_{sol} = Q_{opt} t_{min} \qquad (43)$$

The amount of gas employed in a GC analysis is not usually important, particularly where open tubular columns are used. In LC, however, solvent use presupposes a solvent disposal difficulty if not a toxicity problem and, thus, solvent consumption can be extremely important.

For an LC column, substituting for (t_{min}) from equation (31) and for (Q_{opt}) from equation (42) in equation (43),

$$V_{sol} = \left(\frac{32(1+k'_A)^2}{k'^2_A(\alpha-1)^2}\right)(1+k'_2)\left(\frac{d_{p(opt)}^2}{\pi(1+k'')D_m\sqrt{\frac{\gamma_p}{k''}}}\left(\lambda + \frac{4}{\pi(1+k'')}\sqrt{k''\gamma_p}\right)\right) \times$$

$$\frac{0.0885 k'(\alpha-1)\sigma_E \pi D_m}{(1+k') d_{p(opt)}^2 \left(\lambda + \frac{4}{\pi(1+k'')}\sqrt{k''\gamma_p}\right)}\left(\frac{\gamma_p}{k''}\right)^{\frac{1}{2}}$$

which simplifies to

$$V_{sol} = \frac{5.664(1+k'_2)\sigma_E}{k'_A(\alpha-1)} \qquad (44)$$

Design of Packed Columns

Maximum Sample Volume

The maximum permissible charge that can be placed on an LC column was discussed in chapter 5 and is given by the following equation,

$$V_{sample} = \frac{1.1 V_r}{\sqrt{n}} = \frac{V_o(1+k')}{\sqrt{n}} = \frac{1.1 \varepsilon \pi r_{opt}^2 L_{min}(1+k')}{\sqrt{n}} \qquad (45)$$

Now, substituting for ($\sqrt{n}$) in (45) from $\sqrt{n} = \frac{4(1+k')}{k'(\alpha-1)}$, then

$$V_{sample} = \frac{1.1 \varepsilon \pi r_{opt}^2 L_{min}(1+k')}{\frac{4(1+k')}{k'(\alpha-1)}} = 0.275 \varepsilon \pi r_{opt}^2 L_{min} k'(\alpha-1) \qquad (46)$$

It is seen from equation (46) that, as would be expected, the maximum sample volume decreases as the separation ratio decreases, *i.e.*, with difficulty of the separation.

Synopsis

The column design protocol consists primarily of three data bases, *performance criteria, instrument constraints* and *elective variables*. The performance criteria are derived from the reduced chromatogram which consists of four peaks, the first, the dead volume peak; the second and third (referred to as the *critical pair*) which are the two peaks in the mixture that are eluted closest together; and the fourth, the last peak, which determines when the analysis is complete. Employing the data provided by the reduced chromatogram, the chromatographic resolution is defined and it is specified that the separation must be achieved in the minimum time. The solvent consumption must be minimal for the given separation and the minimum mass sensitivity must be available. The instrument constraints are, basically, the instrument specification, which in particular will be the maximum operating pressure; the extra-column dispersion; and the maximum and minimum available solvent flow rates. Finally, the elective variables that are selected by the operator,will determine the separation ratio of the critical pair, the capacity ratios of the first of the critical pair and the last eluted peak together with the viscosity of the mobile phase, and the diffusivity of the first eluted solute of the critical pair. Firstly. the number of theoretical plates required is calculated form the Purnell equation which can provide an expression for the column length from the product of the column efficiency and the variance per unit length of the column. The function for the column length, so obtained, can be equated to that for the column length derived from the D'Arcy equation and, thus, an expression for the optimum particle diameter obtained. Using the expression for the optimum particle diameter, expressions for the optimum velocity, minimum variance per unit length, minimum column length and minimum analysis time can then be obtained. When deriving the respective equations for a GC column, the compressibility of the mobile phase must be taken into account. The product of the optimum flow rate and the analysis time will give an expression for the total solvent consumption, and a simple arithmetic product of the retention volume times the reciprocal of the square root of the column efficiency will give an expression for the maximum sample volume.

Chapter 13

Chromatography Column Design

The Design of Open Tubular Columns for GC

In a similar manner to the design process for *packed columns,* the physical characteristics and the performance specifications can be calculated theoretically *for open tubular columns.* The same protocol will be observed and again, the procedure involves the use of a number of equations that have been previously derived and/or discussed. However, it will be seen that as a result of the geometric simplicity of the open tubular column, there are no packing factors and no multi-path term and so the equations that result are far less complex and easier to manipulate and to understand.

Starting with the same basic equation of Purnell (chapter 6) which is applicable to all forms of chromatography, and allows the number of theoretical plates required to separate the critical pair of solutes to be calculated,

$$n = \left(\frac{4(1 + k')}{k'(\alpha - 1)} \right)^2 = 16 \frac{(1 + k')^2}{k'^2 (\alpha - 1)^2} \qquad (1)$$

where (α) is the separation ratio of the critical pair
and (k') is the capacity ratio of the first eluted peak of the critical pair.

The next equation of importance is the relationship between the column length (L) and the height of the theoretical plate (H),

$$L = nH \qquad (2)$$

where (n) is the column efficiency as defined in equation (1).

Now, it has been previously shown from the Golay equation (chapter 8) that the value of (H) is given by a function of the form,

$$H = \frac{B}{u} + Cu \qquad (3)$$

where, (u) is the linear velocity of the mobile phase
and (B) and (C) are constants.

Now, from the Golay equation,

$$B = 2D_m \qquad (4)$$

where (D_m) is the diffusivity of the solute in the mobile phase

and it was shown by Ogan and Scott [1] that for a compressible mobile phase such as a gas, if (u_o) is taken as the exit velocity of the mobile phase and the solute diffusivity in the mobile phase is measured at atmospheric pressure ($D_{(o)m}$), then

$$C = \frac{(1+6k'+11k'^2)r^2}{24(1+k')^2 D_{(o)m}} + \frac{k'^3}{6(1+k')^2 K^2 D_S(\gamma+1)} \qquad (5)$$

where (K) is the distribution coefficient of the solute between the two phases,
(D_S) is the diffusivity of the solute in the stationary phase,
(r) is the radius of the tube,
(γ) is the inlet/outlet pressure ratio of the column
and (u_o) is now the gas velocity measured at atmospheric pressure at the outlet.

Restating the second expression in equation (5), which is for the resistance to mass transfer in the stationary phase,

$$\frac{k'^3}{6(1+k')^2 K^2 D_S(\gamma+1)}$$

Now, from the Plate Theory,

$$k' = \frac{KV_S}{V_m} = \frac{K 2\pi r d_f L}{\pi r^2 L} = \frac{2K d_f}{r}$$

where (V_S) is the volume of stationary phase in the open tube column,
(V_m) is the volume of mobile phase in the open tube column,
(d_f) is the film thickness of the stationary phase,
and (L) is the length of the open tube column.

Column Design (Capillary)

Substituting for (k'2) in the expression for the resistance to mass transfer in the stationary phase,

$$\frac{K 4K^2 d_f^2 r^2}{6(1+k')^2 r^2 K^2 D_S(\gamma+1)} = \frac{2k' d_f^2}{3(1+k')D_S(\gamma+1)}$$

Replacing the modified expression for the resistance to mass transfer in the stationary phase in equation (5),

$$C = \frac{(1+6k'+11k'^2)r^2}{24(1+k')^2 D_{(o)m}} + \frac{2 k d_f'^2}{3(1+k')^2 D_S(\gamma+1)} \quad (6)$$

Now when k' = 0, then $C \to \dfrac{r^2}{24 D_{(o)m}}$

and when k' is larg, then $C \to \dfrac{11 r^2}{24 D_{(o)m}} + \dfrac{2 d_f'^2}{3 k' D_S(\gamma+1)}$

Taking (G) as the ratio of the function for the resistance to mass transfer in the mobile phase to that in the stationary phase, it is seen that

$$G = \frac{(1+6k'+11k'^2)}{k'} \frac{D_s(\gamma+1)r^2}{16 D_{(o)m} d_f^2}$$

Now, taking the typical column diameter as 250 μm (i.e., $r^2 = 1.56 \times 10^{-4}$ cm, the maximum film thickness for film stability as 0.25 μm, (i.e., $d_f^2 = 6.25 \times 10^{-10}$ cm), the inlet pressure as 45 psi (i.e, $\gamma = 4$) the diffusivity of the solute in the mobile phase as about 2×10^{-1} cm^2 s^{-1}, the diffusivity of the solute in the stationary phase as about 2×10^{-5} cm^2 s^{-1}, then

$$G = \frac{6.25(1+6k'+11k'^2)}{k'} = 6.25\left(\frac{1}{k'} + 6 + 11k'\right)$$

Differentiating the function of (k') and equating to zero, it is possible to find the minimum value for the function of (k'),

$$\frac{dG}{dk'} = -\frac{1}{k'^2} + 11 \text{ or at the minimum } k' = \sqrt{\frac{1}{11}} = 0.302$$

Thus, when (k'=0.302), the minimum value for the function of (k') will be 12.63.

Therefore, the minimum value for (G) will be 6.25 x 12.63 = 78.9.

It follows that the contribution from the resistance to mass transfer in the stationary phase can be ignored with respect to that in the mobile phase for all values of (k'), and the HETP equation simplifies to

$$H = \frac{B}{u_o} + Cu_o = \frac{2D_{o(m)}}{u_o} + \frac{1+6k'+11k'^2}{24(1+k')^2} \frac{r^2 u_o}{D_m} \tag{7}$$

By differentiating and equating to zero,

$$u_{o(opt)} = \sqrt{\frac{B}{C}} = \left(\frac{2D_{o(m)}}{\frac{1+6k'+11k'^2}{24(1+k')^2} \frac{r^2}{D_{o(m)}}} \right)^{0.5} = \frac{4D_{o(m)}}{r} \left(\frac{3(1+k')^2}{1+6k'+11k'^2} \right)^{0.5} \tag{8}$$

Substituting for (u_{opt}) from equation (8) in equation (7), an expression for the minimum value of (H), (*viz.* H_{min}) is obtained.

$$H_{min} = \frac{B}{\sqrt{\frac{B}{C}}} + C\sqrt{\frac{B}{C}} = 2\sqrt{BC} = 2\sqrt{2D_{o(m)} \frac{1+6k'+11k'^2}{24(1+k')^2} \frac{r^2}{D_{o(m)}}}$$

i.e.,
$$H_{min} = r\sqrt{\frac{1+6k'+11k'^2}{3(1+k')^2}} \tag{9}$$

The Optimum Column Radius

The efficiency obtained from an open tubular column can be increased by reducing the column radius, which, in turn will allow the column length to be decreased and, thus, a shorter analysis time can be realized. However, the smaller diameter column will require more pressure to achieve the optimum velocity and thus the reduction of column diameter can only be continued until the maximum available inlet pressure is needed to achieve the optimum mobile phase velocity.

Now, the column length (L) can be defined as the product of the minimum plate height and the number of theoretical plates required to complete the separation as specified by the Purnell equation.

Thus,
$$L_{min} = nH_{min} \tag{10}$$

Column Design (Capillary)

The length of the column is also defined by the Poiseuille equation that describes the flow of a fluid through an open tube in terms of the tube radius, the pressure applied across the tube (column), the viscosity of the fluid and the linear velocity of the fluid. Thus, for a compressible fluid,

$$L = \frac{\pi r^2 P_o}{\eta_o u_{(o)opt}} (\gamma^2 - 1) \tag{11}$$

where (P_o) is the inlet pressure to the column
and (η) is the viscosity of the mobile phase.

(It should be emphasized that the dimensions of the applied pressure (P) must be appropriate for the dimensions in which the viscosity (η) is measured and also the radius, (r)),

Thus, equating equations (10) and (11), an expression for (r_{opt}) can be developed,

$$\frac{\pi r_{opt}^2 P_o}{\eta_o u_{(o)opt}} (\gamma^2 - 1) = nH_{min} \tag{12}$$

Substituting for (u_{opt}) from equation (8) and for (H_{min}) from equation (9) and for (n) from equation (1),

$$\frac{\pi r_{opt}^2 P_o}{\eta_o \sqrt{\frac{B}{C}}} (\gamma^2 - 1) = n 2\sqrt{BC}$$

Simplifying,
$$\frac{\pi r_{opt}^2 P_o}{\eta_o} (\gamma^2 - 1) = n 2B$$

Substituting for (B) from equation (4), for (n) from equation (1) and rearranging,

$$\frac{\pi r_{opt}^2 P_o}{\eta_o} (\gamma^2 - 1) = \frac{16(1+k')^2}{k'^2 (\alpha - 1)^2} 4 D_{o(m)}$$

$$r_{opt}^2 = \frac{64(1+k')^2 \eta_o D_{(o)m}}{\pi P_o k'^2 (\gamma^2 - 1)(\alpha - 1)^2}$$

Thus,

$$r_{opt} = \frac{8(1+k')}{k'(\alpha - 1)} \sqrt{\frac{\eta_o D_{(o)m}}{\pi P_o (\gamma^2 - 1)}} \tag{13}$$

Equation (13) is the first important equation for open tubular column design. It is seen that the optimum radius, with which the column will operate at the optimum velocity for the given inlet pressure, increases rapidly as an inverse function of the separation ratio ($\alpha-1$) and inversely as the square root of the inlet pressure. Again it must be remembered that, when calculating (r_{opt}), the dimensions of the applied pressure (P) must be appropriate for the dimensions in which the viscosity (η) is measured.

It is now necessary to obtain new equations for (u_{opt}) and (H_{min}). Reiterating equation, (8),

$$u_{opt} = \frac{4D_m}{r}\left(\frac{3(1+k')^2}{1+6k'+11k'^2}\right)^{0.5}$$

Replacing (r) with the function for (r_{opt}) from equation (11),

$$u_{opt} = \frac{4D_m}{\frac{8(1+k')}{k'(\alpha-1)}\sqrt{\frac{\eta_o D_m}{\pi P_o(\gamma^2-1)}}}\left(\frac{3(1+k')^2}{1+6k'+11k'^2}\right)^{0.5}$$

$$u_{opt} = \frac{k'(\alpha-1)}{2}\left(\frac{3\pi P_o(\gamma^2-1)D_m}{\eta_o(1+6k'+11k'^2)}\right)^{0.5} \quad (12)$$

Reiterating equation (9),

$$H_{min} = r\sqrt{\frac{1+6k'+11k'^2}{3(1+k')^2}}$$

Replacing (r) with the function for (r_{opt}) from equation (11),

$$H_{min} = \frac{8(1+k')}{k'(\alpha-1)}\sqrt{\frac{\eta_o D_m}{\pi P_o(\gamma^2-1)}}\sqrt{\frac{1+6k'+11k'^2}{3(1+k')^2}}$$

$$H_{min} = \frac{8}{k'(\alpha-1)}\sqrt{\frac{\eta_o D_m(1+6k'+11k'^2)}{3\pi P_o(\gamma^2-1)}} \quad (13)$$

The Minimum Length of an Open Tubular Column

The column length is given by equation (2), *viz.*,

$$L_{min} = nH_{min}$$

Column Design (Capillary)

Substituting for (n) and (H_{min}) from equations (1) and (13), respectively

$$H_{min} = 16\frac{(1+k')^2}{k'^2(\alpha-1)^2}\frac{8}{k'(\alpha-1)}\sqrt{\frac{\eta_o D_m(1+6k'+11k'^2)}{3\pi P_o(\gamma^2-1)}}$$

$$L_{min} = 128\frac{(1+k')^2}{k'^3(\alpha-1)^3}\sqrt{\frac{\eta_o D_m(1+6k'+11k'^2)}{3\pi P_o(\gamma^2-1)}} \tag{14}$$

Minimum Analysis Time

The analysis time is given by the following equation,

$$t_{min} = (1+k'_2)\frac{L_{min}}{u_{mean}} \tag{15}$$

Now,
$$u_{mean} = u_{opt}\frac{3(\gamma^2-1)}{2(\gamma^3-1)}$$

Thus,
$$t_{min} = (1+k'_2)\frac{2L_{min}(\gamma^3-1)}{3u_{opt}(\gamma^2-1)}$$

Substituting for (L_{min}) and (u_{opt}) from equations (12) and (11) respectively,

$$t_{min} = (1+k'_2)\frac{128\dfrac{(1+k')^2}{k'3(\alpha-1)3}\sqrt{\dfrac{\eta_o D_m(1+6k'+11k'^2)}{3\pi P_o(\gamma^2-1)}}}{\dfrac{k'(\alpha-1)}{2}\left(\dfrac{3\pi P_o(\gamma^2-1)D_m}{\eta_o(1+6k'+11k'^2)}\right)^{0.5}}\frac{3(\gamma^2-1)}{2(\gamma^3-1)}$$

or

$$t_{min} = (1+k'_2)\frac{256\eta_o(1+k')^2(\gamma^3-1)(1+6k'+11k'^2)}{9k'^4(\alpha-1)^4\pi P_o(\gamma^2-1)^2} \tag{16}$$

The Optimum Flow Rate

The optimum flow-rate (Q_{opt}) will be given by

$$Q_{opt} = \pi r_{opt}^2 u_{opt}$$

Substituting for (r_{opt}) and (u_{opt}) from equations (11) and (12),

$$Q_{opt} = \pi \left(\frac{8(1+k')}{k'(\alpha-1)} \sqrt{\frac{\eta_o D_{(o)m}}{\pi P_o(\gamma^2-1)}} \right)^2 \frac{k'(\alpha-1)}{2} \left(\frac{3\pi P_o(\gamma^2-1)D_m}{\eta_o(1+6k'+11k'^2)} \right)^{0.5}$$

$$Q_{opt} = \frac{32(1+k')^2}{k'(\alpha-1)} \left(\frac{3\pi \eta_o D_{(o)m}^3}{P_o(\gamma^2-1)(1+6k'+11k'^2)} \right)^{0.5} \qquad (17)$$

Maximum Sample Volume and Maximum Extra-Column Dispersion

In a packed column the HETP depends on the particle diameter and is not related to the column radius. As a result, an expression for the optimum particle diameter is independently derived, and then the column radius determined from the extra-column dispersion. This is not true for the open tubular column, as the *HETP* is determined by the column radius. It follows that a converse procedure must be employed. Firstly the optimum column radius is determined and then the maximum extra-column dispersion that the column can tolerate calculated. Thus, with open tubular columns, the chromatographic system, in particular the detector dispersion and the maximum sample volume, is dictated by the column design which, in turn, is governed by the nature of the separation.

As already stated, the maximum extra-column dispersion that can be tolerated, such that the resolution is not significantly reduced, will be equivalent to an increase of 10% of the column variance.

Thus, $\qquad \sigma_E^2 = 0.1\sigma_C^2$

where (σ_E^2) is the variance of the extra-column dispersion
and (σ_C^2) is the variance of the peak eluted from the column.

If the total extra-column dispersion is shared equally between the sample volume and the detector,

$$\sigma_S^2 = 0.05\sigma_C^2$$
$$\sigma_D^2 = 0.05\sigma_C^2$$

where (σ_S^2) is the extra-column dispersion resulting from the sample volume
and (σ_D^2) is the extra-column variance resulting from dispersion in the detector.

Column Design (Capillary)

or
$$\sigma_S = 0.22\sigma_C \qquad (18)$$

and
$$\sigma_D = 0.22\sigma_C \qquad (19)$$

Assuming the peak variance is increased by 5% due to the sample volume, (V_i) will then be the maximum sample volume, developing the expression for the maximum permissible sample volume in the usual manner. Consider a volume (V_i), injected onto a column, forming a rectangular distribution at the front of the column. The variance of the final peak will be the sum of the variances of the sample volume, plus the normal variance of a peak for a small sample. Now, the variance of the rectangular distribution of sample volume at the beginning of the column is $V_i^2/12$, and the column variance (from the Plate Theory) is $\left(\dfrac{V_r}{\sqrt{n}}\right)$ where (V_r) is the retention volume of the solute. Thus, with the new limiting conditions,

$$\frac{V_i^2}{12} + \left(\frac{V_r}{\sqrt{n}}\right) = 1.05\left(\frac{V_r}{\sqrt{n}}\right)^2$$

$$\frac{V_i^2}{12} = 0.05\left(\frac{V_r}{\sqrt{n}}\right)^2$$

and
$$V_i = 0.77\frac{V_r}{\sqrt{n}} \qquad (20)$$

Substituting the function for the retention volume for an open tube,

$$V_i = \frac{0.77(1+k')}{\sqrt{n}} \pi r_{opt}^2 L_{min} \qquad (21)$$

Substituting in equation (21) for ($\sqrt{n}$), (r_{opt}) and (L_{opt}) from equations (1), (11) and (12), respectively,

$$V_i = \frac{0.77(1+k')}{\dfrac{4(1+k')}{k'(\alpha-1)}} \pi \left(\frac{8(1+k')}{k'(\alpha-1)}\sqrt{\frac{\eta_o D_m}{\pi P_o(\gamma^2-1)}}\right)^2 \frac{128(1+k')^3}{k'^3(\alpha-1)^3}\left(\frac{\eta_o D_m(1+6k'+11k'^2)}{3\pi P_o(\gamma^2-1)}\right)^{0.5}$$

$$V_i = 6.16(1+k')\pi\left(\frac{\eta_o D_{o(m)}}{\pi P_o(\gamma^2-1)}\right)^{\frac{3}{2}}\left(\frac{4(1+k')}{k'(\alpha-1)}\right)^4\left(\frac{(1+6k'+11k'^2)}{3}\right)^{0.5} \qquad (22)$$

Synopsis

The design process for *open tubular columns* is similar to that for packed columns, and the physical characteristics and performance specifications can be calculated using the same basic principles. An analogous protocol is observed, however, as a result of the geometric simplicity of the open tubular column; there are no packing factors and there is no multi-path term, consequently, the equations that result are far less complex and easier to manipulate and to understand. The first equation that is employed is again that of Purnell, utilizing the data from the reduced chromatogram that gives the efficiency necessary to affect the required separation. The next is the Golay equation, appropriately modified for a compressible mobile phase, which defines the variance per unit length of the column. By differentiating and equating to zero, the Golay equation can provide an expression for the optimum velocity and the minimum variance per unit length of the column when operated at the optimum mobile phase velocity. In addition, by taking the product of the required efficiency and the variance per unit length of the column at the optimum velocity, an expression for the column length can be derived. Equating the expression for the column length so obtained to that developed from the Poiseuille equation an expression for the optimum column diameter can be extracted. Substituting the expression for the optimum column diameter into those describing the optimum velocity and the minimum plate height, the fundamental equations for all the column parameters of the fully optimized column can be educed. These will include the column efficiency, column length and column radius together with the minimum analysis time, the maximum sample volume and the maximum acceptable extra-column dispersion.

Reference
1. K. Ogan and R. P. W. Scott, *J. High Res. Chromatogr.*, **7**(**July**) (1984)382.

Chapter 14

Chromatography Column Design

Application of the Design Equations to Packed Liquid Chromatography Columns and Open Tubular Gas Chromatography Columns

In the previous two chapters, equations were developed to provide the optimum column dimensions and operating conditions to achieve a particular separation in the minimum time for both packed columns and open tubular columns. In practice, the vast majority of LC separations are carried out on packed columns, whereas in GC, the greater part of all analyses are performed with open tubular columns. As a consequence, in this chapter the equations for *packed LC columns* will first be examined and the factors that have the major impact of each optimized parameter discussed. Subsequently *open tubular GC columns* will be considered in a similar manner.

Optimized Packed Columns for LC

In order to obtain numeric values for the optimized parameters, it is necessary to define a given separation and the equipment by which the sample is to be analyzed.

Table 1. Typical Chromatographic Operating Conditions for an LC Packed Column

Separation Ratio (α) (Critical Pair)	1.05
Capacity Ratio (first eluted peak of the Critical Pair) (k')	2.5
Capacity Ratio (first eluted peak of the Critical Pair) (k")	4.8
Viscosity of the Mobile Phase (η)	0.023 Poises
Diffusivity of the Solute (D_m)	3.5×10^{-5} cm^2/sec.
D'Arcy's Constant (j)	35
Maximum Inlet Pressure (P)	3000 p.s.i.
Packing Factor (λ)	0.5
Packing Factor (γ_p)	0.6
Capacity Ratio of the Last Eluted Peak (k'_2)	5.0

The different optimization equations derived in chapter 12 will then be used with these realistic chromatographic conditions in a simple optimization procedure. The conditions chosen are typical and might represent the average LC analysis. The values for (λ) and (γ_p) are those estimated by Giddings [1] for a well-packed column, namely, 0.5 and 0.6, respectively.

The Optimum Particle Diameter

The expression that gives the optimum particle diameter is given by equation (27), chapter 12 and is reiterated here. The optimization of the particle diameter will be considered first, as each of the other operating parameters will, directly or indirectly, be determined by the magnitude of the optimum particle diameter.

$$d_{p(opt)} = \frac{4(1+k'_A)}{k'_A(\alpha-1)} \sqrt{\left(\frac{\eta D_m}{D'_A P}\right)\left(\lambda\sqrt{\frac{\gamma_p \pi^2 (1+k'')^2}{k''}} + 4\gamma\right)} \quad (1)$$

It is clear that the major factor controlling the particle diameter will be the separation ratio (α), which reflects the difficulty of the separation. The more difficult the separation, the more theoretical plates are needed, and thus the column must be longer. However, to use a longer column, the particle diameter must be increased to allow the optimum velocity to be realized without exceeding the maximum system pressure. The effect of the capacity ratio of the first solute of the critical pair on the optimum particle diameter is complex. Extracting the function of the capacity ratio (f(k')) from equation (1),

$$f(k') = \frac{Y(1+k'_A)}{k'_A} \sqrt{\left(\lambda\sqrt{\frac{\gamma_p \pi^2 (1+k'')^2}{k''}} + 4\gamma\right)}$$

where
$$Y = \frac{4}{(\alpha-1)} \sqrt{\left(\frac{\eta D_m}{D'_A P}\right)}$$

The function f(k') is shown plotted against the thermodynamic capacity ratio in Figure 1. It is seen that for peaks having capacity ratios greater than about 2, the magnitude of (k') has only a small effect on the optimum particle diameter because the efficiency required to effect the separation tends to a constant value for strongly retained peaks. From equation (1) it is seen that the optimum particle diameter varies as the square root of the solute diffusivity and the solvent viscosity. As, in

Column Design Applications

practice, these physical properties of the chromatographic system do not change largely with the solvent used, the nature of the solute has only a small effect on the optimum particle diameter.

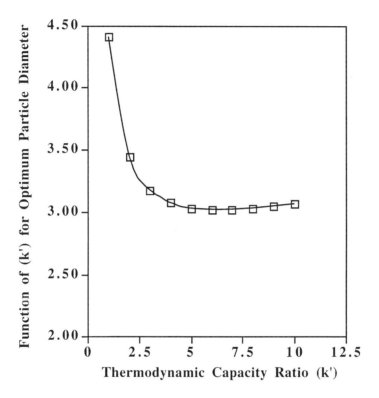

Figure 1. The Effect of Solute Capacity Ratio on the Magnitude of the Optimum Particle Diameter

The optimum particle diameter will decrease as the reciprocal of the available pressure and, thus, pressure will have a very significant effect on the magnitude of $(d_{p(opt)})$. As already stated, the higher the pressure, the smaller the particle diameter, the shorter the column and the faster the analysis. However, the overriding effect on the magnitude of $(d_{p(opt)})$ is the separation ratio (α) of the critical pair. As would be expected, the more difficult the separation, the more plates are needed and, thus, the longer the column. As the pressure is limited, this will demand larger particle diameters and will be accompanied by longer analysis times. The effect of the separation ratio (α) on the optimum particle diameter is shown in Figure 2. The curve was obtained using the data given in Table 1. It is seen that the optimum particle diameter changes dramatically as the separation ratio of the critical pair is reduced. When $\alpha > 1.05$ the optimum particle diameter falls and approaches 1 μm and below. Unfortunately, there are a very limited number of column packings that have diameters of 1 μm or less and, further, there are no well-established and reproducible packing techniques that will provide constant chromatographic properties from columns packed with sub-micron particles. In practice, particle diameters of 2 or 3 μm that give high efficiency reproducible columns are those usually employed for relatively simple separations.

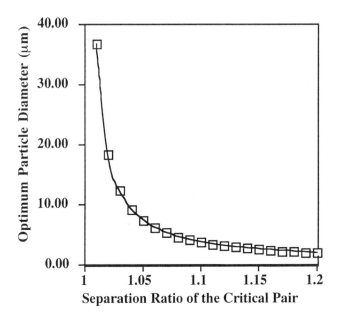

Figure 2. Curve Relating Optimum Particle Diameter to the Separation Ratio of the Critical Pair

The Optimum Velocity

The expression for the optimum mobile phase velocity is given by equation (28) in chapter 12 and is as follows,

$$u_{opt} = \pi(1 + k'') \frac{D_m}{2d_{p(opt)}} \left(\frac{\gamma_p}{k''} \right)^{\frac{1}{2}} \qquad (2)$$

It is seen that the optimum velocity is inversely proportional to the optimum particle diameter and it would be possible to insert the expression for the optimum particle diameter into equation (2) to provide an explicit expression for the optimum velocity. The result would, however, be algebraically cumbersome and it is easier to consider the effects separately. The optimum velocity is inversely

Column Design Applications

proportional to the optimum particle diameter and directly proportional to the solute diffusivity. However, the optimum particle diameter is proportional to the square root of the solute diffusivity and, thus, the optimum velocity is proportional to the square root of the solute diffusivity. The effect of the (k') on the optimum velocity will be largely determined by its effect on ($d_{p(opt)}$) and, thus, at high (k') values, where ($d_{p(opt)}$) changes little, it will become only slightly dependent on the magnitude of (k'). However, at low values of (k'), the value of ($d_{p(opt)}$) increases very rapidly with (k') and so, as the optimum velocity is inversely proportional to ($d_{p(opt)}$), the magnitude of the optimum velocity will begin to fall rapidly. The major factor controlling the optimum velocity will be the same as that controlling ($d_{p(opt)}$), *i.e.*, the separation ratio of the critical pair. Employing the data given in Table 1, values for the optimum velocity can be calculated for different values of the separation ratio of the critical pair. The results are shown in Figure 3.

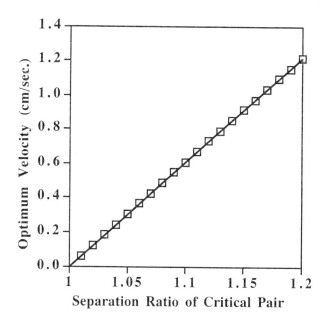

Figure 3. Graph of Optimum Velocity against Separation Ratio of the Critical Pair

It is seen that as the optimum particle diameter is inversely proportional to (α-1), and the optimum velocity is inversely related to the optimum particle diameter, the optimum velocity is linearly related to (α-1). Consequently, as the value (α) gets smaller, the optimum velocity is linearly reduced to accommodate the larger particle diameters that must be used.

The Minimum Plate Height

The expression that gives the minimum plate height is given by equation (29), chapter 12, and is reiterated here.

$$H_{min.} = 2d_{p(opt)}\left(\lambda + \frac{4}{\pi(1+k'')}\sqrt{k''\gamma_p}\right) \tag{3}$$

It is seen that the minimum plate height is linearly related to the optimum particle diameter and, thus, its magnitude will be modified with respect to changes in the capacity ratio of the solute and changes in the separation ratio of the critical pair, in a similar manner to the optimum particle diameter. Its sensitivity to changes in pressure, solvent viscosity and solute diffusivity will also follow the consequent changes in optimum particle diameter. Using the data given in Table 2, given later in this chapter, the expression for $H_{(min)}$ can be simplified to,

$$H_{min.} = 1.74\,d_{p(opt)} \tag{4}$$

It must be emphasized that equation (4) only applies to the separation and conditions defined in Table 1. The effect of the magnitude of the separation ratio on the minimum plate height is shown in Figure 4.

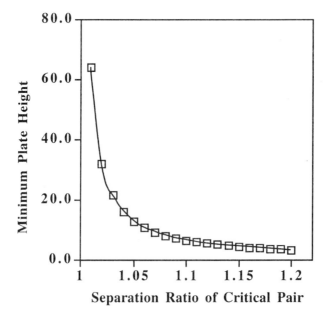

Figure 4. Graph of Minimum Plate Height against Separation Ratio of the Critical Pair

Column Design Applications

It is seen that as a result of using very small particles to separate solute pairs with relatively large separation ratios, the plate height is very significantly reduced and, as the number of plates needed is also small, the columns become very short and exceptionally fast. In contrast, more difficult separations will require longer columns to provide greater efficiency, which, due to the limited inlet pressure, will require larger particles resulting in an increase in the plate height.

The Minimum Column Length

The expression for the optimum column length is given by equation (30) in chapter 12 and is as follows,

$$L_{min} = \left(\frac{16(1+k'_A)^2}{k'^2_A (\alpha-1)^2} \right) \left(2d_{p(opt)} \left(\lambda + \frac{4}{\pi(1+k'')} \sqrt{k''\gamma_p} \right) \right) \tag{5}$$

It is seen that the length (L) will increase as $(1/(\alpha-1)^2)$ and, as (H_{min}) increases linearly with the particle diameter, which also increases as the reciprocal of $(\alpha-1)$.

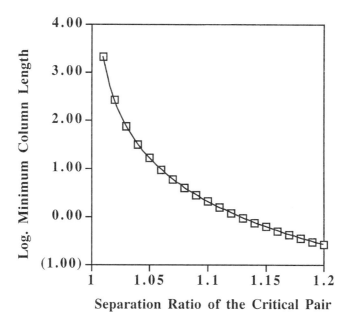

Figure 5. Graph of Log of Column Length against Separation Ratio of the Critical Pair

The column length will actually increase as $(1/(\alpha-1)^3)$. This means that the column length is extremely sensitive to the separation ratio of the critical pair and explains the importance of choosing a phase system that provides as large a value for (α) as possible. The relationship between column length and separation ratio is shown by the curve in Figure 5. The range of optimum column length is very large; to separate a substance having a separation ratio of 1.01 requires a column over 1000 cm in length, whereas to separate solutes which have a separation ratio of 1.2, a column as short as 1 mm is all that is necessary.

The Minimum Analysis Time

The minimum analysis time is given by equation (31) in chapter 12 and takes the following form,

$$t_{min} = \frac{L_{min}}{u_{opt}}(1 + k'_2) \qquad (6)$$

The impact of the value of (k'_2) is obviously linear, but controls the overall analysis time. It emphasizes the importance of carefully selecting the phase system not only to make the value of (α) as large as possible but also to keep (k'_2) to a minimum.

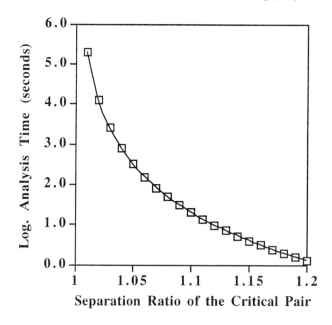

Figure 6. Graph of Log of Analysis Time against the Separation Ratio of the Critical Pair

The minimum column length increases as the value of (α) is reduced and, at the same time, the optimum velocity also falls. This has a double effect on the analysis time and it can be shown that the analysis time increases as $(1/(\alpha-1)^4)$. It is seen that, as the analysis time is a function of $((\alpha-1)^4)$, the range of analysis times is

Column Design Applications

large extending from about 0.5 s. necessary to separate the mixture when the separation ratio of the critical pair was large, *i.e.*, 1.2 to over 2 days when the separation ratio is as low as 1.01. This again emphasizes the critical nature of the choice of phase system and, for repetitive control analyses, considerable work may be well justified to obtain the largest separation ratio that is practical.

The Optimum Column Radius

Rearranging equation (36), chapter 12, and substituting for (n) from the Purnell equation, an expression for the column radius can be obtained,

$$r_{opt} = \left(\frac{1}{(1+k')}\right)\sqrt{\frac{\sigma_E k'(\alpha-1)}{1.28\varepsilon\pi H_{min}}} \quad (7)$$

Employing the data in Table 1 to simplify the function,

$$r_{opt} = 0.291\sqrt{\frac{\sigma_E(\alpha-1)}{H_{min}}} \quad (8)$$

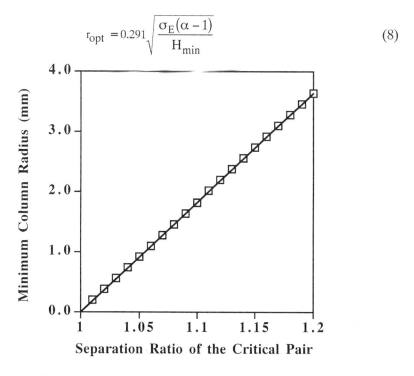

Figure 7. Graph of Optimum Column Radius against Separation Ratio of the Critical Pair

Taking a value for (σ_E) of 2.5 µl (which would be typical for a well-designed column detector system) and using equation (8), values for (r_{opt}) are shown plotted against separation ratio in Figure 7. It is seen that the optimum column radius increases linearly with the separation ratio of the critical pair (ranging from 0.1 mm to nearly 4 mm) and also with the magnitude of the extra-column dispersion (σ_E). Thus, the simpler the separation (the larger (α)), the wider the column and, as a consequence, the mass sensitivity of the system will decrease as the separation ratio of the critical pair increases.

The Optimum Flow Rate

The optimum flow rate will be given by the simple relationship,

$$Q_{opt} = e\pi r_{opt}^2 u_{opt}$$

where all the symbols have the meanings previously ascribed to them.

Substituting for the constants,

$$Q_{opt} = e\pi r_{opt}^2 u_{opt} = 1.88\, r_{opt}^2 u_{opt} \qquad (9)$$

Although the optimum column radius increases linearly with the separation ratio of the critical pair, this simple relationship is moderated by the ratio of the square of the optimum radius to the optimum velocity, both of which are functions of (α). The relationship between the optimum flow rate and the separation ratio of the critical pair is shown in Figure 8.

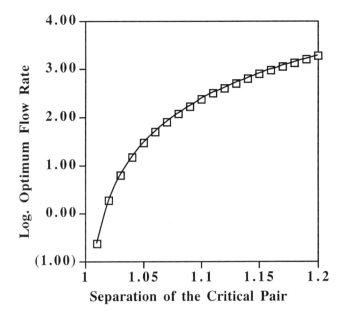

Figure 8. Graph of Log of Optimum Flow Rate against Separation Ratio of the Critical Pair

It is seen that, contrary to what may be commonly thought, the optimum flow rate *increases dramatically* as the *separation becomes more simple* ((α) becomes larger). This is because smaller particle diameters can be used and, thus, the optimum velocity is very high. The interacting effects have an interesting influence on the solvent consumption as seen below.

Solvent Consumption

The solvent consumption is simply obtained by multiplying the analysis time by the optimum flow rate as shown below.

$$V_{sol} = Q_{opt} t_{min} \qquad (10)$$

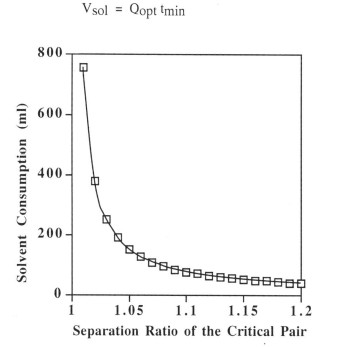

Figure 9. Graph of Solvent Consumption against the Separation Ratio of the Critical Pair

The solvent consumption appears to be in conflict with the corresponding optimum flow rates. Substances with high (α) values have a very high optimum flow rate (over 1l per min. for (α=1.2) and the column diameter is over 6 mm which would indicate a very large solvent consumption. However, because the separation is simple, a very rapid separation is achieved with analysis times of less than a second. As a consequence, only a few ml of solvent is necessary to complete the analysis. The apparatus, however, must be designed with an exceedingly fast response and very special sample valves would be necessary. In contrast, a very difficult separation would require nearly a liter of solvent although the flow rate will be small (little more than 0.1 ml/min.) and the column only about 100-200 μm in diameter. However, the analysis time is now about 2 days, which even at the low flow rates will still involve the use of nearly 800 ml of solvent.

The Maximum Sample Volume

The expression for the maximum sample volume can be found in Chapter 12 equation (45) and is as follows,

$$V_{sample} = \frac{1.1 \varepsilon \pi r_{opt}^2 L_{min}(1+k')}{\sqrt{n}}$$

and, substituting for √n,

$$V_{sample} = 0.275 \varepsilon \pi r_{opt}^2 L_{min} k'(\alpha - 1) \tag{11}$$

Employing the data from Table 1 the values already calculated for (r_{opt}) and (L_{min}) the maximum sample volume was calculated for a range of values for (α) and the results are shown as a graph in Figure 10.

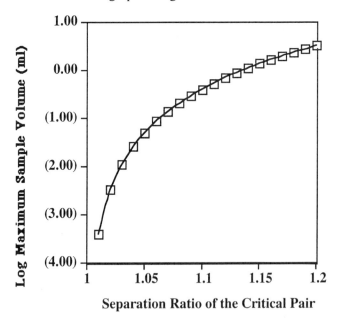

Figure 10. Graph of Log of Maximum Sample Volume against the Separation Ratio of the Critical Pair

It is seen that the maximum sample volume ranges from a fraction of a microliter when (α=1.01) to several ml when (α) = 1.2. The small sample volume results from the very high efficiencies required to separate a pair of peaks having a separation

Column Design Applications

ratio of only 1.01 which entails very small peak volumes (particularly as the column diameter must be only 100-200 µm in diameter). At the same time, such conditions provide very small limiting mass sensitivity which allow the use of very small samples. To obtain the maximum speed for the separation of the solute pair (where $\alpha=1.2$) however, can employ sample volumes of a few ml which may result in very poor mass sensitivity depending on the detector and the nature of the solutes.

It is seen that it is possible to design the optimum packed column for any specific separation. In the next section, a similar procedure will be adopted to develop a regime for designing the optimum open tubular column. The conditions used and given in Table 1 are fairly typical for most LC analyses of relatively small molecular weight substances. It is seen that there is a limited range over which the operation of LC columns would be practical. A long column, packed with relatively large particles, would separate the critical pair with a separation ratio of only 1.01, but it would take days to complete an analysis. It is possible that there might be a sample of sufficient importance to spend this amount of time on the analysis, but it is more likely that the analyst would seek an alternative method. At the other extreme, particles having diameters less than 2-3 microns, although available, are difficult to pack. Assuming that the practical peak capacity is about half the theoretical peak capacity, a column a few cm long packed with particles 3 microns in diameter would separate a mixture containing about 16 components in about 30 seconds. Such columns are commercially available in close-to-optimized form (known colloquially as the 'three by three') and would probably be the most suiTable for the rapid analysis of simple mixtures. The two other common, commercially available, particle sizes, 5 and 10 microns, should be packed into columns 13 and 148 centimeters long, respectively, for optimum performance. Such columns would provide a 'column family' for general use which would encompass the vast majority of LC applications that face the analyst today. It should be emphasized, however, that for an analysis that is repeated many times every day, as in a quality control laboratory, the construction of the fully optimized column that is specific for the particular analysis would always be economically worthwhile.

Gradient Elution

The design procedure described above will, in theory, be applicable only to samples that are separated by isocratic development. Under gradient elution conditions the (k') value of each solute is continually changing, together with the viscosity of the mobile phase and the diffusivity of each solute in the mobile phase. As a consequence the equations derived from the plate and rate theories will only be approximate at best and, in many cases, may give misleading values, particularly for the required column efficiency. As the efficiency required to separate the critical pair is crucial to column design, the optimum column for use in gradient elution development, although not impossible to calculate, is very complex and the equations bulky and involved.

However, gradient elution is often employed as an alternative to isocratic development to *avoid* the design and construction of the optimum column, which is seen as a procedure which can be tedious and time consuming. Samples that contain solutes that cover a wide polarity range, when separated with a solvent mixture that elutes the last component in a reasonable time, often fail to provide adequate resolution for those solutes eluted early in the chromatogram. For the occasional sample, gradient elution provides a satisfactory and immediate solution to this problem. However, for quality control, where there is a high daily throughput of samples, isocratic development, employed in conjunction with an optimized column,. is likely to be more economic. In many applications, the optimized column, operated isocratically, will provide a shorter analysis time than that obtained by gradient elution used with an *ad hoc* column. Furthermore, isocratic development eliminates the time required after gradient elution to bring the column back to equilibrium with the initial solvent mixture before commencing the next analysis. Excluding samples of biological origin, where possible isocratic development is the preferred method for routine LC analysis, as with an optimized column, it is usually faster, utilizes less solvent and requires less expensive apparatus.

Samples of biological origin fall in a class of their own. Many biological samples can *only* be separated by gradient elution, particularly the macromolecules, polypeptides, proteins, etc. However, due to the nature of the samples, the gradient is often very small indeed as a slight change in mobile phase composition can make a dramatic change in elution rate of many macromolecules. Consequently, the solvent viscosity and solute diffusivity does not change significantly during the program and, for the purposes of column design, the chromatogram can be treated as though the separation was developed isocratically. The separation ratios, capacity ratios and efficiencies are calculated in the normal way, assuming the gradient

Column Design Applications

chromatogram to be the result of an isocratic development. Optimizing the LC column can be tedious procedure but, if the analysis is to be a standard that is repeated continuously through the day for many years, it can save much time, reduce solvent costs dramatically and, in particular, reduce the problems of toxic solvent disposal.

Optimized Open Tubular Columns for GC

In a similar manner to the optimization of an LC column, in order to obtain numeric values for the optimized parameters, it is necessary to define a given separation and the equipment and materials by which the sample is to be analyzed. The data given in Table 2 are for a general GC separation using an open tubular column.

Table 2. Typical Chromatographic Operating Conditions for an LC Packed Column

Separation Ratio (α) (Critical Pair)	1.05
Capacity Ratio (first eluted peak of the Critical Pair) (k')	2.5
Viscosity of the Mobile Phase (η) (nitrogen @ 25°C)	178.1 x 10^{-6}
Diffusivity of Solute (D_m)	2.0 x 10^{-1} cm^2/sec.
Inlet Pressure (P) 30 p.s.i. $\equiv$ 3 atmospheres absolute $\equiv$ 3.039 x 10^4 dynes/sq. cm	
Inlet/Outlet Pressure Ratio (γ)	3.0
Capacity Ratio of the Last Eluted Peak (k'_2)	5.0

In this optimization procedure those equations derived in chapter 13 will then be used with the data provided in Table 2.

The Optimum Column Radius

The equation used to calculate the optimum radius is that given in chapter 13, equation (13)

$$r_{opt} = \frac{8(1+k')}{k'(\alpha-1)} \sqrt{\frac{\eta_o D_{(o)m}}{\pi P_o (\gamma^2 - 1)}} \tag{12}$$

It is seen that the magnitude of ($r_{(opt)}$) will vary as the square root of the product of the viscosity and diffusivity. As the solute diffusivity and the viscosity of a gas tend to be inversely proportional, then the optimum radius is largely independent of the gas that is used. Employing the data in Table 2, equation (12) can be reduced to

$$r_{opt} = \frac{7.646 \times 10^{-5}}{(\alpha - 1)} \tag{13}$$

Employing equation (2), the effect of separation ratio on the optimum radius is shown by the graph in Figure 11.

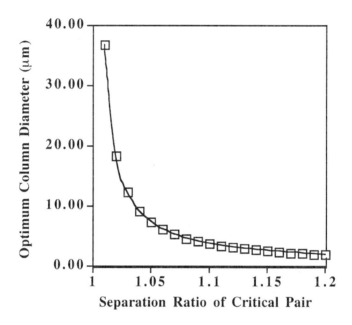

Figure 11. Graph of Optimum Column Radius against the Separation Ratio of the Critical Pair

The first, perhaps somewhat surprising, fact that emerges from the curve shown in Figure 11 is that small diameter columns are best for separating simple mixtures where there is little challenge in the separation. Conversely, for difficult separations wider diameter columns are necessary. This is because as the separation becomes more difficult, a longer column is required. However, as the inlet pressure is limited, then in order to obtain the optimum flow rate, the column radius must be increased to reduce the flow impedance. The second interesting point is that columns 15 to 40 μm in diameter (note the data in Figure 11 are for the column radius) are best for relatively simple mixtures, and contemporary GC equipment is not suiTable for operating columns as small as this although columns 50 μm in diameter have been successfully utilized.

Column Design Applications

The Optimum Mobile Phase Velocity

The optimum mobile phase velocity for an open tubular GC column is given in chapter 13, equation (14). Reiterating this equation,

$$u_{opt} = \frac{4D_m}{r_{opt}} \left(\frac{3(1+k')^2}{1+6k'+11k'^2} \right)^{0.5} \qquad (14)$$

Employing the data in Table 2, equation (14) can be reduced to

$$u_{opt} = \frac{5270}{r_{opt}} \qquad (15)$$

Employing equation (15), curves relating the optimum velocity to the separation ratio of the critical pair were calculated and the results are shown in Figure 12.

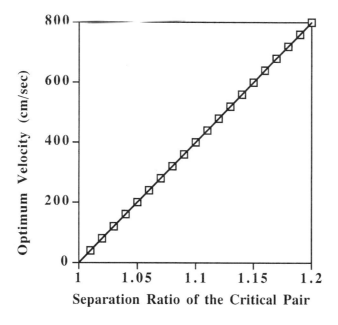

Figure 12. The Linear Curve Relating the Optimum Velocity to the Separation Ratio of the Critical Pair

As the optimum column radius is inversely proportional to $(\alpha-1)$, and (u_{opt}) is inversely proportional to (r_{opt}), the simple linear relationship between optimum velocity and the separation ratio is to be expected. The high velocities employed for the more simple separations are not immediately obvious and appear extremely high. However, the reason for these high velocity values will become clear when (H_{min}) and (L_{min}) are evaluated.

The Calculation of (H_{min})

The expression for (H_{min}) is given in chapter 13 and a simplified form of this equation is as follows,

$$H_{min} = r_{opt}\sqrt{\frac{1+6k'+11k'^2}{3(1+k')^2}}, \qquad (16)$$

The explicit form of the equation being

$$H_{min} = \frac{8}{k'(\alpha-1)}\sqrt{\frac{\eta_o D_m(1+6k'+11k'^2)}{3\pi P_o(\gamma^2-1)}} \qquad (17)$$

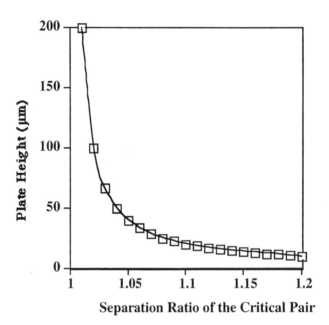

Figure 13. Graph of Minimum Plate Height against Separation Ratio of the Critical Pair

Employing the data in Table 2, equation (16) can be reduced to

$$H_{min} = 1.51 r_{opt} \qquad (18)$$

Column Design Applications

Employing equation (18), the value of (H_{min}) was calculated for a series of separation ratios of the critical pair and the results are shown in Figure 13. It is seen that under optimum conditions the plate height can be only a few microns providing a large number of plates in a relatively small length of column. The reason for the high velocities is beginning to become a little more obvious. The column length will provide the final clue.

The Optimum Column Length

The equation for the column length is give in chapter 13, equation (14) and is reproduced as follows.

$$L_{min} = nH_{min} = \frac{16(1+k')^2}{k'^2(\alpha-1)^2} H_{min} \qquad (19)$$

or

$$L_{min} = 128 \frac{(1+k')^2}{k'^3(\alpha-1)^3} \sqrt{\frac{\eta_o D_m (1+6k'+11k'^2)}{3\pi P_o (\gamma^2-1)}} \qquad (20)$$

The minimum column length was calculated for a series of separation ratios using equation (19) and the results obtained are shown in Figure 14.

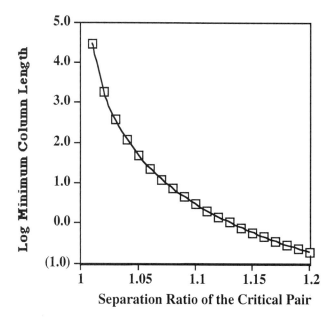

Figure 14. Graph of the Log of the Minimum Column Length against the Separation Ratio of the Critical Pair

It is seen that the minimum column length ranges from 100 meters to 2 or 3 mm. It should be pointed out at this time that the theory may not predict data accurately when the column lengths become less than a few centimeters. It is now of interest to calculate the corresponding analysis times.

The Analysis Time

The equations for the analysis time are given in chapter 13. Reiterating equations (15),

$$t_{min} = (1 + k'_2)\frac{L_{min}}{u_{mean}} \tag{21}$$

Now,
$$u_{mean} = u_{opt}\frac{3(\gamma^2 - 1)}{2(\gamma^3 - 1)}$$

Thus,
$$t_{min} = (1 + k'_2)\frac{2L_{min}(\gamma^3 - 1)}{3u_{opt}(\gamma^2 - 1)} \tag{22}$$

Employing equation (22) the analysis time was calculated for a series of different separation ratios and the results obtained are shown in Figure 15.

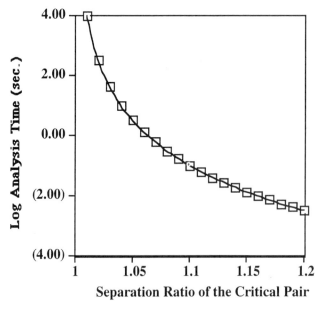

Figure 15. Graph of the Log of the Analysis Time against the Separation Ratio of the Critical Pair

Column Design Applications

It is seen from Figure 15 that the analysis time ranges from about 10,000 seconds (a little less than 3 hr) to about 30 milliseconds. The latter, high speed separation, is achieved on a column about 2 mm long, 12 microns in diameter, operated at a gas velocity of about 800 cm/second. Such speed of elution for a multicomponent mixture is of the same order as that of a scanning mass spectrometer.

The Optimum Flow Rate

The equation for the optimum flow rate is given in chapter 13, which is given here as follows.

$$Q_{opt} = \pi r_{opt}^2 u_{opt} \tag{23}$$

In Figure 16 the relationship between column flow rate (exit flow rate) and the separation ratio of the critical pair is demonstrated, the data being calculated using equation (23).

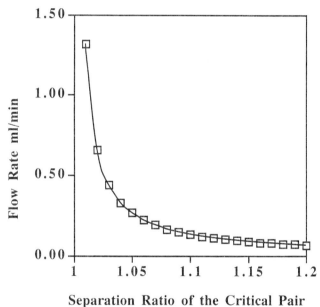

Figure 16. Graph of Optimum Flow Rate against the Separation Ratio of the Critical Pair

It is seen that the optimum flow rate ranges from about 1.3 ml/min to about 65 µl per min. It is noted that once the magnitude of the separation ratio is 1.05 or more, then the flow rate is less than 200 µl/min.

The Maximum Sample Volume

The expression defines the maximum sample volume is given in chapter 13, equation (21) and is reproduced as follows,

$$V_i = \frac{0.77(1+k')}{\sqrt{n}} \pi r_{opt}^2 L_{min} \qquad (24)$$

Simplifying equation (24) using numerical data from Table 2, then equation (2) becomes

$$V_i = 1.511(\alpha - 1) r_{opt}^2 L_{min} \qquad (25)$$

Employing equation (25), the effect of the separation ratio of the critical pair on the maximum sample volume was calculated using equation (25) and the results are shown in Figure 17.

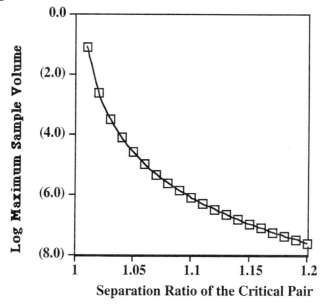

Figure 17. Graph of Log of Maximum Sample Volume against the Separation Ratio of the Critical Pair

It is seen that, in all cases, the maximum sample volume is extremely small ranging from about 100 μl for a very difficult, lengthy separation to less than a tenth of a nanoliter for simple, rapid separations.

From the optimization calculations, it is clear that the open tubular column can have a performance that is far better than that accepted by contemporary GC operators. There are two main reasons for this. Firstly, the remarkable level of performance that can be obtained form the open tubular column in GC is largely not known or appreciated. Consequently, a grossly inferior performance is tolerated in the majority of GC analyses. This problem can be solved only by education. Secondly,

Column Design Applications

to realize the higher performance, the GC apparatus must be greatly improved. Much smaller diameter columns must be made available and used with an apparatus that is designed to ensure they do not become blocked during operation. The overall response of the detecting system must be made much faster and the dispersion that can occur in the apparatus other than the column must be reduced by at least an order of magnitude. Sample volumes must be reduced by two orders of magnitude and, at the same time, they must be constructed to be accurate and precise. Most GC separations employ temperature programming which complicates the optimization procedure. Approximate optimization can be achieved by employing the same technique, but using the reduced temperature chromatogram as an isothermal chromatogram. In the early days of GC, the gas chromatograph was called the poor man's mass spectrometer. Today the apparatus must be designed to emulate the speed and resolution of the contemporary mass spectrometer if the remarkable speed and resolution available from the open tubular column is to be realized.

Synopsis

Employing a set of standard chromatographic conditions to separate a closely defined mixture of solutes, the equations previously developed for column optimization were employed to design a fully optimized chromatographic packed column system. Initially, it was shown that the optimum particle diameter became independent of the magnitude of capacity ratio of the first eluted solute of the critical pair, providing the capacity ratio was above 2.5. Optimum particle diameters ranged from about 40 µm to 1 µm as the separation ratio changed from 1.01 to 1.2. The optimum velocity changed linearly with the separation ratio from about 0.05 cm/sec. to 1.2 cm/sec. The minimum plate height, when employing the optimum particle diameter, and operating at the optimum velocity, ranged from about 60 µm to less than 5 µm the values at separation ratios below 1.05 ranging tightly between about 5 µm and 1 µm. The minimum column length for a separation ratio of 1.01 was over 10 m, but at the other extreme for a separation ratio of 1.2 the column length was only a few millimeters. In a similar manner, the analysis time covered a wide range, taking over 2 days to separate the mixture when the separation ratio was 1.01 and only 0.5 sec. when the separation ratio was 1.2. Over the same span of separation ratios, the optimum flow rate ranged from 0.2 ml/min to 3 ml/min, the solvent consumption from 800 ml to about 3 ml and the maximum sample volume from a fraction of a microliter to a fraction of a milliliter. Optimization involves considerable work but if there are sufficient samples to be analyzed every day, then it is well worthwhile as the throughput will be high and the solvent consumption at a minimum. In general, if gradient elution is employed, the optimization procedure is too complex to employ. If, however, the change in solvent composition does not modify the diffusivity or viscosity significantly, the gradient chromatogram can be treated as an isocratic chromatogram and the system optimized approximately in the same way. Open tubular GC columns can be optimized in a similar manner, the optimum radius ranging from 40 μm to about 1μm (diameters 80 μm to 2 μm) for separation ratios of 1.01 and 1.2, respectively. The optimum velocity changes linearly from about 1 cm/sec. to 800 cm/sec. and the minimum plate height from 200 μm to about 10 μm over the same separation ratio range. The corresponding changes in column length were from 100 m to 1 millimeter, changes in analysis time from 3 hr to 30 milliseconds and flow rate from about 1.3 ml/min to 0.05 ml/min. In order to take full advantage of the open tube column optimization, the chromatographic system needs to be redesigned to provide very small sample volumes, very sharp injection bands and very low column-detector dispersion. The performance of contemporary gas chromatographs is far from optimal and with improved design could provide striking improvements. Most GC separations employ temperature programming which complicates the optimization procedure. Approximate optimization can be achieved by employing the same procedure but using the reduced temperature programmed chromatogram as equivalent to a reduced isothermal chromatogram.

Reference

1. J. C. Giddings, *"Dynamics of Chromatography"*, Marcel Dekker, New York (1965) 56.

Chapter 15

Preparative Chromatography

The term preparative chromatography tends to be automatically associated with large samples. In fact, although most preparative separations do, indeed, involve relatively large samples, the term *preparative* really applies to those separations where some, or all, of the individual constituents are collected for subsequent use. If the collected sample is the final product of a synthesis, or even an intermediate, then the quantity isolated may well be several hundred grams or even kilos. If, however, the sample is merely required for subsequent spectroscopic examination, then the amount of material collected may be only a few milligrams. It follows that preparative chromatography does not necessarily involve the use of large sample loads. Nevertheless, it is when large samples are used to provide significant amounts of material that operating problems arise. Although routine separations for production purposes usually require specially designed columns and equipment, in most preparative separations standard columns are often used which are heavily overloaded to accommodate the necessary sample loading. Thus, sample overload is an important aspect of preparative chromatography and requires some theoretical examination.

Column Overload

A column can be overloaded in basically two different ways.

1. If the sample is relatively insoluble in the mobile phase, then it can be dissolved, as a dilute solution, in a relatively large volume of solvent. A large volume of the solution can then be placed on the column, a procedure that results in *volume overload*.

2. Conversely, if the sample is readily soluble in the mobile phase, then a much stronger solution of the sample can be prepared, but, in this case, a moderately small

420   Chromatography Theory

sample volume can be placed on the column. However, because of the high concentration of solute, the sample mass will be excessive, which will result in *mass overload*.

The two types of overload have quite different effects on the separation process, *i.e.*, the resolution that is obtained from the column and the shape of the resulting peaks. Both sampling techniques can be very effective, but need to be carefully controlled and the procedure well understood if sample purity is to be maintained.

Column Overload Due to Excess Sample Volume

An expression for the maximum charge that can be placed on a column without impairing resolution has already been derived, but the approach, when dealing with an overloaded column for preparative purpose, will be quite different. For preparative purposes the phase system is chosen to provide the maximum separation of the solute of interest from its *nearest neighbor*. It should be pointed out that the separation may, but probably will not, involve the closest eluting pair in the mixture. Consequently, the maximum resolving power of the column will not be required for the purpose of separation and the excess resolution of the solute of interest from its nearest neighbor can be used to increase the column load.

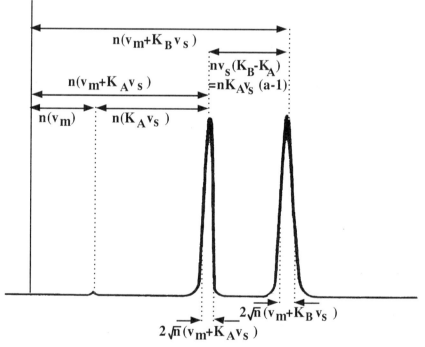

Courtesy of the Journal of Chromatography [ref. 1]

Figure 1. Theory of Volume Column Overload

Preparative Chromatography

Consider the separation depicted in Figure 1. It is assumed that the pair of solutes represent the elution of the solute of interest and its nearest neighbor. Now, when the sample volume becomes extreme, the dispersion that results from column overload, to the first approximation, becomes equivalent to the sample volume itself as the sample volume now contributes to the elution of the solutes. Thus, from Figure 1, the peak separation in milliliters of mobile phase will be equivalent to the volume of sample plus half the sum of the base widths of the respective peaks.

Now, half the peak width at the base is approximately twice the peak standard deviation, and thus, assuming the two peaks have around the same efficiency (n),

$$(\alpha_{B/A} - 1) n K_A v_s = V_L + 2\sqrt{n}(v_m + K_A v_s) + 2\sqrt{n}(v_m + K_B v_s)$$

where (K_A) is the distribution coefficient of solute (A),
(K_B) is the distribution coefficient of solute (B),
(α) is the separation ratio of solute (B) to solute (A),
and (V_L) is the sample overload volume.

Rearranging,

$$V_L = (\alpha_{B/A} - 1) n K_A v_s - 2\sqrt{n}\left[(v_m + K_A v_s) + (v_m + K_B v_s)\right]$$

Noting that

$$n K_A v_s = V'_A \quad \text{and} \quad n K_B v_s = V'_B$$

$$\frac{V'_A}{V_o} = k'_A \quad \text{and} \quad \frac{V'_B}{V_o} = k'_B$$

$$k'_B = \alpha_{B/A} k'_A \quad \text{and} \quad V_o = n v_m$$

Then

$$V_L = V_o\left[(\alpha_{B/A} - 1) k'_A - \frac{2}{\sqrt{n}}\left(2 + k'_A + \alpha_{B/A} k'_A\right)\right] \quad (1)$$

Employing equation (1), curves relating maximum sample volume to the capacity ratio of the first eluted peak for different separation ratios were calculated and constructed and the results are shown in Figure 2.

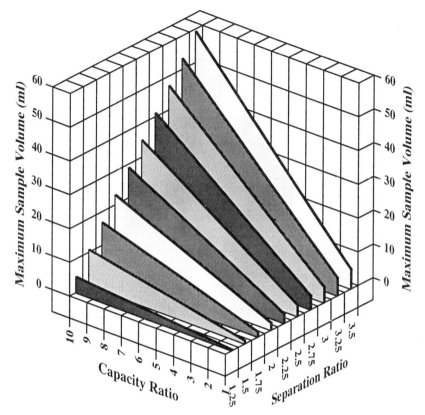

Figure 2. Curves Relating Maximum Sample Volume to the Capacity Ratio of the First Peak for Different Separation Ratios

The curves were calculated for a column 4.6 mm I.D. 25 cm long, packed with particles 20 mm in diameter and having an efficiency of 6,000 theoretical plates. It is seen that the maximum acceptable sample volume increases with both the capacity ratio of the first eluted peak and the separation ratio of the peak of interest to its nearest neighbor as predicted by equation (1). It is also seen that even with an analytical scale column (4.6 mm I.D.), if reasonable retention is ensured and the phase system is selected to provide a large separation ratio, sample volumes of 60 ml or more can be employed. It follows that providing a very large *mass* of material is not required, then for preparative separations, it is often more convenient to overload an analytical column than to construct or purchase a high capacity column designed specifically for the separation in hand. The same principle, however, applies to large columns. Using equation (1), it can be shown that givena column 4 ft long and 1 in. I.D., having an efficiency of 2000 theoretical plates, separating a pair of substances having a separation ratio of 2 and a capacity ratio for the first peak of 6, then the maximum sample volume of sample will be well over 2l. It is clear that the potential preparative capacity of any column is well in excess of that normally accepted for analytical purposes and, whatever the column, overload will always be a viable technique for increasing sample capacity. This does assume that a phase system can be found that will provide a reasonably high capacity ratio for the first peak and a fairly large separation ratio between the solute of interest and its nearest neighbor.

The effective use of column volume overload for preparative separations was experimentally demonstrated by Scott and Kucera [1]. These authors used a column 25 cm long, 4.6 mm I.D. packed with Partisil silica gel 10 mm particle diameter and employed *n*–heptane as the mobile phase. The total mass of sample injected was kept constant at 176 mg, 8 mg and 0.3 mg of benzene, naphthalene and anthracene, respectively, but the sample volumes used which contained the same mixture of solutes were 1 µl, 1 ml, 2 ml and 3 ml. The chromatograms of each separation are shown in Figure 3. The dead volume of the column was found to be 3.48 ml and the retention volumes of benzene, naphthalene and anthracene were found to be 4.22, 8.11 and 15.0 ml, respectively.

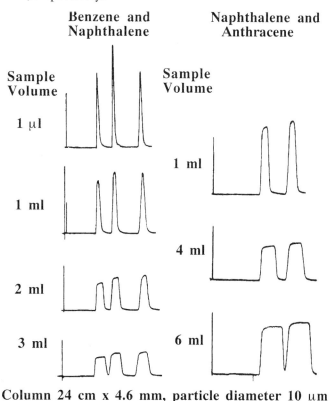

Column 24 cm x 4.6 mm, particle diameter 10 µm
Courtesy of the Journal of Chromatography (ref. 1)

Figure 3. Chromatograms Demonstrating Volume Overload

424   Chromatography Theory

The capacity ratios, efficiencies and separation ratios are given in Table 1, together with the maximum sample volume as calculated from equation (1). Comparing the maximum sample volume used for benzene given in Table 1, with the actual sample volumes for benzene shown in Figure 3, it would appear that equation (1) can be used with confidence for calculating (V_L). In a similar manner, using the same column, 2 ml, 4 ml and 6 ml samples, each containing 9.0 mg and 0.3 mg of naphthalene and anthracene, respectively, were placed on the column and the chromatograms obtained from each sample volume are included in Figure 3.

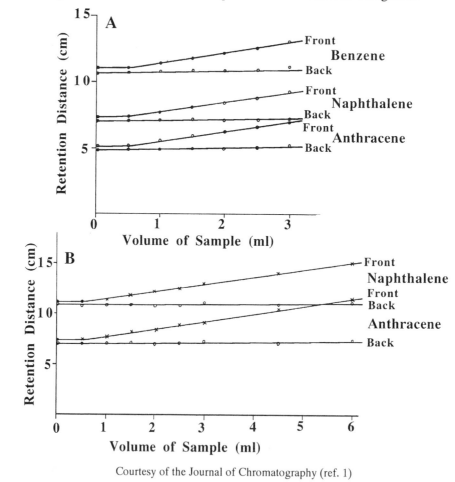

Courtesy of the Journal of Chromatography (ref. 1)

Figure 4. Graphs of the Retention Distance of the Peaks of Benzene, Naphthalene and Anthracene against Sample Volume

Scott and Kucera repeated a series of similar experiments again using 0 to 3 ml samples of the benzene mixture and 0 to 6 ml samples of the naphthalene mixture but

Preparative Chromatography

increased the sample volume in increments of 0.5 ml. It is seen that the maximum volume of charge is again accurately predicted by equation (1).

Table 1. Analytical Column Overload Properties

Parameter	Benzene	Naphthalene	Anthracene
(k')	1.18	2.33	4.31
(n)	1850	4480	5470
(α)	< 1.97	>< 1.85	>
(V_L)	3.1 ml	6.1 ml	

The retention volume of the front and the back of each peak was measured and the curves relating these measurements to the volume of sample added are shown in Figures 4A and 4B. Figure 4 shows the effect of volume overload on the overall peak dispersion more clearly. From Figures 4A and 4B, it is seen that the front of the peak has a constant retention distance irrespective of the volume of the charge. The retention distance of the rear of the peak, however, only remains constant for sample volumes less than 0.5 ml for the three component mixture, and 1 ml for the two component mixture. Subsequently, the retention distance of the rear of the peak increases linearly with sample volume as predicted by equation (1). This increase in retention volume of the rear of the peak is the same for each solute and is independent of the capacity ratio of the individual solutes. It is seen in Figure 4A at the predicted sample volume of 3 ml, the back of the benzene peak has just reached the front of the naphthalene peak. Similarly, Figure 4B shows that at the predicted sample volume of 6 ml, the back of the naphthalene peak has just reached the front of the anthracene peak. Peak dispersion towards greater retention is characteristic of volume overload which is in direct contrast to the peak distortion that arises from mass overload.

Column overload that results from increased sample volume distorts the normal elution process of development towards that of frontal analysis. In fact, if elution development is carried out with progressively increasing sample volume, the distorted elution development actually culminates in a frontal analysis chromatogram. This transition from elution development to frontal analysis is unambiguously demonstrated in Figure 5, where progressively larger volumes are placed on the column, up to a maximum of 16 ml. It is seen that the first chromatogram takes the typical form that arises from elution development. A sample volume of 2 ml produces significant peak dispersion but each peak is still eluted discretely and apart from the others. When the sample volume is increased to 3 ml, the peaks for naphthalene and anthracene begin to coalesce although the three components are still discernible. A sample volume of 6 ml, however, causes the frontal analysis shape of the chromatogram to begin to form.

The peaks for naphthalene and anthracene now exhibit three distinct parts one comprising pure naphthalene, one pure anthracene and the other, in the center of the envelope, a peak containing both naphthalene and anthracene. The back of the triple peak now constitutes the last two steps of the frontal analysis concentration profile.

At this stage the benzene is still just separated from the naphthalene. A 16 ml sample volume demonstrates the complete frontal analysis chromatogram of the mixture. The first and second steps containing pure benzene and benzene + naphthalene, respectively, the center peak containing all three solutes and the last two steps containing naphthalene plus anthracene and pure anthracene, respectively.

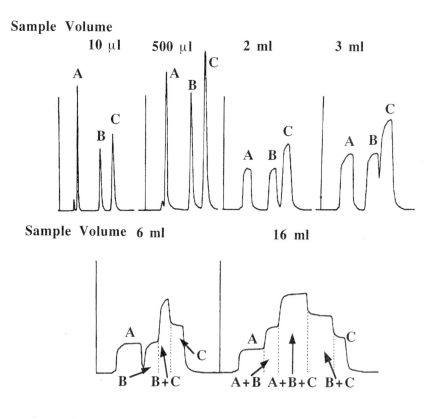

Courtesy of the Journal of Chromatography (ref. 1)

Figure 5. The Transition from Elution Development to Frontal Analysis by Using Large Sample Volumes

Preparative Chromatography

Column Overload Due to Excess Sample Mass

The effect of mass overload on the separation process can be extremely complicated and for this reason it is difficult to examine comprehensively. Specific mass overload effects are sometimes subject to theoretical treatment, but it is laborious to take into account all the secondary effects of high sample masses, simultaneously. Initially, as the mass of solute placed on the column is increased, the adsorption isotherm becomes nonlinear due to the limited capacity of the stationary phase. In many examples the distribution coefficient is determined by the Langmuir adsorption isotherm, which has already been discussed in some detail. The effect of nonlinear adsorption isotherms on peak shape caused by mass overload has been examined by Guiochon and his co-workers [2-4] by using computer simulation which, although somewhat complicated, is now fairly well understood. However, in addition, increasing the concentration of solute in the stationary phase will change the nature of the solute-stationary phase interactions, as the solute itself will now act as a significant part of the stationary phase. At even higher masses, the concentration of the solute in the mobile phase will become high enough to also significantly change the nature and probability of solute interactions with the components of the mobile phase. Thus, the solute in the mobile phase itself will act as an additional component of the mobile phase and aid in eluting the solute and, as a consequence, reduce the retention time.

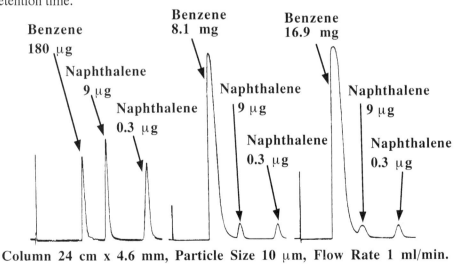

Column 24 cm x 4.6 mm, Particle Size 10 μm, Flow Rate 1 ml/min.

Courtesy of the Journal of Chromatography (ref. 1)

Figure 6. An Experimental Example of the Mass Overload of Benzene

The problem is made more difficult because these different dispersion processes are interactive and the extent to which one process affects the peak shape is modified by the presence of another. It follows if the processes that causes dispersion in mass overload are not random, but interactive, the normal procedures for mathematically analyzing peak dispersion can not be applied. These complex interacting effects can, however, be demonstrated experimentally, if not by rigorous theoretical treatment, and examples of mass overload were included in the work of Scott and Kucera [1]. The authors employed the same chromatographic system that they used to examine volume overload, but they employed two mobile phases of different polarity. In the first experiments, the mobile phase n–heptane was used and the sample volume was kept constant at 200 µl. The masses of naphthalene and anthracene were kept constant at 9 mg and 0.3 mg, respectively, but the mass of benzene was increased from 1.80 mg to 8.1 mg and then to 16.9 mg. The elution curves obtained are shown in Figure 6.

The chromatograms shown in Figure 6 depict the normally loaded column on the left and the column overloaded with 8 and 16 mg of benzene center and right, respectively. The three mass overload effects are clearly demonstrated in both chromatograms. The benzene peaks have significantly broadened and exhibit gross asymmetry with long tails. It is clear that the concentration of benzene in both overload chromatograms has reached a level where the adsorption isotherm is no longer linear. It can also be seen that the retention times have been reduced, which results from reduced interactions with the stationary phase and increased interactions of the solute with the mobile phase. It also appears that the dispersion of the anthracene peak is minimal and its shape remains symmetrical. Due to the benzene peak merging with the naphthalene peak, the effect of benzene overload on the naphthalene peak is obscured.

To demonstrate the effect in more detail a series of experiments was carried out similar to that of volume overload, but in this case, the sample mass was increased in small increments. The retention distance of the front and the back of each peak was measured at the nominal points of inflection (0.6065 of the peak height) and the curves relating the retention data produced to the mass of sample added are shown in Figure 7. In Figure 7 the change in retention time with sample load is more obvious; the maximum effect was to reduce the retention time of anthracene and the minimum effect was to the overloaded solute itself, benzene. Despite the reduction in retention time, the band width of anthracene is still little effected by the overloaded benzene. There is, however, a significant increase in the width of the naphthalene peak which

Preparative Chromatography

is the closest eluted peak to benzene. In a similar manner, and under the same conditions, the effect of the mass overload of naphthalene was demonstrated using a simple binary mixture of naphthalene and anthracene.

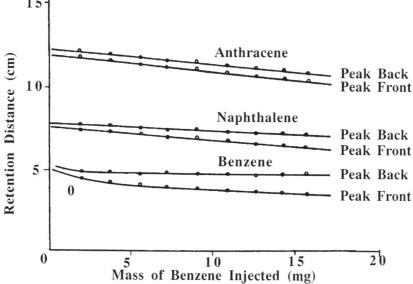

The front and back of the peaks were measured at the points of inflection of the curves (i.e. 0.607 x Peak Height)

Courtesy of the Journal of Chromatography (ref. 1)

Figure 7 The Effect of Mass Overload of Benzene

The mass of naphthalene injected was progressively increased from 1.1 mg to 19.1 mg. The results obtained are shown in Figure 8, as curves relating the retention distance of the front and rear of each peak against sample mass.

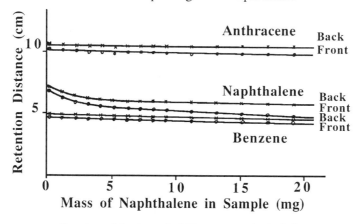

Courtesy of the Journal of Chromatography (ref. 1)

Figure 8. The Effect of Mass Overload of Naphthalene

The overload of the binary mixture shows a significant reduction in retention time but not to the same extent as with benzene overload. As the distribution coefficient of naphthalene with respect to the stationary phase is much larger than that of benzene, the concentration of the solute in the mobile phase will be smaller, but the effect on the shape of the adsorption isotherm would be greater. It would therefore appear that the reduction in retention distance from mass overload was more due to increased interactions in the mobile phase than the effect of changes in the shape of the adsorption isotherm. Overload effects were also examined employing a different mobile phase and three different solutes. The solutes selected were anisole, benzyl acetate and acetophenone, and the mobile phase 5%v/v ether in n-heptane. The composition of the initial mixture was 19.6 mg of anisole, 44.6 mg benzyl acetate and 20.5 mg of acetophenone in 200 µl of mobile phase. The sample volume was maintained constant at 200 µl and the mass of benzyl acetate progressively increased from 1.1 to 30 mg. Graphs relating the distance of the front and back inflection points from the point of injection to the mass of benzyl acetate injected is shown in Figure 9.

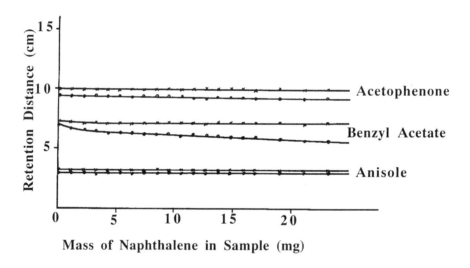

Courtesy of the Journal of Chromatography (ref. 1)

Figure 9. The Effect of Mass Overload of Benzyl Acetate

The results from the overload of the more polar solute are similar to that for the aromatic hydrocarbons, but the effect of the overloaded peak on the other two appears to be somewhat less. It is seen that there is little change in the retention of anisole and acetophenone, although the band width of acetophenone shows a slight increase. The band width of benzyl acetate shows the expected band broadening

Preparative Chromatography

towards shorter retention but the retention distance of the back of the peak stays sensibly constant.

It is clear that column overload can be very effective in preparative chromatography and it is an attractive alternative to the design and use of larger columns. It is also seen that the general effect of overload can be qualitatively predicted but would be extremely difficult and time consuming to evaluate quantitatively. Nevertheless, the approach of Guiochon and his co-workers would help to quantitatively access the effect of column overload but, to be successful, would also demand a considerable amount of basic physical chemical measurements to be made (*i.e.*, carefully determined adsorption isotherms at high solute concentrations for each of the pertinent solutes). This work would be both time consuming and costly and would need to be economically justified by the nature of the separation and the value of the product involved.

The Control of Sample Size for Normal Preparative Column Operation

The technique of column overloading is only feasible if more than adequate resolution is possible between the solute of interest and its nearest neighbor. Many samples require a column to be constructed that will only just separate the solutes of interest and under these circumstances the loading capacity must be increased without overloading the column. It has been shown in earlier chapters that the maximum sample (mass or volume) that can be placed on a column is proportional to the *plate volume* of the column and the square root of its efficiency. Thus, the maximum sample mass (M) will be given by

$$M = A\sqrt{n}(v_m + Kv_s)$$

where (n) is the efficiency of the column,
(v$_m$) is the volume of mobile phase per plate,
(v$_s$) is the volume of stationary phase per plate,
(K) is the distribution coefficient of the solute,
and (A) is a constant.

Now, it has been shown that $V_r = n(v_m + Kv_s)$, where (V$_r$) is the retention volume of the solute. Thus,

$$M = A\frac{V_r}{\sqrt{n}}$$

Now, if $V_m \ll KV_s$ (that is, the solute is well retained) and it is recalled that $k' = \frac{KV_s}{v_m} = \frac{KV_s}{V_m}$ where (V_m) and (V_s) are total volumes of mobile phase and stationary phase in the column, respectively, and (k') the capacity ratio of the solute,

then as
$$V_r \rightarrow nKv_s \rightarrow KV_s \rightarrow V_m k' \rightarrow 0.6\pi r^2 l k'$$

It follows that
$$M = A \frac{0.6\pi r^2 l k'}{\sqrt{n}}$$

Now, for a well packed column, (n) will be directly proportional to the column length, thus,

$$M = A' \, 0.6\pi r^2 \sqrt{l} \, k' \tag{2}$$

where (A') is another constant.

It follows from equation (2) that the sample load will increase as the square of the column radius and thus the column radius is the major factor that controls productivity. Unfortunately, increasing the column radius will also increase the volume flow rate and thus the consumption of solvent. However, both the sample load and the mobile phase flow rate increases as the square of the radius, and so the solvent consumption per unit mass of product will remain the same.

Sample load only increases as the square root of the column length, but the separation time increases linearly with column length. Thus, increasing the sample load by increasing the column length is less effective than increasing the column radius and will significantly increase the separation time. However, under some special circumstances, increasing the column length can be a very effective means of increasing productivity. If the mixture contains only two components, *e.g.*, in the separation of a pair of enantiomers, the column can be increased in length to allow several separations to proceed along the column sequentially. In this way several pairs of enantiomers will be present in the column being separated at one time. This technique increases the column load for a longer column very significantly while actually reducing the separation time.

It is also clear from equation (2) that the sample mass can also be extended by increasing the capacity ratio (k') of the eluted solutes (*i.e*, by careful phase selection). In this case the maximum load will increase linearly with the value of (k') but so will

Preparative Chromatography

the separation time. Again in the special case of a simple binary mixture, multiple injections can be placed on the column which would compensate for the increased separation time that results from an extended value of (k').

The Moving Bed Continuous Chromatography System

The concept of a continuous gas solid chromatographic system was first reported by Freund *et al.* [5], as early as 1956, for the separation of acetylene from an acetylene–methane mixture.

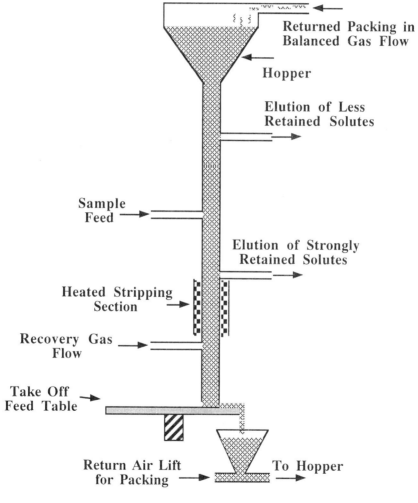

Figure 10 A Continuous Chromatography Separation System

The system proposed by Freund *et al.* contained all the essential properties of the modern moving bed, or pseudo moving bed chromatographic systems. The procedure was extended by Scott [6], in 1958, to gas/liquid chromatography and employed the moving bed technique to isolate pure benzene from coal gas. Domestic gas at that time was derived from the pyrolysis of coal during the production of coke and contained a number of aromatic hydrocarbons including significant quantities of benzene and toluene and some xylenes. However, despite early successes, the technique has not been as widely applied as first thought. Nevertheless, the methodology is still being actively developed, but not in the original form. Continuous elution chromatography for preparative purposes can be experimentally very difficult and this will be better appreciated when the basic principles of the technique are discussed and understood. The basic system is depicted diagramatically in Figure 10. The stationary phase coated on a support is arranged to freefall down a tower. This, in effect, provides a moving stationary phase. A feed of gas containing a mixture of solutes enters the tower at the center and by the careful adjustment of pressures is arranged to pass up the tower and out through an exit at the top. If the band velocity of a solute is greater than the velocity of stationary phase descent, then the solute will pass up the tower. If the solute velocity is less than the velocity of stationary phase descent, then it will be carried down the tower into the lower part of the tower.

The stationary phase in the lower part of the tower meets a second upward stream of gas and a portion of the lower part of the tower is heated. This results in the distribution coefficients of the solutes being reduced and their velocity up the tower increased so that they are now greater than the velocity of stationary phase descent. The stripping gas exits at another port at the top of the lower section of the tower. The packing passes out of the tower at the base, the amount of fall being controlled by the rate of rotation of a feed Table. The packing is then gas-lifted back to the top hopper for recycling. This procedure will split the solutes into two fractions and if the lower part of the tower contains a similar splitting system, then a specific solute can be selected from the mixture and obtained pure. The system used for the isolation of pure benzene from coal gas is shown diagramatically in Figure 11. If $(V'_{r(A)})$ and $(t'_{r(A)})$ are the corrected retention volume and the corrected retention time of a solute passing through unit length of packing respectively, and $K_{(A)}$ the distribution coefficient of solute (A) then it has been established that

$$V'_{r(A)} = K_{(A)}V_L = Q_{(G)}t_{r(A)}$$

Now, if the band velocity is (χ), then

$$\chi = \frac{1}{t_{r(A)}} = \frac{Q_{(G)}}{K_{(A)}V_L}$$

Preparative Chromatography

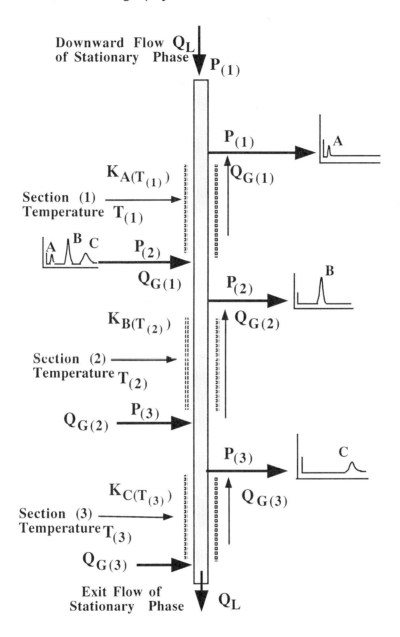

Figure 11. Theoretical Diagram of the Moving Bed System

Consider the system shown in Figure 11. Let the gas flow carrying the sample through the upper half of the tower be ($Q_{G(1)}$). Let the volume of mobile phase passing down the tower per unit time be (Q_L) and the distribution of solute (A) between the stationary phase and the mobile phase at temperature ($T_{(1)}$) be ($K_{A(T)}$).

Assuming the down velocity of the stationary phase is (Q_L), then if

$$\frac{Q_{(G)}}{K_{(A)}V_L} > Q_L$$

then solute (A) will pass up the column.

Conversely, if
$$\frac{Q_{(G)}}{K_{(A)}V_L} < Q_L$$

then solute (A) will pass down the column.

The conditions for the continuous separation of substances (A), (B) and (C) are now clear.

In section (1), solute (A) will be separated from solutes (B) and (C) when,

$$\frac{Q_{G(1)}}{K_{A(T_{(1)})}V_L} > Q_L \qquad (3)$$

$$\frac{Q_{G(1)}}{K_{B(T_{(1)})}V_L} < Q_L \qquad (4)$$

$$\frac{Q_{G(1)}}{K_{C(T_{(1)})}V_L} > Q_L \qquad (5)$$

These conditions will ensure that solute (A) passes up the tower and solutes (B) and (C) pass down the tower.

Thus the downflow of stationary phase on the support and the gas flow rate carrying the sample into section (1) must be adjusted such that at temperature $(T_{(1)})$ the conditions defined in equations (3), (4) and (5) are met. By the use of appropriate restrictions it must also be arranged that the pressure at the top of the tower must be made equal or close to that of the outlet for solute (A) to prevent cross-flow.

Now to separate solute (B) form (C) the following conditions must be met,

$$\frac{Q_{G(2)}}{K_{B(T_{(2)})}V_L} > Q_L \qquad (6)$$

and
$$\frac{Q_{G(2)}}{K_{C(T_{(2)})}V_L} > Q_L \qquad (7)$$

Preparative Chromatography

Now, the downflow of stationary phase has already been established and so the conditions defined in equations (6) and (7) must be met by adjustment of ($Q_{G(2)}$) and ($T_{(2)}$). Again, by the use of appropriate restrictions, it must also be arranged that the pressures at the sample inlet and that of the outlet for solute (B) must be made equal, or close to equal, to prevent cross-flow.

Finally, solute (A) can be recovered under the following condition,

$$\frac{Q_{G(3)}}{K_{C(T_{(3)})} V_L} > Q_L \qquad (8)$$

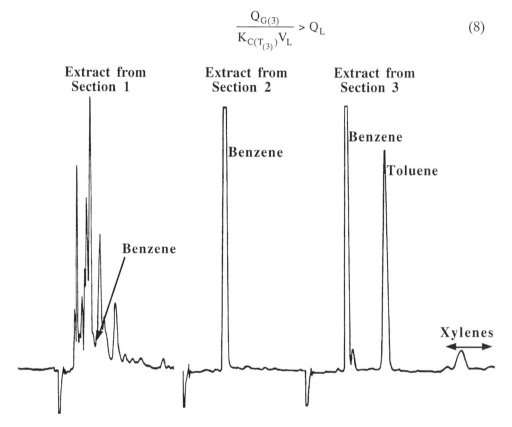

Courtesy of Benzole Producers Ltd. (ref.7)

Figure 12 The Extraction of Pure Benzene from Coal Gas by Continuous Extraction Using a Moving Bed Technique

The condition defined by equation (8) is met by adjustment of ($Q_{G(3)}$) and ($T_{(3)}$). The pressures at the second stripping flow inlet and that of the outlet for solute (C) must be made equal, or close to equal, to prevent cross-flow. Scott and Maggs [7] designed a three stage moving bed system, similar to that described above, to extract pure benzene from coal gas. Coal gas contains a range of saturated aliphatic hydrocarbons, alkenes, naphthenes and aromatics. In the above theory the saturated aliphatic hydrocarbons, alkenes and naphthenes are represented by solute (A).

The main product, benzene, is represented by solute (B), and the high boiling aromatics are represented by solute (C) (toluene and xylenes). The analysis of the products they obtained are shown in Figure 12. The material stripped form the top section (section (1)) is seen to contain the alkanes, alkenes and naphthenes and very little benzene. The product stripped from the center section appears to be virtually pure benzene. The product from section (3) contained toluene, the xylenes and thiophen which elutes close to benzene. The thiophen, however, was only eliminated at the expense of some loss of benzene to the lower stripping section. Although the system works well it proved experimentally difficult to set up and maintain under constant operating conditions. The problems arose largely from the need to adjust the pressures that must prevent cross-flow. The system as described would be virtually impossible to operate with a liquid mobile phase.

A pseudo moving bed system was described by Barker [8,9] who simulated the process by employing a column in circular form. A diagram depicting this wheel concept is shown in Figure 13.

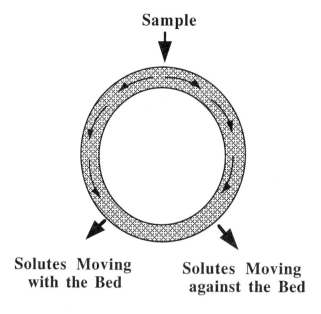

Figure 13. The Circular Simulated Moving Bed Column

The column takes the form of a circular hole round the wheel circumference and there are regular apertures in the rim allowing gas access to the packing. The apertures are closed until they are connected to one of the ports. The ports occupy fixed positions

Preparative Chromatography

at the periphery. The packed bed moved continuously in one direction relative to the ports and the mobile phase moved in the opposite direction. Those that traveled at a rate slower than the wheel velocity continued in the direction of the wheel movement, those that migrated faster than the wheel velocity moved with the mobile phase and in the opposite direction to the rotation of the wheel. The two fractions were collected from the take off ports and the mobile phase returned for recirculation. Unfortunately the wheel system only achieved moderate success due to problems with leaks past the sliding port surfaces on the wheel periphery.

The first practical simulated moving bed system was described by Hurrel in the late 1960s [10], and was an ingenious extension of the large circular column devised by Barker. There are two main advantages to the *simulated* moving bed over the *actual* moving bed. It is, first, technically simple and, second, as the mobile phase is recirculated, less solvent is used and the stationary phase is conserved making the system more economic. In the system devised by Hurrel the circular column is replaced by a number of sections joined in the form of a circle, each section consisting of a short preparative column. The columns were U shaped and packed individually by standard packing techniques. The columns are all joined by means of a large rotary disc valve fitted with appropriate ports. The fabrication and operation of the disc valve was relatively simple compared with that of the massive rotating wheel and could be made leak proof even at the high pressures. The valve consisted of two discs, the upper containing connections to the different ports and the lower disc connections to the columns. The upper disc of the valve was mounted on a steel frame and the lower disc supported the columns and could be rotated simulating the moving bed. Contemporary simulated moving bed systems used for preparative chromatography are mostly based on the Hurrel system. Nevertheless, they all work on the same principle and the simple theory given for the actual moving bed system will apply, albeit in some cases in a slightly modified form.

Synopsis

Preparative chromatography involves the collection of individual solutes as they are eluted from the column for further use, but does not necessarily entail the separation of large samples. Special columns can be designed and fabricated for preparative use, but for small samples the analytical column can often be overloaded for preparative purposes. Columns can be either *volume* overloaded or *mass* overloaded. Volume overload causes the peak to broaden, but the retention time of the front of the peak remains constant, while the retention time of the rear of the peak increases linearly with sample volume. The sample volume can be increased until the rear of the peak meets its nearest neighbor or *vice versa*. An explicit equation can be developed to permit the maximum overload sample volume to be calculated. Excessive sample volume overload culminates in frontal analysis. Excess mass overload causes peak dispersion and gross peak asymmetry and tends to reduce the retention time of later eluting solutes. This is due to the excess sample in the mobile phase acting as part of the mobile phase and thus increases the elution rate of later peaks. The front of the overloaded peak extends to shorter analysis times whereas retention time of the rear of the peak often remains sensibly constant. In general, an explicit equation for the maximum sample mass can not be developed and, unless a significant amount of distribution data for each solute is obtained, the chromatogram cannot be simulated by the computer. Consequently, the maximum sample mass is best obtained by experiment. The mass of the sample is increased until the overloaded peak just reaches its nearest neighbor. The maximum sample mass and solvent consumption for a large scale column increases as the square of the column radius and so the solvent consumption per unit mass of product remains sensibly constant. The maximum sample mass only increases as the square root of the column length, but the separation time increases linearly with the column length. Thus, increasing the column throughput by increasing the column length is only successful at the expense of time. A longer column, however, may allow more than one separation to be carried concurrently, if only two closely eluting solutes (such as an enantiomeric pair) are to be separated. The moving bed system, an alternative procedure for handling large sample throughput, has been used for continuous preparative separations. In principle, the column packing is made to move counter-current to the mobile phase. If the band velocity is greater than the bed velocity, it will be carried in the direction of the mobile phase. If the band velocity is less than the bed velocity, it will be carried in the same direction as the bed. Thus, the direction of movement of a particular solute will be determined by its distribution coefficient. By using a column having sections heated to different temperatures, and a number of stripping and collection ports, a single substance can be selectively and continuously separated form a mixture in a very pure state. The falling bed system is experimentally difficult to operate in a constant manner but the simulated moving bed column is more practical. The simulated moving bed system consists of a number of columns joined in series to form a circle by means of a rotating disc valve. The moving bed is simulated by rotating the valve so that the inlet and outlet ports move round the column instead of the packing moving past the ports. This system has been successfully used for both GC and LC separations.

References

1. R. P. W. Scott and P. Kucera, *J. Chromatogr.*, **119**(1976)467.
2. S. Golshan-Shirazi, A. Jaulmes and G. Guiochon, *Anal. Chem.*, **60**(1988)1856.
3. S. Golshan-Shirazi, A. Jaulmes and G. Guiochon, *Anal. Chem.*, **61**(1989)1276.
4. S. Golshan-Shirazi, A. Jaulmes and G. Guiochon, *Anal. Chem.*, **61**(1989)1368.
5. M. Freund, P. Benedek and L. Szepesy, *"Vapour Phase Chromatogrphy"*, (Ed. D. H. Desty) Butterworths Scientific Publications, London (1957)359.
6. R. P. W. Scott, *"Gas Chromatogrphy 1958"*, (Ed. D. H. Desty) Butterworths Scientific Publications, London (1958)287.
7. R. P. W. Scott and R. J. Maggs, *Benzole Producers Research Paper*, 5–1960.
8. P.E. Barker *"Preparative Gas Chromatography"*, (Ed. A. Zlakiz and V. Pretorious) Wiley Interscience, London (1971)325.
9. P. E. Barker and R. E. Deeble, *Anal. Chem.*, **45**(1973)1121.
10. R. Hurrel, US patent No. 3,747,630(1972).

Chapter 16

Thin Layer Chromatography Theory

The thin layer distribution system does not lend itself easily to theoretical examination due to the complex inhomegeneity of both the film of moving mobile phase on the gel surface and the stationary phase proper (the layer of silica or bonded silica on the plate surface). The complex nature of the elution profile of mobile phase and the consequent adsorbed layer on the stationary phase surface, has been discussed in detail in chapter 1, page 13. In practice, as the sample is separated between the two phases and the mobile phase generally contains two or more solvents, there is a concurrent mobile phase gradient system formed which, in turn, produces a corresponding complex stationary phase gradient. Thus, the distribution system on which the separation is carried out is continually changing from the point of sampling to the solvent front. It is also seen that such a system is very difficult to investigate theoretically and in detail, although the process can be examined qualitatively and the major effects may be deduced and measured. Such TLC plate parameters can provide measurements that can be used for spot identification, quantitative evaluation and, to a limited extent, TLC plate efficiency.

The interactions between solute and the phases are exactly the same as those present in LC separations, namely, dispersive, polar and ionic interactions. At one extreme, the plate coating might be silica gel, which would offer predominately polar and induced polar interactions with the solute and, consequently, the separation order would follow that of the solute polarity. To confine the polar selectivity to the stationary phase, the mobile phase might be n-hexane which would offer only dispersive interactions to the solute. The separation of aromatic hydrocarbons by induced polar selectivity could be achieved, for example, with such a system.

On the other hand, if the stationary phase consists of bonded silica containing hydrocarbon chains, then the stationary phase interactions with the solute would be predominantly dispersive and, consequently, the separation order would be determined by the dispersive character of the solute. To ensure that dispersive interactions dominate in the stationary phase, the mobile phase would, ideally, be water but, as the plate must be wetted by the solvent, a methanol/water mixture would be necessary which would offer predominantly, but not exclusively, polar interactions with the solute. Such a system could, for example, be used for the separation of a homologous series of fatty acids.

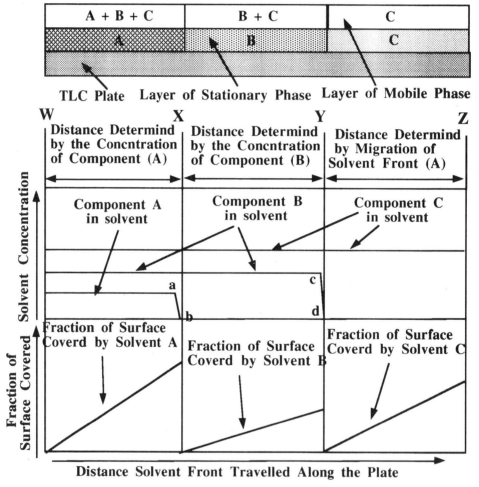

Figure 1. Distribution of Three Components of a TLC Mobile Phase After Traveling Some Way Along a TLC Plate

The two examples that have been given are simple and basic, and illustrate the principles of a TLC separation. Ion exchange material can also be bonded to the silica, allowing ionic interactions to be dominant in the stationary phase and, thus,

Thin Layer Chromatography

also provide ionic selectivities. Nevertheless, the simple thin layer plate employed with a very simple developing procedure can provide extremely high and subtle selectivity's that results from the frontal analysis of the solvent as it passes over the plate surface (chapter 1 page 13). It should be pointed out that, by using an ionic moderator in the solvent mixture, ionic interactions can be easily introduced into the mobile phase. Furthermore, if the moderator is adsorbed onto the surface of the stationary phase, then an effective ion exchanger can also be located on the stationary phase. Consider the thin layer plate shown in figure 1, which depicts the separation of a sample using a three component solvent mixture. The number of solutes in the mixture is not limited in practice, but the factors that provide the extraordinary wide range of selectivity can be readily understood by considering a simple three component solvent mixture.

Consider a silica coated plate that is employed with a mobile phase containing ethanol, ethyl acetate and n-hexane to separate a mixture of fatty acids. The condition of the plate, after the solvent front has traveled from (W) to point (Z), is depicted in figure 1. As the mobile phase percolates along the plate, the most polar solvent, ethanol, is extracted from the mobile phase and held on the surface of the silica plate in the section marked WX. Thus, any fatty acid that is eluted through this section will be distributed between the original solvent and the silica gel surface covered with ethanol. In the next section, XY, the solutes will distribute themselves between a mobile phase consisting of n-hexane and ethyl acetate and a silica surface covered with ethyl acetate. At the boundaries there will not be abrupt changes in mobile phase composition but more gradual as shown in figure 1. These more gentle changes at the boundaries add even more subtle selectivities to the distribution system. Finally, in the last stage, YZ, the distribution takes place between pure n-hexane and silica covered with n-hexane and thus the interactions with the stationary phase will be almost exclusively polar with the hydroxyl groups on the silica gel surface and exclusively dispersive with the mobile phase. However, the solvent that left section WX and entered the next section XY was depleted of ethanol and only contained ethyl acetate and n-hexane. Likewise, that which left section XY and entered YZ would be depleted of ethyl acetate and only contain n-hexane.

Consider the effect of the abstraction of individual solvents on consequent solute concentration. Let the volume fraction of ethanol, ethyl acetate and n-hexane in the original mobile phase mixture be (α_E), (α_{EA}) and (α_H), respectively. If a given solute is placed on the TLC plate at a concentration (c_s) then, assuming for the moment, there is no band dispersion, the concentration of the peak leaving section (WX) and entering (XY), (c_X), will be given by.

$$c_x = \frac{c_S}{\alpha_{EA} + \alpha_H}$$

During passage through the section XY, the mobile phase was depleted of component (B) the ethyl acetate thus; again assuming there is no band dispersion, the concentration of the peak leaving section (XY) and entering (YZ), (c_y), will be given by.

$$c_y = \frac{c_x}{\alpha_H} = \frac{c_S}{\alpha_H(\alpha_{EA} + \alpha_H)}$$

Taking the example where,

$$\alpha_{ET} = \alpha_{EA} = \alpha_H = \frac{1}{3}$$

$$c_y = \frac{c_S}{\frac{1}{3}\left(\frac{1}{3} + \frac{1}{3}\right)} = \frac{c_S}{\frac{2}{9}} = \frac{9c_S}{2} = 4.5c_S$$

This is an oversimplified treatment of the concentration effect that can occur on a thin layer plate when using mixed solvents. Nevertheless, despite the complex nature of the surface that is considered, the treatment is sufficiently representative to disclose that a concentration effect does, indeed, take place. The concentration effect arises from the frontal analysis of the mobile phase which not only provides unique and complex modes of solute interaction and, thus, enhanced selectivity, but also causes the solutes to be concentrated as they pass along the TLC plate. This concentration process will oppose the dilution that results from band dispersion and thus, provides greater sensitivity to the spots close to the solvent front. This concealed concentration process, often not recognized, is another property of TLC development that helps make it so practical and generally useful and often provides unexpected sensitivities.

TLC Measurements.

TLC measurements used for the identification of unknown solutes depends on two basic parameters. Firstly, the distance traveled by the solvent front, measured from the sampling point or sampling boundary, and secondly, on the distance traveled by the spot from the sampling point or sampling boundary. These are the sole

Thin Layer Chromatography

measurements that are made on the thin layer plate for *qualitative* analysis. The basic ratio that is used in qualitative analysis is the (Rf) factor or simple derivatives of this ratio. Consider the elution of solute (a) along a thin layer plate as depicted in figure 2.

The (Rf) factor of solute (a), (*i.e.* R_{f_a}) is defined as

$$R_{f_a} = \frac{\text{Distance Traveled by Solute (a)}}{\text{Distance Traveled by Solvent Front}} = \frac{x}{y} \qquad (1)$$

The distances can be measured in any convenient unit of length and scanning techniques are generally found to provide more accurate and precise data. The point of greatest density is taken as the spot center. (R_{f_a}) values can range from zero to unity but can never be greater than unity. If the separation of two components is considered as depicted in figure 3, the ratio of the (Rf) values of each component will be

$$\frac{\dfrac{\text{Distance Traveled by Solute (a)}}{\text{Distance Traveled by Solvent Front}}}{\dfrac{\text{Distance Traveled by Solute (b)}}{\text{Distance Traveled by Solvent Front}}} = \frac{R_{f_a}}{R_{f_b}}$$

$$= \frac{\text{Distance Traveled by Solute (a)}}{\text{Distance Traveled by Solute (b)}} = \frac{x}{z} = R_{x_a} \qquad (2)$$

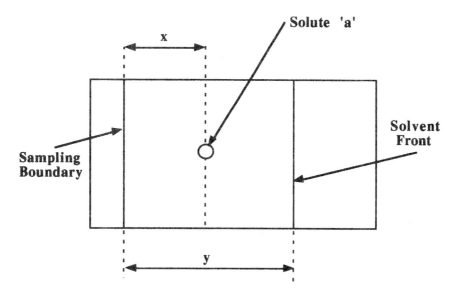

Figure 2. The Elution of Solute (a) Along a Thin Layer Plate

where solute (a) would be the unknown substance and solute (b) the reference substance placed in the mixture to aid in the identification of solute (a). It is seen later that (R_{x_a}) is thermodynamically equivalent to the separation ratio in GC and LC. In a similar manner to separation ratios, and in contrast to the (Rf) values, (R_{x_a}) values are not restricted to less than unity. It has been shown that, in GC and LC, separation ratios depend solely on the thermodynamic properties of the phase system; in TLC, however, (R_{x_a}) values will depend *largely* on the phase system selected but are not entirely independent of other factors, particularly the physical characteristics of the thin layer plate.

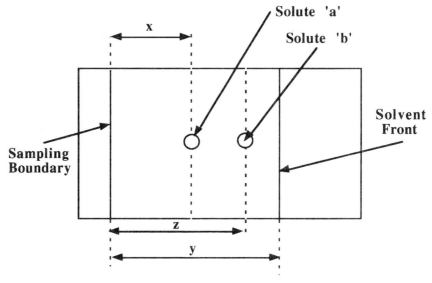

Figure 3. The Elution of Two Solutes, (a) and (b) Along a Thin Layer Plate

Referring again to the retention of a solute on a thin layer plate depicted in figure 2 and comparing this with a normal GC or LC chromatogram, then the distance $(y - x)$ represents the total retention of the solute relative to the solvent front and the distance (x) represents the difference between the retention of the solute and the retention of the solvent, *i.e.*, the dead volume. Thus, the equivalent capacity ratio (k'), for a solute on a thin layer plate is given by

$$k' = \frac{y-x}{x} = \frac{\frac{y-x}{y}}{\frac{x}{y}} = \frac{1-\frac{x}{y}}{\frac{x}{y}} = \frac{1-R_f}{R_f} \quad (3)$$

or, alternatively, (Rf) can be expressed in terms of the capacity ratio (k'), *i.e.*,

$$R_f = \frac{1}{k'+1} \quad (4)$$

Thin Layer Chromatography

Thus, employing the relationship for (k') given in equation (4), the various expressions used to define some of the other parameters of the classical chromatograms given by GC and LC can be expressed for TLC in terms of the (R_f) parameter.

Resolution

Resolution in TLC is defined in a similar manner to that in GC and LC, as the ratio of the mean spot width (peak width at the base) to the separation distance between the centers of the two spots (see figure 4). However, this will only be acceptable if the spots are not skewed under which circumstances the point of highest concentration within the spot is used.

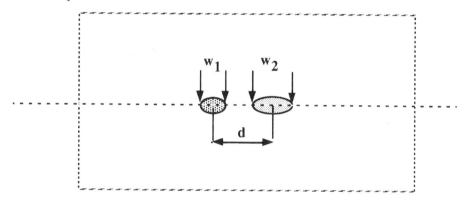

Figure 4. Measurement of TLC Resolution

In TLC, the spot diameter, taken as the width between the observable limits of the spot, is considered equivalent to four standard deviations of the spot dispersion, whatever manner the chromatogram was developed. This assumption can, at best, be only considered as approximate but has survived due to the lack of a better alternative. If the spots are scanned (which is becoming more common as accurate scanning devices become increasingly available at reasonable cost), the profile of the elution curves can be constructed. Then, by measuring the distance between the intersection of the tangents to the sides of the scanned peak to the base line produced, a more accurate value for (w_1) and (w_2), and consequently (R), can be obtained. The separation of two spots is depicted in figure 4 and, thus, the resolution (R) is given by,

$$R = \frac{d}{\frac{(w_1 + w_2)}{2}} = \frac{2d}{(w_1 + w_2)} \quad (5)$$

where (w_1) and (w_2) are the spot diameters of solutes (1) and (2),

An alternative expression for resolution was put foreword by Karger, Snyder and Horvath (1), *viz*,

$$R = \frac{1}{4}\left(\frac{\alpha-1}{\alpha}\right)N^{1/2}\left(\frac{k'}{1+k'}\right) \quad (6)$$

Now, from (3) as $k' = \frac{1-R_{f(a)}}{R_{f(a)}}$, it can be easily shown that, $\frac{k'}{1+k'} = 1 - R_{f(a)}$

and, from (2) as $\alpha = \frac{R_{f(b)}}{R_{f(a)}}$, it can easily be shown that $\frac{\alpha-1}{\alpha} = \frac{R_{f(b)} - R_{f(a)}}{R_{f(b)}}$

Then substituting for $\frac{k'}{1+k'}$ and $\frac{\alpha-1}{\alpha}$ in (8)

$$R = (1 - R_{f(a)})\left(1 - \frac{R_{f(a)}}{R_{f(b)}}\right)N^{1/2} \quad (7)$$

The alternative expression for resolution given in equation (7) demonstrates that the plate resolution, as in other forms of chromatography, depends on the number of theoretical plates, the selectivity and the capacity ratio of the solute for the particular plate concerned. In practice, however, the expression given in equation (7) appears to be the more practically useful for TLC separations.

Measurement of TLC Efficiency

In TLC, plate efficiency is measured in a similar manner to that in GC and LC employing a similar expression. However, the retention distance from the sampling point to the position of the spot center is substituted for the retention volume, time or distance and the spot width as an alternative to the peak base width, *viz.*,

$$N = 16\left(\frac{x}{w_b}\right)^2$$

where (N) is the column efficiency in theoretical plates,
  (x) is the distance moved by the spot,
and (w_b) is the peak width at the base.

The measurement of plate efficiency is depicted in figure 5.

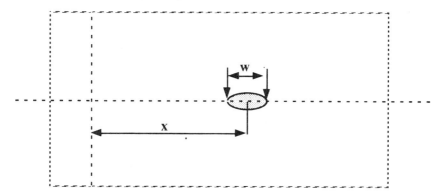

Figure 5. Measurement of Plate Efficiency

This equation, although originating from the plate theory, must again be considered as largely empirical when employed for TLC. This is because, in its derivation, the distribution coefficient of the solute between the two phases is considered constant throughout the development process. In practice, due to the nature of the development as already discussed for TLC, the distribution coefficient does not remain constant and, thus, the expression for column efficiency must be considered, at best, only approximate. The same errors would be involved if the equation was used to calculate the efficiency of a GC column when the solute was eluted by temperature programming or in LC where the solute was eluted by gradient elution. If the solute could be eluted by a pure solvent such as n-heptane on a plate that had been presaturated with the solvent vapor, then the distribution coefficient would remain sensibly constant over the development process. Under such circumstances the efficiency value would be more accurate and more likely to represent a true plate efficiency.

In addition, as mentioned before, the visual identification of the spot width is also open to considerable error but, if the plate is scanned, and the elution curve obtained, then both the peak width (w) or the peak width at the base (w_b) can be obtained fairly accurately (assuming appropriate software is available).

Employing the peak width, the efficiency would be given as

$$N = 4\left(\frac{x}{w}\right)^2$$

Employing the value (N) obtained for the plate efficiency, again, in an analogous manner to GC and LC the plate height of a TLC plate (H_{TLC}) is given by

$$H_{TLC} = \frac{x}{N}$$

where (x), the distance traveled by the solute along the plate, is taken as the equivalent length of the plate over which the measured value of (N) is pertinent.

Dispersion on a Thin Layer Plate

Dispersion equations, typically the van Deemter equation (2), have been often applied to the TLC plate. Qualitatively, this use of dispersion equations derived for GC and LC can be useful, but any quantitative relationship between such equations and the actual thin layer plate are likely to be fraught with error. In general, there will be the three similar dispersion terms representing the main sources of spot dispersion, namely, multipath dispersion, longitudinal diffusion and dispersion due to resistance to mass transfer between the two phases.

The multipath dispersion on a thin layer plate is the process most likely to be described by a function similar to that in the van Deemter equation. However, the actual mobile phase velocity is likely to enter that range where the Giddings function (3) applies. In addition, as the solvent composition is continually changing (at least in the vast majority of practical applications) the solute diffusivity is also altered and thus, the mobile phase velocity at which the Giddings function applies will vary.

The longitudinal diffusion on a thin layer plate will not be a constant function of the mobile phase velocity, as the diffusivity of the solute is continually changing. This means that the optimum velocity (if such a concept can be considered under these circumstances) will also vary continuously along the plate. It should likewise be pointed out that the actual mobile phase velocity will also be changing as a direct result of the selective solvent extraction which generates the concentration effect that has already been discussed. Indeed, as the solute is eluted along the plate, the solute diffusivity changes, its optimum velocity changes and, in addition, the *actual* mobile phase velocity changes. Thus, the conditions assumed in the development of the traditional dispersion equations are widely contravened.

In an exactly similar manner the true functions that describe dispersion resulting from the resistance to mass transfer of the solutes between the phases in TLC will be far

Thin Layer Chromatography

more complex than those developed for GC and LC. This increased complexity will result from changes in solute diffusivity, changes in the distribution coefficient of the solute between the phases and changes in the *actual* mobile phase velocity itself, all of which arise from the displacement effects of the solvent as it passes along the plate. Nevertheless, dispersion resulting from solute resistance to mass transfer between the phases will be a significant factor controlling the size of the TLC spot. At present, however, a precise theoretical treatment is not available and the best that can be concluded is that the minimum dispersion will be obtained from plates coated *very evenly* with *small* diameter particles used with solvents having minimum viscosity and, therefore, high solute diffusivities.

Despite the uncertain nature of the distribution system, the theoretical examination of the processes involved in a TLC separations has received considerable attention. Most of the work, however, tend to give the impression of being somewhat esoteric, and of uncertain value in the practice of thin layer chromatography. A more detailed discussion of TLC theory can be found elsewhere. (4)

Synopsis

The complex distribution system that results from the frontal analysis of a multicomponent solvent mixture on a thin layer plate makes the theoretical treatment of the TLC process exceedingly difficult. Although specific expressions for the important parameters can be obtained for a simple, particular, application, a general set of expressions that can help with all types of TLC analyses has not yet been developed. One advantage of the frontal analysis of the solvent, however, is to produce a concentration effect that improves the overall sensitivity of the technique. The primary parameter used in TLC is the (R_f) factor which is a simple ratio of the distance traveled by the solute to the distance traveled by the solvent front. The (R_f) factor will always be less than unity. If a standard is added to the mixture, then the ratio of the (R_f) factors of the solute to that of the standard is termed the (R_x) factor and is thermodynamically equivalent to the separation ratio (α) in GC or LC. In a similar manner, the capacity ratio (k') of a solute can be calculated for TLC from its (R_f) factor. Resolution is measured as the distance between the centers of two spots to the mean spot width. Alternative expressions for the resolution can be given in terms of the (R_f) factor and the plate efficiency. The plate efficiency is taken (by analogy to GC and LC) as sixteen times the square of the ratio of the retention distance of the spot to the spot width, but the analogy between TLC and the techniques of GC and LC can only be used with extreme caution. The so called

HETP of a TLC plate is taken as the ratio of the distance traveled by the spot to the plate efficiency. The same three processes cause spot dispersion in TLC as do cause band dispersion in GC and LC. Namely, they are multipath dispersion, longitudinal diffusion and resistance to mass transfer between the two phases. Due to the aforementioned solvent frontal analysis, however, neither the capacity ratio, the solute diffusivity or the solvent velocity are constant throughout the elution of the solute along the plate and thus the conventional dispersion equations used in GC and LC have no pertinence to the thin layer plate.

References

1. B. L. Karger, L. R. Snyder and C. Horvath, *An Introduction to Separation Science,* John Wiley and Sons. Inc., New York, London(1973) 150.
2. J. J. Van Deemter, F. Zuiderweg and A. Klinkenberg, *Chem. Eng. Sci.,* **5**(1956)271.
3. J. C. Giddings, *Anal Chem.*, **35**(1963)1338.
4. F. Geiss, Fundamentals of Thin Layer Chromatography, A. Hüthig Verlag, Heidelberg-Basel New York (1987)

Appendix 1

The integration of the differential equation that describes the rate of change of solute concentration within a plate to the volume flow of mobile phase through it. The integral of this equation will be the equation for the elution curve of a solute through a chromatographic column.

The differential equation is as follows:

$$\frac{dX_{m(p)}}{dv} = X_{m(p-1)} - X_{m(p)}$$

First consider the conditions of the above equation when an initial charge of concentration $X_{o(m)}$ has been placed on the first plate of the column, but chromatographic development has not commenced, *i.e.*, v=0.

Then, $X_{m(p)} = X_{o(m)}$ when p=0 (*i.e.*, the first plate)

and $X_{m(p)} = 0$ when p>0 (*i.e.*, for any other plate in the column)

The first condition merely states that before the chromatographic development commences, the concentration in plate p=0 is that resulting from the injection of the sample onto the column. The second condition states that the remainder of the column is free of solute.

Thus for plate p=0, and as there is no plate (p-1),

$$\frac{dX_{m(0)}}{dv} = X_{m(0)}$$

$$\frac{dX_{m(0)}}{X_{m(0)}} = dv$$

Integrating, $\log_e X_{o(m)} = -v + \text{constant}$

Now, when v = 0, $X_{o(m)} = X_{o(m)}$

Consequently, the constant $= \log_e X_o$, and $\log_e X_{o(m)} = -v + \log_e X_o$

or
$$X_{(m)o} = X_o e^{-v} \tag{1}$$

For Plate 1,
$$\frac{dX_{m(1)}}{dv} = X_{m(0)} - X_{m(1)} \tag{2}$$

Substituting for $X_{m(0)}$ from (11) in (12),

$$\frac{dX_{m(1)}}{dv} = X_o e^{-v} - X_{m(1)}$$

$$\frac{dX_{m(1)}}{dv} + X_{m(1)} = X_o e^{-v}$$

Multiplying throughout by e^v,

$$\frac{dX_{m(1)}}{dv} e^v + X_{m(1)} e^v = X_o$$

Now, this equation can be recognized as the differential of a product.

Hence,
$$\frac{d(X_{m(1)} e^v)}{dv} = X_o$$

Integrating,
$$X_{m(1)} e^v = X_o v + k$$

Now, when $v = 0$, $X_{m(1)} = 0$, thus, $k = 0$. Furthermore,

$$X_{m(1)} = X_o e^{-v} v$$

In a similar way it can be shown that

for Plate 2,
$$X_{m(2)} = X_o \frac{e^{-v} v^2}{1.2}$$

for Plate 3,
$$X_{m(3)} = X_o \frac{e^{-v} v^3}{1.2.3}$$

and for the nth Plate,
$$X_{m(n)} = X_o \frac{e^{-v} v^n}{n!}$$

Appendix 2

Accurate and Precise Dispersion Data for LC Columns

To confirm the pertinence of a particular dispersion equation, it is necessary to use extremely precise and accurate data. Such data can only be obtained from carefully designed apparatus that provides minimum extra-column dispersion. In addition, it is necessary to employ columns that have intrinsically large peak volumes so that any residual extra-column dispersion that will contribute to the overall variance is not significant. Such conditions were employed by Katz *et al.* (E. D. Katz, K. L. Ogan and R. P. W. Scott, *J. Chromatogr.*, **270**(1983)51) to determine a large quantity of column dispersion data that overall had an accuracy of better than 3%. The data they obtained are as follows and can be used confidently to evaluate other dispersion equations should they appear in the literature.

Reproduced by the kind permission of the Elsevier Publishing Company and the courtesy of the Journal of Chromatography.

TABLE I

PROPERTY DATA FOR MOBILE PHASES AND SOLUTES

Mobile phase	Benzyl acetate			Hexamethylbenzene D_m $(10^{-5}$ $cm^2/sec)$
	k'	k_e	D_m $(10^{-5}$ $cm^2/sec)$	
1 4.58% (w/v) ethyl acetate in *n*-pentane	2.05	4.07	3.61	3.51
2 4.86% (w/v) ethyl acetate in *n*-hexane	1.97	3.94	3.06	2.73
3 4.32% (w/v) ethyl acetate in *n*-heptane	2.04	4.06	2.45	2.23
4 4.50% (w/v) ethyl acetate in *n*-octane	2.01	4.01	2.01	1.71
5 4.41% (w/v) ethyl acetate in *n*-nonane	2.12	4.20	1.65	1.35
6 4.82% (w/v) ethyl acetate in *n*-decane	2.01	4.01	1.46	1.17

Table II

PEAK DISPERSION DATA FOR BENZYL ACETATE ($k_e = 4.0$)

u (cm/sec)	u_e (cm/sec)	H (cm)	Van Deemter coefficients
4.68% (w/v) ethyl acetate in n-pentane			
0.01819	0.03027	0.004788	
0.02721	0.04527	0.003704	
0.03911	0.06507	0.003116	$A = 0.001189$ cm
0.06024	0.10023	0.002526	$B = 0.0001079$ cm^2/sec
0.07847	0.1306	0.002292	$C = 0.002525$ sec
0.09937	0.1653	0.002176	
0.1495	0.2488	0.002246	($r = 0.999844$)
0.1914	0.3185	0.002360	(std. error = $5.9 \cdot 10^{-5}$)
0.2880	0.4792	0.002678	
0.3623	0.6028	0.002856	
4.86% (w/v) ethyl acetate in n-hexane			
0.01825	0.03037	0.004182	
0.02711	0.04521	0.003352	
0.03923	0.06527	0.002731	
0.06112	0.1017	0.002293	$A = 0.001144$ cm
0.07784	0.1295	0.002246	$B = 0.00009045$ cm^2/sec
0.09968	0.1659	0.002110	$C = 0.003008$ sec
0.1500	0.2496	0.002305	
0.1852	0.3082	0.002407	($r = 0.999885$)
0.2908	0.4839	0.002778	(std. error = $4.8 \cdot 10^{-5}$)
0.3631	0.6042	0.003098	
4.32% (w/v) ethyl acetate in n-heptane			
0.01763	0.02933	0.003982	
0.02692	0.04479	0.003109	
0.03817	0.06351	0.002605	
0.05631	0.09369	0.002251	$A = 0.001114$ cm
0.07167	0.1193	0.002191	$B = 0.00008141$ cm^2/sec
0.08596	0.1430	0.002208	$C = 0.003362$ sec
0.1294	0.2153	0.002199	
0.1729	0.2877	0.002365	($r = 0.999955$)
0.2718	0.4523	0.002849	(std. error = $2.9 \cdot 10^{-5}$)
0.3623	0.6028	0.003253	
4.50% (w/v) ethyl acetate in n-octane			
0.01821	0.03030	0.003439	
0.02741	0.04561	0.002829	
0.03954	0.06579	0.002428	
0.06059	0.1008	0.002163	$A = 0.001210$ cm
0.07862	0.1308	0.002195	$B = 0.00006460$ cm^2/sec
0.09992	0.1663	0.002252	$C = 0.003661$ sec
0.1506	0.2506	0.002356	
0.1956	0.3255	0.002602	($r = 0.999946$)
0.2894	0.4815	0.003120	(std. error = $3.2 \cdot 10^{-5}$)
0.3771	0.6275	0.003602	

Table II (continued)

u (cm/sec)	u_e (cm/sec)	H (cm)	Van Deemter coefficients
4.41% (w/v) ethyl acetate in n-nonane			
0.01813	0.03017	0.002985	
0.03037	0.05053	0.002385	
0.03945	0.06564	0.002186	
0.05447	0.09063	0.002119	$A = 0.001208$ cm
0.07230	0.1203	0.002143	$B = 0.00004893$ cm^2/sec
0.08735	0.1453	0.002184	$C = 0.004298$ sec
0.1316	0.2190	0.002384	
0.1753	0.2917	0.002640	($r = 0.999934$)
0.2706	0.4503	0.003307	(std. error = $3.4 \cdot 10^{-5}$)
0.3561	0.5925	0.003789	
4.82% (w/v) ethyl acetate in n-decane			
0.01813	0.03017	0.002888	
0.02720	0.04526	0.002464	
0.03932	0.06542	0.002226	
0.06053	0.10072	0.002123	$A = 0.001237$ cm
0.07862	0.1308	0.002256	$B = 0.00004587$ cm^2/sec
0.09937	0.1653	0.002324	$C = 0.004786$ sec
0.1490	0.2479	0.002625	
0.1883	0.3133	0.002864	($r = 0.999935$)
0.2848	0.4739	0.003544	(std. error = $3.6 \cdot 10^{-5}$)
0.3570	0.5940	0.004194	

TABLE III

PEAK DISPERSION DATA FOR HEXAMETHYLBENZENE ($k_e = 0.67$)

u (cm/sec)	u_e (cm/sec)	H (cm)	Van Deemter coefficients
4.68% (w/v) ethyl acetate in n-pentane			
0.01819	0.03027	0.003676	
0.02721	0.04527	0.002849	
0.03911	0.06507	0.002208	$A = 0.0009097$ cm
0.06024	0.10023	0.001828	$B = 0.00008309$ cm^2/sec
0.07847	0.1306	0.001656	$C = 0.001021$ sec
0.09937	0.1653	0.001531	
0.1459	0.2488	0.001542	($r = 0.999803$)
0.1914	0.3185	0.001564	(std. error = $4.7 \cdot 10^{-5}$)
0.2880	0.4792	0.001590	
0.3623	0.6028	0.001617	
4.86% (w/v) ethyl acetate in n-hexane			
0.01825	0.03037	0.003132	
0.02711	0.04521	0.002408	
0.03923	0.06527	0.001968	$A = 0.0009713$ cm
0.06112	0.1017	0.001655	$B = 0.00006317$ cm^2/sec
0.07784	0.1295	0.001659	$C = 0.001260$ sec
0.09968	0.1659	0.001529	
0.1500	0.2496	0.001615	($r = 0.999693$)

TABLE III *(continued)*

u (cm/sec)	u_e (cm/sec)	H (cm)	Van Deemter coefficients
0.1852	0.3082	0.001632	(std. error = $5.5 \cdot 10^{-5}$)
0.2908	0.4839	0.001699	
0.3631	0.6042	0.001798	
4.32% (w/v) ethyl acetate in n-heptane			
0.01763	0.02933	0.002800	
0.02692	0.04479	0.002078	
0.03817	0.06351	0.001777	A = 0.0010014 cm
0.05631	0.09369	0.001620	B = 0.00004870 cm^2/sec
0.07167	0.1193	0.001561	C = 0.001536 sec
0.08596	0.1430	0.001579	
0.1294	0.2153	0.001606	(r = 0.999011)
0.1729	0.2877	0.001719	(std. error = $9.4 \cdot 10^{-5}$)
0.2718	0.4523	0.001917	
0.3623	0.6028	0.001880	
4.50% (w/v) ethyl acetate in n-octane			
0.01821	0.03030	0.002363	
0.02741	0.04561	0.001892	
0.03954	0.06579	0.001677	A = 0.001036 cm
0.06059	0.10082	0.001496	B = 0.00003673 cm^2/sec
0.07862	0.1308	0.001605	C = 0.001952 sec
0.09992	0.1663	0.001602	
0.1506	0.2506	0.001733	(r = 0.999576)
0.1956	0.3255	0.001844	(std. error = $6.1 \cdot 10^{-5}$)
0.2894	0.4815	0.002073	
0.3771	0.6275	0.002259	
4.41% (w/v) ethyl acetate in n-nonane			
0.01813	0.03017	0.002101	
0.03037	0.05053	0.001647	
0.03945	0.06564	0.001559	A = 0.001081 cm
0.05447	0.09063	0.001565	B = 0.00002651 cm^2/sec
0.07230	0.1203	0.001586	C = 0.002353 sec
0.08735	0.1453	0.001652	
0.1316	0.2190	0.001756	(r = 0.999607)
0.1753	0.2917	0.001920	(std. error = $5.8 \cdot 10^{-5}$)
0.2706	0.4503	0.002206	
0.3561	0.5925	0.002474	
4.82% (w/v) ethyl acetate in n-decane			
0.01813	0.03017	0.001899	
0.02720	0.04526	0.001613	
0.03932	0.06542	0.001518	A = 0.001174 cm
0.06053	0.10072	0.001592	B = 0.00001751 cm^2/sec
0.07862	0.1308	0.001658	C = 0.002501 sec
0.09937	0.1653	0.001695	
0.1490	0.2479	0.001906	(r = 0.999550)
0.1883	0.3133	0.002116	(std. error = $6.4 \cdot 10^{-5}$)
0.2848	0.4739	0.002328	
0.3570	0.5940	0.002680	

Appendix 2

EXPERIMENTAL VALUES FOR DISPERSION EQUATION COEFFICIENTS OBTAINED BY CURVE FITTING PROCEDURES TO THE DATA GIVEN IN TABLE II

Mobile phase						
	n-Pentane 4.68% (w/v)	n-Hexane 4.86% (w/v)	n-Heptane 4.32% (w/v)	n-Octane 4.50% (w/v)	n-Nonane 4.41% (w/v)	n-Decane 4.82% (w/v)

Van Deemter equation $\quad H = A + \dfrac{B}{u} + Cu$

A	0.001189	0.001144	0.001114	0.001210	0.001208	0.001237
B	0.0001079	0.00009045	0.00008141	0.00006460	0.00004893	0.00004539
C	0.002525	0.003008	0.003362	0.003661	0.004298	0.004786

Giddings equation $\quad H = \dfrac{A}{1 + E/u} + \dfrac{B}{u} + Cu$

A	0.001189	0.001144	0.001123	0.001210	0.001407	0.001257
B	0.0001079	0.00009045	0.00008622	0.00006460	0.00006651	0.00005297
C	0.002525	0.003008	0.003348	0.003661	0.004001	0.004754
E	0	0	0.005243	0	0.03370	0.008100

Huber equation $\quad H = \dfrac{A}{1 + E/u^{1/2}} + \dfrac{B}{u} + Cu + Du^{1/2}$

A	0.001455	0.001048	0.0009864	0.001196	0.0007022	0.1612
B	0.0001035	0.00009204	0.00008345	0.00006483	0.00005710	0.00005624
C	0.003302	0.002728	0.002979	0.003622	0.002769	0.02310
D	−0.0009161	0.0003309	0.0004472	0.00004591	0.001775	−0.06804
E	0	0	0	0	0	2.131

Knox equation $\quad H = Au^{1/3} + \dfrac{B}{u} + Cu$

A	0.002509	0.002422	0.002390	0.002545	0.002608	0.002626
B	0.0001232	0.0001051	0.00009518	0.00008025	0.00006389	0.006111
C	0.0008720	0.001407	0.001754	0.002003	0.002518	0.0023035

Horváth equation $\quad H = \dfrac{A}{1 + E/u^{1/3}} + \dfrac{B}{u} + Cu + Du^{2/3}$

A	0.001366	0.001057	0.001013	0.001197	0.0008252	0.005583*
B	0.0001044	0.00009215	0.00008332	0.00006486	0.00005625	0.00005139
C	0.003572	0.002495	0.002744	0.003585	0.001948	0.009169
D	−0.001104	0.0005411	0.0006474	0.00007991	0.002454	−0.008577
E	0	0	0	0	0	*

* The best fit obtained for E sufficiently large that the first term reduced to $A/E \, u^{-1/3}$. The value given is for A/E.

Appendix 3   List of Symbols

C_s	concentration of an adsorbed layer of a moderator
D_m	diffusivity of a solute in a fluid
d_p	particle diameter
H	height equivalent to a theoretical plate (HETP)
h	reduced plate height
H_{min}	minimum plate height
K	distribution coefficient of a solute between the two phases
k'	capacity ratio of a solute
k''	dynamic capacity ratio of a solute
K_A	distribution coefficient of solute (A)
K_B	distribution coefficient of solute (B)
L	column length
l	distance between molecules
l	distance traveled axially by the solute band
M	molecular weight
n	refractive index of a substance
N_A	Avogadro's number
N_E	effective plate number
P_i	column inlet pressure
P_M	molar polarizability
P_o	column outlet pressure
$p^o{}_1$	solute bulk vapor pressure
P_r	average pressure in the column
P_x	pressure at a distance (x) from the front of the column
Q	flow rate, ml/min. If the mobile phase is compressible, then Q is the mean flow rate
Q_i	volume flow of mobile phase into the column at (P_i)
Q_o	flow of mobile phase from the column at (P_o)
r	column radius
R	gas constant
R	molar refractivity of a substance
R_{MS}	ratio of resistance to mass transfer in the mobile phase to that in the stationary phase
S_A	area covered by an adsorbed moderator molecule
T	absolute temperature
t''_0	retention time of a non permeating, non-absorbed solute
t_0	dead time, i.e., the time elapsed between the injection point and the dead point.
t_r	retention time

u_i	column inlet velocity
u_o	column oulet velocity
U_p	energy that arises during interaction between two dipolar molecules
u_x	mobile phase linear velocity at a distance (x) from the front of the column
V'_r	corrected retention volume
V_0	dead volume, i.e., the volume of mobile phase passed through the column between the injection point and the dead point
V^0_g	specific retention volume of a solute
V_c	column volume
V_E	extra-column volume from valves, tubing, unions, etc.
V_I	interstitial volume that is actually moving
V_L	sample overload volume
V_m	volume of mobile phase in the column
V_p	that portion of V_m that is contained in the pores
V_r	retention volume
V_s	volume of stationary phase in the column
$w_{0.5}$	peak width at half height
w_B	peak width at the base
$X_{(n)}$	concentration of solute in the mobile phase leaving the (n)th plate
X_0	initial concentration of solute placed on the first plate of the column
X_m	solute concentration in the mobile phase
$X_{m(p)}$	concentration of solute in the mobile phase leaving the (p)th plate
X_s	solute concentration in the stationary phase
Z	velocity of a solute band along a column
η	viscosity of a gas
η_{Ti}	viscosity of a gas at temperature (T) after a temperature program has progressed for time (t)
φ	total chromatographically available surface area
ΔG^0	standard free energy change
ΔH^0	standard enthalpy change
ΔS^0	standard entropy change
Ψ	fraction of the static interstitial volume accessible to the solute
Ω	fraction of the pore volume accessible to the solute
α	separation ratio (selectivity)
α	response of a detector to solute (A)
α_p	polarizability of a molecule
α_Q	rate of change of flow rate per unit time
β	response of a detector to solute (B)

Appendix 3

$\bar{\gamma}_1$	fully corrected solute activity coefficients
μp	dipole moment of a molecule
ϕ	coil aspect ratio of a coiled tube, i.e., the ratio of the tube radius to the coil radius
ρ	density of the medium
σ^2_C	dispersion (variance) in the column
σ^2_{CF}	dispersion (variance) in the sensor volume resulting from Newtonian flow
σ^2_{CM}	dispersion (variance) in the sensor volume from band merging
σ^2_E	extra-column dispersion (variance)
σ^2_s	dispersion (variance) due to sample volume
σ^2_T	dispersion (variance) in the valve-column and the column-detector connecting tubing
σ^2	variance of a chromatographic band
ξ	fraction of the stationary phase accessible to the solute

Author Index

Adlard, E. R., 18
Alhedai, A., 45
Amundson, N. R., 17
Arnold, J. H., 359
Ashley, J. W., 233
Asshauer, J., 286
Atwood, J. G., 286, 314
Barker, P. E., 442
Beesley, T. E., 85, 142, 164
Benedek, P., 442
Bird, R. B., 165
Bophemen, J., 233
Chilton, C. H., 234
Claxton, G. C., 234
Curtis, C. F., 164
Davis, J. M., 234
Deeble, R. E., 442
Desty, D. H., 233
Done, J. N., 286
Dreyer, B., 45
Endele, R., 286
Engelhardt, H., 45
Eon, C., 313
Eyring, H., 359
Feller, W., 260
Freund, M., 442
Geiss, F., 455
Gerlach, H. O., 313
Giddings, J. C., 18, 233, 234, 260, 285, 286, 419, 455
Gilbert, M. T., 313
Glasstone, S., 85
Gluekauf, E., 17
Golay, M. J. E., 18, 45, 261, 286, 314
Goldstein, J., 286
Goldup, A., 233
Golshan-Shirazi, S., 18, 442
Groszek, A. J., 234
Grushka, E., 359
Guiochon, G., 18, 313, 442
Gutierrez, J. E. N., 142
Gutlich, K. F., 313
Halasz, I., 286, 313
Herington, E. F. G., 18
Hildebrand, C. P., 233
Hofmann, 313
Horne, D. S., 260

Horvath, Cs., 18, 286, 455
Huber, J. F. K., 18, 285, 286
Hulsman, A. R. J., 18, 285
Hirschfelder, J. O., 164
Hurrel, R., 442
Hurtubise, R. J., 142
Hussain, A., 142
Jaulmes, A., 18, 442
Kahn, M. A., 18
Kaliszan, R. J., 45
Karger, B. L., 45, 455
Katti, A., 18
Katz, E., 18, 142, 286, 314, 334, 359
Kennedy, G. J., 18, 286
Keulemans, A. I. M., 17
Kikta, Jr., E. J., 359
Kirkpatrick, S. D., 234
Klinkenberg, A., 45, 233 234, 313, 334, 455
Knox, J. H., 18, 45, 260, 286, 313
Kraak, J. C., 286
Kucera, P., 45, 142, 233 234, 313, 314, 442
Kwantes, A., 18
Langer, S. H., 85
Lapidus, L., 17
Laub, R. J., 85, 142
Lawrence, J. G., 85
Liao, H. L., 18, 85
Lin, H. J., 18, 286
Lochmuller, C. H., 142
London, F., 85
Madden, S., 142
Maggs, R. J., 442
Martin, A. J. P., 17, 45, 85
Martin, M., 313
Martire, D. E., 18, 45, 85
Mayer, S. W., 17
McCann, M., 142
McCormick, R. M., 45
McElroy, S. C., 142
Mclaren, E., 260
McTaggart, N. G., 18
Muller, H., 45
Nieass, C. S., 45
Nir, S., 359

Ogan, K. L., 18, 142, 286, 334
Pantazopolos, G., 85
Pecsok, R. L., 85
Perry, R. H., 234
Poppe, H., 286
Pretorius, V. J., 234
Primevesi, R. G., 18
Purnell, J. H., 142, 233, 334
Ray, N. H., 236
Reese, C. E., 233, 313
Reidl, P., 85
Reidt, P., 18
Reilley, C. N., 233
Richter, P. W., 234
Rijnders, G. W. A., 18
Robbins, W. K., 142
Said, A. S., 17, 45
Scott, C. G., 18, 234
Scott, R. P. W., 18, 45, 85, 142, 143, 164, 233, 234, 286, 313, 314, 334, 359, 442
Sheriden, J. P., 18
Silver, H. F., 142
Simpson, C. F., 45, 142
Smith, J., 234
Smith, R. J., 45
Smuts, T. W., 234
Snyder, L. R., 455
Stein, W. D., 359
Synge, R. L. M., 17, 45
Szepesy, L., 442
Tewari, Y. B., 18, 85
Tijsen, R., 314
Tompkins, E. R., 17
Tunitski, N. N., 17
Trukelítaub, A., 233
van Deemter, J. J., 18, 45, 334, 455
van den Berg, J. H. M., 286
Walking, P., 313
Wellington, C. A., 142
Whitham, B. T., 18
Zahn, C., 85
Zhu, P. L., 45
Zhukovitski, A. A., 233
Zuiderweg, F. J., 45, 334, 455

Subject Index

adsorption isotherm
  bi-layer 93, 95
  bi-layer,function for 96
  chloroform,butylchloride 94
  effect on peak shape 176
  for alcohols 90
  Freundlich 177
  Langmuir 88
  single layer 88
analysis,quantitative,from retention measurements 171
associated
  methanol-water 125
  solvents, solute interactions with 135
association
  methanol water
    thermodynamic properties 133
  methanol-water mixtures
    standard enthalpy of 133
    standard entropy of 133
assymetry,peak,causes 175
avaerage mobile phase velocity 272
available stationary phase 36
base line,definition of 14
bi-layer adsorption 95
  isotherm for ethyl acetate 97
  isotherm,function for 96
bonded phase
  brush type 92
  bulk type 92
bonding,stationary phase,effect on retention 92
butyl chloride,adsorption isotherm 94
capacity
  peak
    as function of capacity ratio 207
    equation for 206
  plate 23
  ratio
    equation for 27
    function of temperature 49
    limited by detector sensitivity 208
    of a solute 26
cell,detector, low dispersion 308
chloroform,adsorption isotherm 94
chromatogram
  alternative axes 238
  reduced 362
chromatography
  column design
    applications 396
    capillary 386
    packed 360
  column,expression for peak capacity 206
  continuous
    moving bed 434
    simulated moving bed 439
  definition of 4
  development 7
  preparative 420
  terminology 14
  theory
    history of 4
    introduction to 3
  thin layer 444
    effective capacity ratio 450
    efficiency,measurement of 452
    plate dispersion 453
    resolution 450
    Rf factor 448, 449
    Rx factor 448
  vacancy 195
    elution curve equation 197
    multicomponent mixture 200
co-elution temperature, expression for 82
column
  capillary
    design 386
    maximum sample volume 393
    minimum elution time 392
    minimum H 391
    minimum length 391
    optimum flow rate 392
    optimum radius 389
    optimum velocity 391
  dead volume 34
  design
    applications 396
    packed column 360
    protocol 360
  dispersion,accurate measurement of 317
  efficiency 179
    expression for 181

plates necessary for separation 186
expression for peak capacity 206
moving phase 35
optimized specifications 366
optimized, minimum elution time 376
packed 368
   maximum sample volume 384
   minimum H 376
   minimum length 376
   minimum solvent consumption 383
   optimum flow rate 382
   optimum particle diameter 371
   optimum radius 380
   optimum velocity 375
   sources of dispersion 245
peak capacity 202
permitted extra column volume 290
physical properties of 43
preparative
   mass overload 428
   optimum sample size 432
   overload 420
   volume overload 421
resolving power 183
static phase 35
volume, distribution of 35
column design
   data bases
      analytical specifications 367
      column specifications 366
      elective variables 365
      instrument constraints 364
      performance criteria 362
   optimum particle diameter 371
   process for packed columns 368
compressibility, mobile phase 28
conduits, tubular, dispersion in 296
continuous chromatography, moving bed 434
D'Arcy's equation 29
dead
   point, definition of 14
   time, definition of 15
   volume
      definition of 15
      measurement of 39
density, change on mixing, methanol-water 128
design protocol, column 360
detector
   heat of adsorption, theory of 218
   sensor volume
      apparent dispersion 307

dispersion from Newtonian flow 306
detector cell low dispersion 308
development
   chromatographic 7
   displacement 7
   elution 9
   elution in TLC 12
   frontal analysis 8
   solvent profile in TLC 13
diffusion
   dispersion equation 245
   eddy 246
   effect of column pressure 275
   effect of pressure on dispersion in LC 276
   Ficks law 243
   liquid, effect of pressure 274
   process 243
diffusivity, function of molecular weight 339
dipole moment, effect on dipole interactions 66
dipole-dipole interactions 66
dipole-induced-dipole interactions 67
dispersion 317
   detector sensor volume
      apparant 307
      Newtonian flow 306
   dynamics of 237
   effect of detector time constant 311
   extra column 288
      maximum for a capillary column 393
      sources 291
   forces
      effect of polarizability 64
      interaction energy of 64
   from Newtonian flow 297
   from sample volume 291
   in a packed bed 245
   in frits 295
   in sample valves 294
   in tubular conduits 296
   in unions 295
   interactive force 63
   longitudinal diffusion 248
      expression for mobile phase 249
      expression for stationary phase 250
   low, connecting tube 301
   low, serpentine tubes 303
   mobile phase, HETP equation corrected for gas compressibility 271
   multipath
      effect 246

Index

expression for 247
on a thin layer plate 453
packed column
   mobile phase compressibility effect on HETP 268
   relationship to molecular weight 344
   radial 240
      expression for 242
   resistance to mass transfer
      mobile phase 251
      stationary phase 252
      stationary phase expression 256
displacement, development 7
distribution coefficient
   effect of temperature 49
   function of free energy 47
   in plate equilibrium 21
   retention control 47
   thermodynamic treatment 47
dynamics of peak dispersion 237
eddy diffusion 246
effective plate number 187
   equation for 188
   relationship to efficiency 189, 190
efficiency
   column 179
      expression for 181
      plates necessary for resolution 186
   relationship to effective plate number 189
elective variables 365
electron acceptor sites 76
electron doner sites 76
elution
   curve, equation for 23
   development in TLC 12
   equation, Gaussian form 165
   of a finite sample 190
      expression for 192
   process of 9
enantiomeric separations 81
energy driven distribution 50
enthalpy
   control of distribution 50
   from Vant Hoff curves 49
   physical significance 48
   positive effect on distribution 53
   standard
      free 48
      relationship with standard entropy 61
entropy
   control of distribution 50
   from Vant Hoff curves 49
   negative effect on distribution 53
   physical significance 48
   standard
      for different group types 58
      free 48
      relationship with standard enthalpy 61
entropy driven distribution 51
equilibrium
   established between phases 11
   in a theoretical plate 21
exit velocity, mobile phase 273
extensions of the HETP equation 277
extra column dispersion 288
   permitted for different columns 290
   sources of 291
extra column volume 25
Ficks law 244
flow
   optimum for capillary column 392
   optimum for maximized column 382
   programming 144
      compressible fluid, GC 146
      effect on retention time GC 149
      non compressible fluid, LC 144
free energy
   control of distribution 47
   distribution between groups 57
   distribution between interaction types 75
   standard 48
Freundlich adsorption isotherm 177
frits, dispersion in 295
frontal analysis 8
fudge factors 258
Giddings equation 262
Golay equation 267
gradient elution 157
heat of adsorption detector, theory of 218
height
   of a theoretical plate 239
   peak
      definition of 16
      measurement, effect of assymetry 170
HETP, minimum 279
history of chromatography theory 4
homologous series, thermodynamic analysis of 54
Horvath and Lin equation 266
Huber equation 263
hydrophilic interactions 71
hydrophobic interactions 71
inflexion point
   expression for 183
   position of 182

injection point, definition of 14
instrument constraints 364
interstitial volume 35
   measurement of 40
   moving 37
   static 37
ionic
   forces 69
   iteractions 69
isotherm adsorption
   bi-layer 95
   bi-layer for ethyl acetate 97
   bi-layer, function for 96
   chloroform, butyl chloride 94
   effect on peak shape 176
   for alcohols 90
   Freundlich 177
   Langmuir 88
   mono-layer and bi-layer 98
   single layer 88
kinetic dead volume 38
Knox equation 265
longitudinal diffusion 248
   dispersion
      expression for mobile phase 249
      expression for stationary phase 250
low dispersion
   connecting tubes 301
   serpentine tubes 303
lyophilic interactions 72
lyophobic interactions 72
mass overload
   effect on peak shape 176
   on preparative columns 428
maximum sample volume 194
methanol association with water 125
methanol-water
   change in refractivity on mixing 130
   diagram of ternary system 132
   equilibrium constant 131
   standard enthalpy of mixtures 133
   thermodynamic properties of mixtures 133
methyl group, thermodynamic properties of 58
methylene group, thermodynamic properties of 58
minimum HETP 279
mixed phases, theory of 87
mobile phase
   aqueous solvents 124
   binary mixtures 106
   compressible 28
      corrected HETP equation 271
   effect on HETP of a packed column 268
   effect of pressure on LC columns 276
   effect of volume fraction on interaction probability 107
   mixed solvents 109
   ternary solvent mixtures 115
   velocity
      avarage 272
      exit 273
moderator
   effect at high concentrations 106
   effect at low concentrations 88
   solvent 87
molar polarizability, effect on dispersion forces 65
molecular
   forces, dispersion 63
molecular interactions 62
   by displacement from a suface 99
   by displacement, experimental support 102
   by sorption on a surface 99
   by sorption, experimental support 102
   complex formation 77
   dipole-dipole 66
      effect of dipole moment 66
   dipole-induced dipole 67
   hydrophilic 71
   hydrophobic 71
   ionic 69
   lyophilic 72
   lyophobic 72
   polar interactions 65
   polar, differing strengths 75
   solute with stationary phase surface 98
   with associated solvents 135
molecular weight
   from dispersion in a packed column 344
   from dispersion measurements 336
   function of diffusivity 339
monolayer isotherms, of butyl chloride, chloroform 94
moving phase 35
multipath
   dispersion 246
      expression for 247
   process 247
optimum velocity 278
overload, mass, effect on peak shape 176
particle diameter, optimum 371
peak

Index 473

assymetry, causes 175
capacity 202
   as function of capacity ratio 207
   equation for 206
composite, measurement errors 168
distortion from mass overload 430
distortion from volume overload 425
height
   definition of 16
   measurement,effect of assymetry 170
   maximum definition of 15
measurement of close eluting 167
profiles,from different cell volumes 309
width at base,definition of 16
width, at half height,definition of 16
width,definition of 16
phase
   mixed,theory of 87
   mobile
      binary mixtures 106
      mixed solvents 109
      ternary mixtures 115
plate
   capacity 23
   equilibrium in 21
   temperature,change due to solute 209
Plate Theory 4, 20
   differential equation for the plate 22
   extensions of 165
   plate concept 5
   the Poisson function 24
point
   dead,definition of 14
   inflexion
      expression for 183
      position of 182
   injection,definition of 14
polarizability,effect on dispersion force strength 64
pore volume 35
   measurement of 42
pressure
   correction factor for a gas 32
   effect on diffusivity 274
probability of interaction
   control by volume fraction,experimental evidence 109
   effect of volume fraction of solvent 107
programming
   flow 144
      compressible fluid, GC 146
      effect on elution time GC 149
      effect on elution time LC 145
      non compressible fluid,LC 144
   solvent 157
   techniques 143
   temperature 149
      effect on elution time GC 154
      effect on separaration ratio GC 156
pseudo gradient,along a TLC plate 445
quantitative analysis from retention measurements 171
radial dispersion 240
   expression for 242
random walk model 240
Rate Theory 4
   equations of 262
ratio,capacity,limited by detector sensitivity 208
reduced chromatogram 362
references
   chapter 1 17
   chapter 2 45
   chapter 3 85
   chapter 4 142
   chapter 5 164
   chapter 6 233
   chapter 7 260
   chapter 8 285
   chapter 9 313
   chapter 10 334
   chapter 11 359
   chapter 14 419
   chapter 15 442
   chapter 16 455
refractivity, change on mixing,water-methanol 130
resistance to mass transfer dispersion in stationary phase 256
resistance to mass transfer, dispersion in mobile phase 251
resolution
   column 183
   conditions for 25
retention
   corrected volume, function of group type and number 55
   measurements, close eluting peaks 167
   relative contribution of methyl and methylene groups 59
   time
      corrected,definition of 16
      definition of 16
   volume
      corrected,definition of 16
      definition of 16

equation for 24
sample
  finite, equation for elution curve 192
  maximmum volume 384
  valves, dispersion in 294
  volume
    dispersion from 291
    expression for maximum 195
    maximum for a capillary column 393
    maximum permissible 194
separation
  conditions for 25
  ratio 27
    effect of solvent program rate 161
    equation for 27
    function of effective plate number 190
serpentine tubes 303
simulated moving bed chromatography 439
solute
  interaction with stationary phase surface 98
  transfer, at back and front of peak 10
solvent
  aqueous 124
  conditions for minimum consumption 383
  elution development in TLC 13
  mixture, composition and temperature effects on separation 118
  moderator 87
  programming 157
standard
  free energy 48
  free enthalpy 48
  free entropy 48
static phase 35
stationary phase
  available 36
  surface solute interaction with 98
  unavailable 36
summation of variances 193
synopsis
  chapter 1 16
  chapter 2 45
  chapter 3 83
  chapter 4 140
  chapter 5 163
  chapter 6 231
  chapter 7 259
  chapter 8 284
  chapter 9 312
  chapter 10 334
  chapter 11 357
  chapter 12 384
  chapter 13 394
  chapter 14 418
  chapter 15 440
  chapter 16 454
temperature
  and solvent composition, efect on separation 118
  change in plate due to solute 209
  coelution, determination of 82
  effect on capacity ratio 49
  effect on inlet/outlet pressure ratio 150
  effect on visocosity in temperature programming GC 153
  elution, effect of program rate GC 155
  of co-elution 80
  of co-elution, expression for 82
  programming 149
    effect on elution time GC 154
    effect on separation ratio GC 156
terminology, chromatography 14
ternary mobile phases 115
  experimental support of elution equation 116
thermodynamics
  analysis of homologous series 54
  applied to separation 80
  dead volume 38
  of water methanol association 133
  treatment of distribution coefficient 47
thin layer chromatography 444
  effective capacity ratio 450
  efficiency, measurement of 452
  plate dispersion 453
  resolution 450
  Rf factor 448
  Rx factor 448, 449
  solute concentration on plate 447
time
  constant, effect on peak dispersion 311
  dead, definition of 15
  retention
    definition of 16
    function of solvent program rate 161
TLC, elution development 12
tubes, low dispersion 301
unavailable stationary phase 36
unions, dispersion in 295
vacancy chromatography 195

Index

elution curve equation 197
multicomponent mixture 200
Van Deemter equation
  validation 316
    correct form of capacity ratio 331
    longitudinal dispersion reciprocally related to diffusivity 325
    mass transfer dispersion function of particle diameter 330
    mass transfer dispersion linearly related to diffusion 329
    Minimum (H) independant of diffusivity 327
    optimum velocity linearly related to diffusivity 328
Vant Hoff
  normal curves 49
  pseudo curves 52
variance
  HETP of a column 239
  per unit length of column 238
  summation of 193
velocity, optimum 278
viscosity, effect of temperature programming GC 153
volume
  change on mixing, water-methanol 127
  column, distribution of 35
  corrected retention
    equation for 25
    function of group type and number 55
  dead
    column 34
    definition of 15
    kinetic 38
    measurement of 39
    thermodynamic 38
  extra column 25
    dispersion for different columns 290
  fraction in mobile phase, effect on on reaction probability 107
  interstitial 35
    measurement of 40
    moving 37
    static 37
  overload, preparative columns 421
  pore 35
    measurement of 42
  retention
    corrected
      definition of 16
      linear with 1/volume fraction of solvent 159
    definition of 16
    equation for 24
  sample
    dispersion from 291
    expression for maximum sample volume 195
    maximum permissible 194, 384
  variance expression for 180
water association with methanol 125
water-methanol
  change in refractivity, on mixing 130
  diagram of ternary system 132
  effect of temperature on association 135
  equilibrium constant 131
  standard enthalpy of mixtures 133
  standard entropy of mixtures 133
width
  peak at base, definition of 16
  peak, definition of 16